Springer Texts in Education

Springer Texts in Education delivers high-quality instructional content for graduates and advanced graduates in all areas of Education and Educational Research. The textbook series is comprised of self-contained books with a broad and comprehensive coverage that are suitable for class as well as for individual self-study. All texts are authored by established experts in their fields and offer a solid methodological background, accompanied by pedagogical materials to serve students such as practical examples, exercises, case studies etc. Textbooks published in the Springer Texts in Education series are addressed to graduate and advanced graduate students, but also to researchers as important resources for their education, knowledge and teaching. Please contact Yoka Janssen at Yoka.Janssen@springer.com or your regular editorial contact person for queries or to submit your book proposal.

Peter Wulff · Marcus Kubsch · Christina Krist
Editors

Applying Machine Learning in Science Education Research

When, How, and Why?

 Springer

Editors
Peter Wulff
Heidelberg University of Education
Heidelberg, Baden-Württemberg, Germany

Marcus Kubsch
Freie Universität Berlin
Berlin, Germany

Christina Krist
Graduate School of Education
Stanford University
Stanford, CA, USA

ISSN 2366-7672   ISSN 2366-7680 (electronic)
Springer Texts in Education
ISBN 978-3-031-74226-2   ISBN 978-3-031-74227-9 (eBook)
https://doi.org/10.1007/978-3-031-74227-9

This work was supported by Peter Wulff, Marcus Kubsch and Christina Krist.

The publication was made possible through open-access funds from the Heidelberg University of Education and co-financing for open-access monographs and edited volumes by Freie Universität Berlin.

What I cannot create, I do not understand.

—*Richard Feynman*

Foreword

In the last three years, AI and machine learning have found their way into our daily lives, including research activities. How can AI, particularly machine learning, support our research activities in science education? Peter Wulff, Marcus Kubsch, and Christina Krist in "Applying Machine Learning in Science Education Research: When, how, and why?" take you on an essential journey on how you can use machine learning to analyze and categorize trends in large data sets to explore important science education questions or reduce the work we need to do.

As they argue, as science educators, it is not our research focus to perform fundamental research in AI, but as researchers, we do need to know enough about AI and machine learning to use ML in our research. The various chapters in the book will help you develop the technical skills and principles for making informed decisions on when, how, and why to use ML in science education research.

The fusion of machine learning (ML) with science education research stands as a beacon of innovation, heralding a new era of research that will help us to improve learning by tailoring science learning environments to the needs and backgrounds of teachers and learners throughout the globe using data-driven support. Peter Wulff, Marcus Kubsch, and Christina Krist offer a comprehensive exploration of how ML can help reshape the way we teach and learn science through research using ML and how we understand the processes underpinning science education.

Machine learning, a subset of artificial intelligence focused on building systems that learn from data, has transformed how individuals engage in work across the globe, and education is no exception. In science education, where the complexity of concepts and the diversity of learner needs present unique challenges, ML offers possible solutions that were unimaginable just a decade ago. ML holds the potential to support us in the development of automated assessment tools that provide instant and meaningful feedback to students, the creation of dynamic learning environments that adjust to individual learning trajectories, and the use of predictive analytics to support at-risk students. As such, ML empowers educators and learners alike, making science education more accessible, engaging, and effective.

The book takes you on a journey through the complex world of ML, helping you understand ideas such as supervised and unsupervised machine learning and large language models, and provides rich case studies that guide you along the path. However, using ML does take intellectual work and prior knowledge, as does the book. To get the most out of the book, knowledge of either Python or R is a prerequisite. One critical aspect of the book that I found valuable is its warning regarding whether we need to use ML and AI in our research. ML and AI are only tools, but like any tool, ML is designed for a purpose, and learning how to use it effectively takes time.

Machine learning can have a significant influence on science education research. Through the meticulous collection and analysis of data, ML algorithms uncover insights into learning behaviors, educational outcomes, and the efficacy of teaching methodologies. This book delves into how you can use machine learning in your research. Peter Wulff, Marcus Kubsch, and Christina Krist provide the theoretical background and case studies to start you on your journey of using ML in science education research. The case studies they present illuminate the transformative power of ML. Beyond its immediate impact on teaching and learning, integrating ML into science education research offers the potential to gleam new insights into how students learn challenging ideas and practices in science. Researchers can identify broader trends and patterns by analyzing vast datasets, contributing to a more nuanced understanding of how educational theories and practices intersect with learner outcomes. This book explores these possibilities, offering readers a glimpse into a future where education is not only informed by data but is also continually adapted and improved based on empirical evidence.

Two additional chapters by scholars in the field also explore the unique possibilities of ML in gleaming insights from extensive data sets. A chapter by Hall and Krist builds on ideas in the book to demonstrate how researchers can use various unsupervised pattern recognition approaches to analyze text data to answer research questions. They also discuss using exploratory data analysis tools to explore large text-based datasets before using unsupervised natural language processing techniques to understand the data better. Rosenberg and his colleagues, Bhidya and Pritchard, explain how researchers can use quantitative and qualitative approaches to study complex constructs present in large data sets. As other chapters in the book, these chapters provide valuable insights into how ML can support researchers as a tool for exploring large data sets to find patterns and analyze data.

The book provides a testament to the critical pioneering contributions the authors have made in using ML to improve science education. It challenges other educators and researchers to use these emerging technologies to improve science teaching and learning. Through their insights, readers will gain an appreciation of the complexity and potential of using ML in science education and practical knowledge on implementing ML solutions in their own teaching and research practices.

As we stand on the brink of a new paradigm in education, this book serves as both a guide and an inspiration. It challenges us to rethink our approaches to research, teaching, and learning, advocating for a future where education is more tailored to students' needs to help foster inclusive and effective science education environments than what currently exists. The journey of integrating ML into science education is fraught with challenges, including ethical considerations, the need for robust data privacy measures, and the imperative to ensure equity in access to technology. Yet, the opportunities it presents for enhancing science education are boundless. The authors touch upon these challenges. Peter Wulff, Marcus Kubsch, and Christina Krist in "Applying Machine Learning in Science Education Research: When, How, and Why?" takes you on a journey that requires cognitive engagement, but that journey is essential for anyone interested in the future of science education and the use of machine learning. The book offers a vision of how machine learning and science education research can work together to create more informed, effective, and engaging learning experiences for learners. As we move forward, we need to embrace ML and AI to change education and transform it for the better, making science learning more engaging, accessible, and impactful for students worldwide. Applying Machine Learning in Science Education Research presents a giant step in this direction.

May 2024   Joseph Krajcik
                                                                  Michigan State University
                                                                     East Lansing, USA

Contents

Acronyms

1D, 2D, 3D, ...	One-, two-, three- or Higher Dimensional Data
AGI	Artificial General Intelligence
AI	Artificial Intelligence
ANN(s)	Artificial Neural Network(s)
HDBSCAN	Hierarchical Density-Based Spatial Clustering of Applications with Noise
LLM(s)	Large Language Model(s)
ML	Machine Learning
MLP	Multi-layer Perceptron
NLP	Natural Language Processing
PCA	Principal Components Analysis
RVM	Relevance Vector Machine
SVD	Singular Value Decomposition
t-SNE	T-distributed Stochastic Neighbor Embedding
UMAP	Uniform Manifold Approximation and Projection for Dimension Reduction

Chapter 1
Introduction

Christina Krist, Marcus Kubsch, and Peter Wulff

Abstract This chapter introduces the purpose and goals of this book. It motivates why applying machine learning in science education could offer novel opportunities for data-driven modeling of learning processes and answering novel research questions. It also introduces the grand themes of this textbook: providing basics of machine learning and natural language processing, arguing for the importance of augmenting human analytic capabilities with what machine learning has to offer, and introducing the challenges that might occur when applying machine learning in science education.

1.1 Purpose and Goals of This Book

This book is meant to familiarize science education researchers with introductory machine learning (ML) principles and techniques in a way that allows you to apply them to research projects right away. It is written by science education scholars, for science education scholars.

We, the three authors, have each come to appreciate ML through different paths, and it takes up different spaces in our work and research agendas. Dr. Wulff has come to appreciate ML and AI as tools to (partly) tame the complexity of language data, extract patterns in it, and potentially use ML and AI to enhance pre-service science teacher education and science learning more generally. Dr. Kubsch is interested in the potential of ML as (a) a method that allows the integration of qualitative and quantitative data to better understand science learning and (b) the potential of ML to provide more adaptive learning at scale. For Dr. Krist, it has become an invaluable

C. Krist (✉)
Graduate School of Education, Stanford University, Stanford, CA, USA
e-mail: stinakrist@stanford.edu

M. Kubsch
Freie Universität Berlin, Berlin, Germany

P. Wulff
Heidelberg University of Education, Heidelberg, Baden-Württemberg, Germany

© The Author(s) 2025
P. Wulff et al. (eds.), *Applying Machine Learning in Science Education Research*,
Springer Texts in Education, https://doi.org/10.1007/978-3-031-74227-9_1

tool for augmenting (and at times restructuring) the qualitative research she does using classroom video and audio data–though she will surely caution you that using ML, or any AI-based tool, should actually make your research harder and more complicated rather than easier!

Our experiences have collectively shown us how ML could advance our research agendas in exciting and innovative ways. At the same time, we have noticed that people who have started to take up ML approaches in their work are either largely self-taught or come from a handful of specialized lab groups. In other words, there is a systematic lack of educational opportunity in science education research. This is both an access issue, and a potential ethics issue: when focused primarily on learning-by-doing, there is often not space for considering or developing a broader framework for how one should (or should not) go about integrating ML into science education research.

It is with both issues of access and ethics in mind that we write this book. In terms of access, we intend for this book to be well suited for science education researchers who are new to machine learning but curious about whether and how it might be used. Often we have found that this curiosity is driven by the fear that one is missing out on some "magic bullet" technique that can suddenly make one's research much easier. If this is you, we have some bad news. There is no magic bullet. But we hope that you will continue reading, and we think you will find a tool or application that will make you think about your research question in a new way, explore a new facet of your data, or propose a new set of questions altogether that you may not otherwise have considered.

In terms of ethics, we aim to provide sufficient technical background for conceptual understanding of the underlying algorithms at play (without getting too bogged down in the computational weeds); a broad framework for questions one should ask when considering when, how, and why to apply ML in your own science education research; and concrete examples that walk you through using various ML techniques along with us. We hope this background will equip to you be able to decide when NOT to use ML as expertly as you can decide when and how to use it in your work.

The concrete examples in this book take two forms: "toy models," or simplified examples that illustrate key concepts (Part I), and worked examples in which we provide you with both datasets and code to run yourself as you read (Part II). In both types of examples, we intentionally walk through code excerpts in the body of each chapter, using them as the basis to explain various mechanisms, concepts, and principles of ML in context. We encourage you to try and make sense of the code yourself (as tempting as it might be to just hit "run"!).

A related note of caution regarding the case studies and analyses presented in this book: they are motivated primarily by pedagogy. Thus, the case studies and analyses are explicitly not meant to answer a question or address a problem in the *best* possible way. If you are someone who already has some understanding of machine learning, or if you come back to this book some time after you worked through it, you may consider some analyses too simple, some issues not debated with enough mathematical rigor, or have questions about why we did not use advanced technique X,Y or Z. This is intentional! We have written this book for researchers that are just starting out with using machine learning to serve as an initial map of the realm of machine learning that empowers you to explore that realm–not to be the *entire* territory. Once you begin raising questions and critiques, we congratulate you. You have learned a lot.

To that end, this book is organized into three main parts. In Part I we outline theoretical underpinnings (i.e., the "basics") that are important to reflect upon when applying ML in one's research projects. In Chap. 2 we outline the basics for different forms of ML and what particular challenges researchers might encounter in their ML projects. In Chap. 3 we then motivate using ML for complex data analysis and show in what ways educational data can be considered complex. In Chaps. 4, 5, and 6 we present illustrative examples of how ML can be employed to analyze data and outline workflows for different use cases of ML. In Chap. 7 we introduce natural language processing and large language models which are crucial for processing language data. Part I ends with Chap. 8 on considerations for human-machine tandems (Sherin, 2013), and how ML might augment human analytical powers (or limit it).

In Part II we provide hands-on case studies that demonstrate how the ML approaches introduced in Part I can be realized in science education research projects with science education-specific data sets. In Chap. 9 we first introduce readers to how they can implement the case studies using open-source, freely available software tools. In Chap. 10, a case study is presented with the goal to automate annotation and coding of numerical research data. In Chap. 11 the important goal of pattern recognition for ML is applied to a science education research example. In Chap. 12 we show how annotation of textual data can be automated with ML. In Chaps. 13 and 14, we extend the goal of pattern recognition to unstructured language data. Finally, Chap. 15 utilizes computational tools in conjunction with qualitative methods to triangulate evidence in a science education research problem (measuring scientific uncertainty).

Finally, Part III outlines future perspectives for ML in science education. Chapter 16 features a critical examination of risks and ethical considerations for applying ML in science education research. Chapter 17 assembles ideas for how this rapidly evolving field of applying ML in science education might proceed. Concluding remarks on this textbook as well as its situatedness in the evolving field of applying ML in science education research are presented in Chap. 18.

1.2 Further Reading

As mentioned above, we will not mathematically derive any of the concepts, nor will we engage much with the technical details of the ML algorithms that are used throughout. We refer interested readers to these other resources:

- A classic introduction with advanced mathematics in deep learning is the book by Goodfellow et al. (2016).
- A more recent book called *Understanding Deep Learning* is highly recommended as it provides thorough mathematical as well as precise conceptual explanations

of relevant neural network architectures, focusing on ideas behind important architectures rather than proofs (Prince, 2023).

- Marsland (2015) provides a thorough introduction of many ML algorithms with Python code to implement the ML algorithms on a rather technical level.
- Similar to Marsland (2015), Bishop (2006) and Bishop and Bishop (2024) present in-depth surveys of ML (focus on deep learning) algorithms with Python code to implement the examples.
- There are other ML-related textbooks that introduce specific libraries such as `keras` (Chollet, 2018).
- ML as particularly applied in physics contexts is provided by Rauf (2021).
- Both an introduction to ML and advanced topics (alongside Python code) can be found in Murphy (2022) and Murphy (2023) as well as Murphy (2012). These books provide depth in terms of mathematical background to the concepts.

Readers who are interested in data science, data mining, and mathematical details of ML are referred to the following resources:

- Hastie et al. (2008) provide a reference introduction for statistical learning, which undergirds the success of ML.
- Brunton and Kutz (2019) relate ML to dynamical systems and introduce many underlying concepts for ML with mathematical rigor.
- Also, Kuhn and Johnson (2016) introduce predictive modeling where they outline and detail many important concepts in ML that we also reference in this textbook.
- Similarly, Nisbet et al. (2009) present a thorough survey of mathematical details for implementing data mining and statistical data analysis.

1.3 Why Apply Machine Learning in Science Education?

The current global vision for science education engaging learners in incrementally developing explanatory models for natural phenomena, drawing together core science ideas through participation in science practices (see for example the US *Next Generation Science Standards* (NGSS, 2013), the German *Bildungsstandards* (in physics) (KMK, 2020), or Korean science standards (Song et al., 2019)). Assessing science practices and competencies requires assessment formats where learners can use their knowledge to solve problems and display competencies. In addition, facilitating learning environments that support students' participation in scientific practices requires that teachers provide feedback that is responsive to student thinking. Providing such feedback is resource-intensive. It requires substantial resources from instructors in terms of both time and attention to a broad range of individuals and their specific needs. Compounding this challenge are the social and cultural dimensions of learning: building and sustaining classroom environments in which students interact with each other to develop ideas, and which recognize and value students' culturally diverse ways of knowing, requires both deep pedagogical flexibility and deep knowledge of students.

In the following sections, we describe how various ML tools could be well-suited to assisting scholars in studying the messy and complex learning processes in alignment with this global vision for science education.

Complex constructs and real-world learning processes

A key goal of science education research is to improve science teaching and learning across contexts, including both in-school and out-of-school settings (Abell and Lederman, 2007). However, processes of teaching and learning are complex and dynamic. They are complex because they are multidimensional and hierarchically determined. For example, in formal schooling, individuals are grouped into classrooms, which are embedded in schools, which are often part of larger districts, etc. These levels influence and shape each other, as districts or countries might have different curricula, and individual classrooms can develop social dynamics that largely impact learning and teaching processes.

Moreover, the multidimensionality of learning processes refers to the many factors that influence learning such as prior knowledge and prior experiences as well as other cognitive and motivational constructs such as intelligence or self-efficacy beliefs. In fact, most constructs in educational research are considered to be complex. For example, intelligence and competencies are multidimensional and require complex measurement instruments to validly assess them. Moreover, motivational constructs such as self-efficacy, expectancy of success, sense of social belonging, or task-related values are intricately influenced by situative and culturally focused factors. Also, teaching and learning processes are inherently time-bound, i.e., dynamic. For example, a teacher's curriculum implementation might be adaptively changed in the course of the school year based on the needs of specific learners. Also within learners, knowledge becomes refined as more or different connections between concepts are formed. Moreover, experience and deliberate practice enable effective chunking of knowledge that then changes the engagement with learning materials and requires sensitive measurement tools to be detected.

As an empirically-focused science, science education researchers strive to develop theories on teaching and learning processes in science based on empirical data, which encapsulates the complexity of the aforementioned constructs and processes to various degrees. Empirical data can enable science education researchers to understand, explain, and predict complex science-related teaching and learning processes and their outcomes. Science education researchers seek to gather empirical evidence through multiple sources of evidence. These sources of evidence include qualitative observations, surveys, interviews, digital sensor data, and content-based assessments. Patterns and structures in the collected data can then be used to inform instruction, and improve teaching and learning in science education.

Digitally-enhanced learning environments have added additional possibilities for gathering evidence on teaching and learning processes. For example, computer-based learning environments provide not only information about students' learning products such as their answers to an open ended question but also information about students' learning processes, e.g., which elements in the environment students interact with and for how long. In addition, other modalities of data such as video, audio, or sensor data become increasingly easy to gather and can provide new insights into teaching and learning. Also, digitally-enhanced learning environments can provide support to learners, e.g., providing just-in-time and adaptive guidance for solving problems. In sum, digitally-enhanced learning environments can provide science education researchers with increasingly big and diverse data sets to further understanding and predicting complex science-related teaching and learning processes (we will bracket out privacy and ethical issues that might come up in these environments for the moment).

However, making sense of these multiple types and amounts of empirical data (e.g., extracting patterns and structures in it, efficiently utilizing large amounts of data, or enhancing teaching and learning through extracted patterns) can be daunting, given the size, complexity, and limited resources available. Sophisticated data-driven discovery tools and approaches can often help to make sense of these rich and complex data sets. This calls for science education researchers to rethink, reconsider, and expand their methodological toolset.

Artificial Intelligence and Machine Learning

Emerging technologies and methods in the realm of artificial intelligence (AI) can provide novel means to enhance science education research through guiding researchers in making sense of complex data sets, and implement reliable and valid analytics for teaching and learning process analytics. Technologies and methods in the realm of AI can relate to speech-to-text applications, automated translation, generation of text and images with conversational AI, or complex data analysis. Applications of AI infuse most areas of business, industry, and research. In one form or another almost everyone today uses AI applications in professional or non-professional contexts such as speech-to-text translation, machine translation, movie or song rankings, image classification, object detection, mastering games, proving mathematical theorems, painting pictures, composing music, or solving science problems. And as with most new technologies, applicability in various fields has to be critically and reflectively examined because many challenges and pitfalls (both pragmatically and fundamentally) await.

"Most of what is now classified as AI is really a case of machine learning" (Krauss, 2023, p. 223). Machine learning (ML) is a form of inductive learning to enable computer programs certain capabilities (e.g., pattern identification or classification). As such, ML can be considered a go-to method (sometimes referred to as a workhorse of AI) to process and analyze complex data, and it enables researchers to perform data-driven discovery. As many scientific disciplines increasingly rely on data-driven discovery, it has been described as a fourth paradigm of scientific discovery (see Box in Sect. 1.3).

ML algorithms inductively solve problems based on various amounts of data. For example, ML can be used to extract patterns from complex unstructured data sets that can be "large, noisy, and messy" (Nisbet et al., 2009, p. 17) such as natural language data, or train computers to automatically solve well-defined or ill-defined tasks like single-digit addition (see Chap. 4), or image classification. With the advent of large language models and foundation models many novel generative capabilities such as textual summarization, image creation from text, or all sorts of writing tasks became targets for AI. AI tools powered by ML also enter(ed) many educational software applications such as automated feedback for students.

Science paradigms

Computer scientist Jim Gray outlined a broad-brush, high-level picture of scientific inquiry capturing broad paradigms of discovery. For thousands of years, according to Gray, mostly natural phenomena were described (Hey et al., 2009). This empirical period was then extended for the last few hundred years with a theoretical modeling branch, where models and generalizations were produced (Hey et al., 2009). Examples might include Johannes Kepler's derivation of laws of planetary motion, Isaac Newton's mathematical description of classical mechanics, or James Clarke Maxwell's derivation of fundamental equations governing electromagnetism. These equations and models grew too complicated for many problems to be solved analytically, hence, people started to simulate processes either digitally or analogically(!). Ada Lovelace can be considered the first person to anticipate the importance of simulations (Pontzen, 2023). Simulations then became an important approach for science. Finally, Jim Gray proposes a new, fourth, paradigm called eScience, where IT meets scientism and theory, experiment, and simulations can be unified: "The techniques and technologies for such data-intensive science are so different that it is worth distinguishing data-intensive science from computational science as a new, fourth paradigm for scientific exploration" (Hey et al., 2009, p. xix). While in the fourth paradigm discovery is driven by data science and machine learning, already a fifth paradigm has been proposed "where cognitive systems seamlessly integrate information from human experts, experimental data, physics-based models, and data-driven models to speed discovery" (Zubarev and Pitera, 2019, p. 103).

1.4   Current Applications and Challenges of Machine Learning in Science Education Research

In science education scholarship, many researchers already use AI-based methods such as ML to help them answer their research questions. A vast variety of different ML approaches have been utilized and the recent (i.e., approximately since 2022) evolution of generative large language models such as GPT-4 provides opportunities for science education researchers to answer novel research questions and potentially enhance instruction. Research questions can be qualitative, or quantitative, as well as confirmatory, or exploratory. For all alternatives, ML offers the potential to leverage larger data sets and find patterns in a rather systematic way. It also poses novel challenges for each. Let us dive into some of the applications.

Among the early implementations was Wang et al. (2008) who utilized ML to evaluate problem solving with open-ended questions. They found high agreement of human ratings and machine scores for the tasks and concluded that ML could support the use of open-ended questions for assessment. Following this research, science education researchers have used ML to automate assessments (Donnelly et al., 2016; Zhai et al., 2020, 2022; Wulff et al., 2021; Nehm and Ha, 2011; Graesser et al., 2004).

In contrast to these confirmatory research goals, others have used ML to explore patterns in large data, including shifts in those patterns over time (Odden et al., 2019; Sherin, 2013). Odden et al. (2020) utilized a pattern-seeking approach to extract structures (time trends) from conference abstracts of the Physics Education Research Conference (PERC) between the years 2001 and 2018 in order to analyze waxing and waning topics in science education. Kubsch et al. (2023) used a similar approach to investigate what were the main topics covered in the journal "Unterrichtswissenschaft" (a German journal covering instructional science) and how they changed over time. Still others have begun to leverage ML to develop more accurate qualitative rubrics for assessing complex constructs such as scientific argumentation, explanations, or reflections (Tschisgale et al., 2023; Martin et al., 2023; Wulff et al., 2022). While these applications mostly relate to textual data, also image data was analyzed using ML. Zhai et al. successfully analyzed students' drawings with ML methods (Zhai et al., 2022). Moreover, Rosenberg and Krist (2020) combined confirmatory and exploratory approaches and used ML for pattern-seeking in an assessment context, leveraging ML to develop more accurate qualitative rubrics for scoring, and a classification model that can then be used to automatically classify unseen responses.

In most of these studies, it is also apparent that novel challenges have to be addressed when applying ML in science education research. For example, validation of an ML model is an intricate process where substantial human expertise and involvement is required. We bring in a cautionary tale from a study utilizing ML for image recognition in medicine. This highly cited study sought to identify COVID-19 cases via chest X-rays (a task human raters are not able to perform reliably). The ML algorithm performed well above chance. Some years later, another group re-analyzed the same data and found that the algorithm also performed well above chance even when trained on parts of the images that did not include any sections of the chest in them! This highlights that applying ML in research can be tricky: algorithms might "learn" patterns in the data that you are not aware of, and researchers need to critically examine what pitfalls might occur if, for example, the basis for detecting COVID-19 is something other than the clinically-determined radiological markers for COVID-19. There are several valuable insights from this story, including that data sets should be intentionally curated and that both data sets and details on the ML algorithms used should be open sourced; and that scholars should LOOK at their datasets and data relevant to results with the assumption that the ML algorithm is doing something other than they intended.

Another set of challenges is methodological. For instance, assessing model validity is often not a straightforward task. Liu et al. (2016) point out the importance of evaluating whether a given ML model provides valid scores for relevant subgroups, e.g., gender. Additionally, a more general challenge with inductive learning (extracting patterns from examples) is the fact that it relies upon existing (training) data sets. These training data sets are almost certainly biased in some way; ML algorithms can propagate (and sometimes amplify) these biases. Carefully evaluating which biases matter and in what ways they might impact scientific claims for a given research question is an essential task for researchers. This is particularly difficult in ML research, because it is not always clear how the complex ML algorithms reach their decisions. Even the supposed versatile large language models exhibit many flaws after closer inspection. They might hallucinate[1] knowledge even in innocuous tasks such as textual summarization. Ecological, ethical, and epistemological issues related to large language models have to be critically examined by researchers when using these tools for research, which then might impact practice.

Finally, there are issues of technical access and ethics. Tools for performing ML-based analyses might be proprietary and closed-source, and thus not accessible to anyone without sufficient amounts of money (or other resources such as computer clusters) available. In addition, the processing power required to run ML algorithms comes at a significant environmental and human labor cost, utilizing large amounts of fresh water and often (at least partly) relying on fossil fuels.

1.5 Conclusion

Applying ML in science education research can enhance the formative assessment of competencies in knowledge-in-use assessments, and help automate assessment which can facilitate scaling high-quality learning opportunities. We expect ML and AI more generally to continue to push the frontiers of learning environment design and analysis of complex educational data sets. However, if applied under wrong pretenses, in unsuitable circumstances, or without sufficient thoughtfulness and care, learning and teaching in science education might suffer.

In light of both the potential benefits and challenges, this book aims to equip science education researchers with the background knowledge, concrete tools, and technical acumen needed to effectively and responsibly apply ML to their own research projects. Importantly, we contend that the application of existing ML methods and tools is possible without a deep background in statistics and computer science. We aim to maintain the global vision for science teaching and learning described above as we explore applications of ML to support teaching and learning throughout this book, resisting both purely didactic forms of instruction and purely individualized

---

[1] This refers to making up false information by composing various knowledge fragments.

ones (Biesta, 2016)—each insufficient in some way for achieving the vision for science teaching and learning.

The remainder of this book will give interested readers (you!) an accessible introduction on why, when, and how to apply ML in your (science) education projects, enable you to conduct ML-based analyses, and provide you with a conceptual compass that empowers you to navigate the ML landscape when learning about new methods, evaluating ML methods used by others, and devising your own ML applications. We hope you enjoy the journey!

References

Abell, S. K., & Lederman, N. G. (2007). Preface. In S. K. Abell & N. Lederman (Eds.), *Handbook of research on science education*. Mawhah, New Jersey: Lawrence Erlbaum Associates Publishers.

Biesta, G. (2016). Ict and education beyond learning: A framework for analysis, development and critique. In E. Elstad (Ed.), *Digital expectations and experiences in education* (pp. 29–43). Dordrecht: Sense Publishers.

Bishop, C. M. (2006). *Pattern recognition and machine learning*. Information science and statistics. New York, NY: Springer Science+Business Media LLC.

Bishop, C. M., & Bishop, H. (2024). *Deep learning: Foundations and concepts*. Cham: Springer.

Brunton, S. L., & Kutz, J. N. (2019). *Data-driven science and engineering*. Cambridge University Press.

Chollet, F. (2018). *Deep learning with Python*. Safari Tech Books Online Manning, Shelter Island, NY.

Donnelly, P. J., Blanchard, N., Samei, B., Olney, A. M., Sun, X., Ward, B., Kelly, S., Nystran, M., & D'Mello, S. K. (2016). Automatic teacher modeling from live classroom audio. In *Umap '16: Proceedings of the 2016 conference on user modeling adaptation and personalization* (pp. 45–53).

Goodfellow, I., Bengio, Y., & Courville, A. (2016). *Deep learning*. Cambridge, Massachusetts and London, England: MIT Press.

Graesser, A. C., McNamara, D., Louwerse, M. M., & Cai, Z. (2004). Coh-metrix: Analysis of text on cohesion and language. *Behavior Research Methods, Instruments, & Computers, 36*(2), 193–202.

Hastie, T., Tibshirani, R., & Friedman, J. (2008). *The elements of statistical learning: Data mining, inference, and prediction*. Springer.

Hey, T., Tansley, S., Tolle, K., & Gray, J. (2009). *The fourth paradigm: Data-intensive scientific discovery*. Microsoft Research.

KMK. (2020). Bildungsstandards im fach physik für die allgemeine hochschulreife: Beschluss der kultusministerkonferenz vom 18.06.2020.

Krauss, L. M. (2023). *The known unknowns: The unsolved mysteries of the cosmos*. London: Head of Zeus.

Kubsch, M., Sorge, S., & Wulff, P. (2023). Emotionen beim reflektieren in der lehrkräftebildung. In L. Mientus, C. Klempin, & A. Nowak (Eds.), *Reflexion in der Lehrkräftebildung: Empirisch – Phasenübergreifend – Interdisziplinär* (pp. 261–270). Potsdam: Universitätsverlag Potsdam.

Kuhn, M., & Johnson, K. (2016). *Applied predictive modeling*. New York: Springer. Corrected at 5th printing edition.

Liu, O. L., Rios, J. A., Heilman, M., Gerard, L., & Linn, M. C. (2016). Validation of automated scoring of science assessments. *Journal of Research in Science Teaching, 53*(2), 215–233.

Marsland, S. (2015). *Machine learning: An algorithmic perspective.* Chapman & Hall/CRC machine learning & pattern recognition series. CRC Press, Boca Raton, FL, second edition edition.

Martin, P. P., Kranz, D., Wulff, P., & Graulich, N. (2023). Exploring new depths: Applying machine learning for the analysis of student argumentation in chemistry. *Journal of Research in Science Teaching*, 1–36.

Murphy, K. P. (2012). *Machine learning: A probabilistic perspective.* Adaptive computation and machine learning series. Cambridge MA: MIT Press.

Murphy, K. P. (2022). *Probabilistic machine learning: An introduction.* MIT Press.

Murphy, K. P. (2023). *Probabilistic machine learning: Advanced topics.* MIT Press.

Nehm, R. H., & Ha, M. (2011). Item feature effects in evolution assessment. *Journal of Research in Science Teaching, 48*(3), 237–256.

NGSS. (2013). *Next generation science standards: For states, by states.* Washington: National Academies Press.

Nisbet, R., Elder, J. F., & Miner, G. (2009). *Handbook of statistical analysis and data mining applications.* Amsterdam and Boston: Academic Press/Elsevier.

Odden, T. O. B., Lockwood, E., & Caballero, M. D. (2019). Physics computational literacy: An exploratory case study using computational essays. *Physical Review Physics Education Research, 15*(2).

Odden, T. O. B., Marin, A., & Caballero, M. D. (2020). Thematic analysis of 18 years of physics education research conference proceedings using natural language processing. *Physical Review Physics Education Research, 16*(1), 1–25.

Pontzen, A. (2023). *The universe in a box: Simulations and the quest to code the cosmos.* New York: Penguin Publishing Group.

Prince, S. J. D. (2023). *Understanding deep learning.* MIT Press.

Rauf, I. A. (2021). *Physics of data science and machine learning.* Boca Raton: CRC Press.

Rosenberg, J. M., & Krist, C. (2020). Combining machine learning and qualitative methods to elaborate students' ideas about the generality of their model-based explanations. *Journal of Science Education and Technology.*

Sherin, B. (2013). A computational study of commonsense science: An exploration in the automated analysis of clinical interview data. *Journal of the Learning Sciences, 22*(4), 600–638.

Song, J., Kang, S.-J., Kwak, Y., Kim, D., Kim, S., Na, J., Do, J.-H., Min, B.-G., Park, S. C., Bae, S.-M., Son, Y.-A., Son, J. W., Oh, P. S., Lee, J.-K., Lee, H. J., Ihm, H., Jeong, D. H., Jung, J. H., Kim, J., & Joung, Y. J. (2019). Contents and features of 'korean science education standards (kses)' for the next generation. *Journal of the Korean Association for Science Education, 39*(3), 465–478.

Tschisgale, P., Wulff, P., & Kubsch, M. (2023). Integrating artificial intelligence-based methods into qualitative research in physics education research: A case for computational grounded theory. *Physical Review Physics Education Research, 19*(020123), 1–24.

Wang, H.-C., Chang, C.-Y., & Li, T.-Y. (2008). Assessing creative problem-solving with automated text grading. *Computers & Education, 51*(4), 1450–1466.

Wulff, P., Mientus, L., Nowak, A., & Borowski, A. (2021). Stärkung praxisorientierter hochschullehre durch computerbasierte rückmeldung zu reflexionstexten. *die hochschullehre, 11.*

Wulff, P., Buschhüter, D., Westphal, A., Mientus, L., Nowak, A., & Borowski, A. (2022). Bridging the gap between qualitative and quantitative assessment in science education research with machine learning – a case for pretrained language models-based clustering. *Journal of Science Education and Technology, 31*, 490–513.

Zhai, X., Haudek, K. C., & Ma, W. (2022). Assessing argumentation using machine learning and cognitive diagnostic modeling. *Research in Science Education.*

Zhai, X., Haudek, K., Shi, L., Nehm, R., & Urban-Lurain, M. (2020). From substitution to redefinition: A framework of machine learning-based science assessment. *Journal of Research in Science Teaching, 57*(9), 1430–1459.

Zubarev, D. Y., & Pitera, J. W. (2019). Cognitive materials discovery and onset of the 5th discovery paradigm. ACS symposium series. In E. O. Pyzer-Knapp & T. Laino (Eds.), *Machine learning in chemistry* (Vol. 1326, pp. 103–120). Washington, DC: American Chemical Society.

Part I
Theoretical Background

Chapter 2
Basics of Machine Learning

Peter Wulff, Marcus Kubsch, and Christina Krist

Abstract This chapter presents a historical brief of artificial intelligence and machine learning as well as an overview of conceptual basics of how ML works, alongside examples. Different approaches to ML are reviewed and the challenges of applying ML in research are addressed.

2.1 The Inception of Artificial Intelligence and Machine Learning

Today the term 'artificial intelligence' is regularly used in public parlance and can be found in standard repositories such as the Oxford dictionary or the DUDEN (in Germany). The terms 'artificial intelligence' and 'machine learning' were only coined as recently as the fifties and sixties, primarily in the US. In this chapter, we seek to provide background knowledge on where the historical roots of these terms can be found, how machines are enabled to learn, and what pitfalls occur when applying artificial intelligence and machine learning. If you are more interested in applying ML in science education research, you can skip the following chapters to part II of this book.

Multi-disciplinary origins of AI and ML

Many disciplines have contributed valuable analytical tools and theoretical arguments that have enabled and justified the study of AI and ML. Early (natural) philosophy engaged with questions about the nature of algorithms and methods used

P. Wulff (✉)
Heidelberg University of Education, Heidelberg, Baden-Württemberg, Germany
e-mail: peter.wulff@ph-heidelberg.de

M. Kubsch
Freie Universität Berlin, Berlin, Germany

C. Krist
Graduate School of Education, Stanford University, Stanford, CA, USA

© The Author(s) 2025
P. Wulff et al. (eds.), *Applying Machine Learning in Science Education Research*,
Springer Texts in Education, https://doi.org/10.1007/978-3-031-74227-9_2

15

to discover them. Moreover, the limits of computability and rule extraction from data were outlined. Another important contribution from mathematics came from the field of statistics, where quantifying possible outcomes of events and degrees of subjective belief helped address uncertain situations and intractable problems. Cognitive psychology contributed to the understanding of human actions, among others, by stressing the importance of information processing and transformation of internal (mental) representations. Computer engineering enabled the building of digital electronic computers as promising devices to mimic intelligent behaviors and reasoning, and computer engineers developed important hardware and software that enable large-scale implementation of AI.[1] In linguistics, theories that could explain the creative and productive use of language were developed, and, with insights from pragmatics, it became increasingly appreciated that language and meaning are largely ambiguous, context-dependent, and grounded in usage. Above all, language serves as an important medium of knowledge representation and communication, however, much is left unsaid either in spoken and written communication. Systematically processing language, analyzing it, and even generating human-like language became a testbed for both human and computer (artificial) intelligence.

Coining of artificial intelligence and machine learning

The first recognized work on AI was the development of an artificial neural network (ANN) in 1943 drawing on physiological insights, propositional logic, and the theory of computation. It was discovered that an updating rule for the connection strengths between artificial neurons enabled the ANN to learn, e.g., a dividing hyperplane for multi-dimensional data. This work led to the development of early chess programs and theorem provers (e.g., "Logic Theorist"), but it also encountered significant skepticism. Fundamental questions arose related to its place within science and its potential contributions to solving practical, real-world problems. The Dartmouth workshop in 1956 is recognized as the beginning event of the field "artificial intelligence." Two traditions were often distinguished: logicist and connectionist. Where programs such as "Logic Theorist" (and later the "General Problem Solver") fell into the logicist (symbolist) group, the ANNs were categorized within the connectionist group.[2]

In the 1980s, the logicist group pursued the goal of engineering relevant knowledge into machines (i.e., instructing machines) that could then assist humans. Projects such as Cyc tried to represent all knowledge in machine code and thus enable machines to reason. These efforts, however, did not really pan out. The reasoning capabilities of Cyc were found to be rather shallow and brittle. Eventually, researchers had to

---

[1] Artificial intelligence is widely recognized as an unsuitable term for the phenomena and issues under study. Valiant (2024) proposed the concept of educability, which would eventually better suit the issues at stake. Later on, we will also highlight the misuse of cognitive language in the field of AI and beyond related to computer programs.

[2] As with many dichotomies, there is much gray area and overlap between the two groups, partly because it is not unambiguously agreed upon what constitutes as a symbol and other complexities.

embrace data mining to scrape the web and incorporate knowledge into Cyc. However, logistic programs achieved some fame. The rule-based conversational program called ELIZA simulated (in a very restricted sense) a psychotherapist. It basically responded to statements with generic questions, a process which does not require world-knowledge. Providing this program with knowledge was found to be an intricate problem (we will revisit this example in a background box). Engineering knowledge explicitly into machines eventually requires a large amount of specification that is hardly feasible with finite resources. Moreover, specifying the knowledge that goes into classifying a cat on an image is nearly impossible.

The connectionist group was successful when conceptual basics, hardware, and software were available to update parameters (weights) in an ANN in a way that enabled the ANN to classify previously unseen samples. Different approaches to perform the training were called machine learning (ML). ML was first applied when a computer (a.k.a., a machine) not only played the game of checkers, but also improved upon its gameplay. A novel algorithmic approach was used to optimize this computer-based player. Traditional approaches provided indexed lookup tables that identify correct moves for each possible board configuration. However, for checkers, an indexed lookup table is both ineffective and inefficient: the number of possible board configurations in checkers is astronomical. Instead, an early version of reinforcement learning was used to enable the computer to use an evaluation function while playing. This allowed the computer to improve (i.e., learn) over time.

Moving goalposts?

From the 1960s onward, ML algorithms primarily solved specific problems in well-defined environments. Some researchers (especially outside the field of AI) remained skeptical regarding AI in general. They adopted the belief that "computers could never do X," where X represented any conceivable task. A related phenomenon in the history of public perception of AI and ML research is termed the "AI effect". The AI effect refers to the tendency for the public to perceive that once AI and ML methods can solve a complex problem, such as recognizing hand-written digits, AI is no longer seen as part of the solution: "Every time we figure out a piece of it [a problem deemed intractable by ML], it stops being magical; we say, 'Oh, that's just a computation.'"[3]

ML and games

ML researchers demonstrated that ML algorithms could in fact be used to solve many of these Xs. In addition to novel theorem-provers (e.g., for geometry), performance in games (e.g., Atari games, checkers, Othello, Go, chess, StarCraft, Jeopardy) played an important role in raising awareness of ML and its capabilities. Ever since checkers, ML has outperformed human players in many other games. This theme also found its way into 1980s movies such as WarGames, where computers were programmed

---

[3] https://www.wired.com/2002/03/everywhere/, last access: Nov 2023.

to learn from errors in games. In the movie, the machine WOPR learned through self-play the most important lesson about global thermonuclear war: not to play, since there cannot be a winner. Notably, the IBM chess program Deep Blue defeated a world chess champion in 1996 (though it was not until 1997, with improved program size, that Deep Blue won an entire competition). Deep Blue learned certain chess parameters by systematically analyzing chess books and past games. In 2011, IBM developed Watson, which successfully defeated a Jeopardy champion. Watson is an AI-based program designed to parse natural language, retrieve information, and provide answers to questions. In 2016, the AI company DeepMind trained a machine called AlphaGo to outperform the Go world champion. In game 2 on move 37, the system shocked the Go-community with a seemingly nonsensical move that later proved to be essential fifty moves later. In 2019, AlphaStar defeated professional StarCraft players, now world champions, in all games. AlphaStar was restricted to human-like click-rates, however, it could see the entire map (i.e., the virtual playing ground) simultaneously, which is not possible for human players. Expert commentators dubbed AlphaStar's gameplay as phenomenal and superhuman, with particular advantages in micromanagement (i.e., controlling individual units) and multitasking.

Games have always been an important testbed and popularizing engine for AI and ML algorithms: First, the general public could more easily appreciate performance in games compared to, say, proving mathematical theorems. Second, games require substantial expertise and relate in different ways to real-word human actions. For example, many games, such as StarCraft, require rapid decision-making in complex, uncertain environments. Impressive as these successes in various games might be, these systems also fail glaringly in situations they have not encountered properly during their training. The research firm FAR AI developed a program to systematically track weaknesses in the Go program KataGo, enabling humans to exploit a specific (fairly simple) strategy that helped them win 14 of the 15 games against the AI. Similarly, unusual actions by human players in StarCraft could also trick AlphaStar, revealing that some learned strategies are rather brittle (Vincent, 2019). As such, these ML programs like Watson, AlphaGo, and others also point to important (and fundamental) limitations associated with data-driven discovery and inductive inference, which we will introduce more thoroughly.

ML, computer vision and natural language processing

The connectionist tradition in ML, based on ANNs, experienced a difficult time after its early successes. Computing resources were rather scarce, and training deep artificial neural networks did not really solve any relevant problems that garnered public interest and investment. This led to an "AI winter," where funding was scarce, and working on artificial neural networks was not a field where researchers could thrive. The AI winter ended in 2006 when it was demonstrated that an ANN could be trained to recognize hand-written digits with state-of-the-art precision, a feat previously considered impossible. Essentially, Moravec's paradox states that tasks that are simple (i.e., deeply rooted, sensorimotor) for humans (e.g., recognizing hand-written digits) are particularly difficult for machines, whereas more involved reasoning tasks (e.g., chess) are comparably easy. Hence, recognizing hand-written digits was a genuine breakthrough. The term "deep learning" was coined in this context and the interest of the scientific community in ML rose. It was also recognized that deep learning applications could improve performance for many language-related problems, such as language translation, making deep learning architectures the de-facto standard for many language processing tasks.

The development of large language models (LLMs) was pivotal in language and image processing. Researchers leveraged large data sets such as the Common Crawl of the Internet, book corpora, or Wikipedia, which are (for the most part) publicly available sources of natural language data (note that privacy issues and proprietary questions are at times still unresolved). Researchers trained LLMs based on these large data repositories. As such, LLMs are potential sources of collective opinion, knowledge and ideas, reflecting the fact that they systematically incorporate and combine different sources of knowledge-related data such as Wikipedia (impossible for a human being). Once LLMs have been trained on these data sets they could then be fine-tuned and used in specific tasks such as classifying student responses into predefined categories. Even more so, LLMs could accurately solve tasks without specific training, and with the advent of generative, conversational LLMs, these capabilities could be utilized on a larger scale. Generative LLMs have already sparked significant research activity across various disciplines, including science education, where high performance in certain tasks is often coupled with flaws in other (domain-specific) tasks. While some flaws have diminished with advances in the underlying LLMs, others that relate to biases in training data, explaining model decisions, or reliably presenting correct information seem much more difficult to overcome. Some researchers argue that LLMs are essentially "stochastic parrots." A concern is that LLMs "only" interpolate (or mimic) the training data and regurgitate patterns that they have seen during training. While this may be true to some extent, researchers also showed that these LLMs can generalize beyond the seen training examples, so that to some degree the models are more versatile than presenting efficient storage of humongous amounts of data. This also references the grand questions to what extent LLMs truly understand the world and form world models that they can reason with. We will engage with this question later on, in Chap. 7.

ML and scientific research

AI and ML have also found entrance into the natural sciences, especially in disciplines that tackle particularly complex problems such as biology. An often cited example of ML's success is protein folding, which exemplifies how ML can achieve remarkable capabilities. The problem, in essence, is to predict the 3D protein structure based on an amino acid input sequence, i.e., a 1D sequence of amino acids. Experimentally determining the protein structure is time and cost intensive, and simulating the structure only achieved modest accuracy. ML was found to be an effective and efficient means to help tackle this problem. DeepMind's deep-learning-based model AlphaFold learned to accurately predict 3D protein structure based on an amino acid input sequence (protein folding) by being given sample amino acid sequences with the respective, experimentally determined 3D structures as the training data. At this time, a set of about 100,000 3D structures (of the billions of possible structures) was experimentally determined. AlphaFold outperformed all computer competitors in predicting 3D folding of novel (i.e., unseen to the model) amino acid sequences with nearly experimental accuracy—doing so in a fraction of the time needed by traditional methods. The 2024 Nobel Prize in Chemistry went in fact to AlphaFold's leading researchers. Similar ML models helped to develop COVID-19 drugs and vaccines. AlphaFold 3 is allegedly capable of "predicting the structure of proteins, DNA, RNA, ligands and more, and how they interact," eventually transforming "our understanding of the biological world and drug discovery."[4] However, this does not exempt researchers from performing actual laboratory experiments to corroborate their understanding.

ML also excelled in other tasks where it is deemed intricately difficult to outline a reasonable set of instructions for a machine to solve a problem. For example, DeepMind used ML to improve weather forecasting in 2023. They motivate their ML approach, deep learning in particular, as follows:

> Forecasts typically rely on Numerical Weather Prediction [..], which begins with carefully defined physics equations, which are then translated into computer algorithms run on super-computers. While this traditional approach has been a triumph of science and engineering, designing the equations and algorithms is time-consuming and requires deep expertise, as well as costly compute resources to make accurate predictions.

> Deep learning offers a different approach: using data instead of physical equations to create a weather forecast system. GraphCast is trained on decades of historical weather data to learn a model of the cause and effect relationships that govern how Earth's weather evolves, from the present into the future.[5]

Moreover, researchers showed that magnetic confinement in a nuclear fusion reactor could be enhanced with the help of ML. As such, ML might complement first-principles based research approaches and present a novel form of validating relationships and models that researchers build from the real world.

The extent to which AI can expand true scientific understanding is highly contested (as well as the question what true scientific understanding actually is). Some claimed that AI cannot surpass pure numerics ("To predict is not to explain."). While this is an extreme position, barely seen in the literature, a more productive outline was presented by Krenn et al. (2022, p. 761): "First, AI can act as an instrument revealing properties of a physical system that are otherwise difficult or even impossible to probe. Humans then lift these insights to scientific understanding. Second, AI can act as a source of inspiration for new concepts and ideas that are subsequently understood and generalized by human scientists. Third, AI acts as an agent of understanding. AI reaches new scientific insight and — importantly — can transfer it to human researchers." We notice the crucial links to humans in this outline, and will emphasize

---

[4] See: https://blog.google/technology/ai/google-deepmind-isomorphic-alphafold-3-ai-model/, last access: May 2024.

[5] See: https://deepmind.google/discover/blog/graphcast-ai-model-for-faster-and-more-accurate-global-weather-forecasting/, last access: Nov 2023.

these links for science education research as well, e.g., in Chap. 8. The kinds of insights that AI will provide and the fields in which it will be useful will probably be determined post facto as it is rather challenging to anticipate specific capabilities of AI and ML as outlined above.

2.2 ML as a Data-driven Discovery Procedure

Definition of ML

Let us seek a better understanding of how ML acquires the capabilities reviewed above. As outlined, ML is constituted by truly integrating many different fields of study, drawing from statistics, artificial intelligence, philosophy, information theory, biology, cognitive science, computational complexity, and control theory. ML refers to computer programs (software); however, it has nowadays become intricately linked to hardware such as accelerating computations with specialized processor design, or moving to quantum computing to perform computations. Given the above examples, ML refers to performing well in problems or tasks such as classifying hand-written digits, categorizing images, summarizing and producing text, or playing games. To acquire the capabilities to perform well in these tasks, data related to the tasks is typically necessary as inputs to the machine. Moreover, traditional paradigms of simulation and modeling where primitives, laws, or rules are provided as instruction to transform inputs are transcended and complemented. ML departs from providing the machine instructions (rules) with inputs (data) to providing input-output (data-answer) samples.

A widely recognized (high-level) definition of ML reads: "[Machine learning is the] field of study that gives computers the ability to learn without being explicitly programmed" (as cited in Géron (2017)). A more concrete definition is:

Definition 2.1 A computer program is said to learn from experience E with respect to some classes of task T and performance measure P if its performance can improve with E on T measured by P. (Mitchell, 1997, p. 2).

Experience refers to a set of training data, and classes of tasks can be as diverse as the classification or clustering of the data. As such, ML is in essence a form of inductive learning: Patterns are directly learned from the given training data. This highlights the importance of readily available data sets of appropriate size to cover a reasonable amount of cases to generalize to unseen cases. This also poses novel challenges for researchers to posit what kind of generalizability they expect and in what ways the available data is suitable to cover relevant cases. Generalizing to unseen examples is associated with an empirical risk, because mostly not all cases have been observed. Especially in empirical research contexts it is literally not possible to observe all cases. Say, you want to train an ML model to predict students' exam performance. This raises the question of what the target population is. Even though you could sample all of them at one time, students with different characteristics (e.g., prior knowledge) might enter your course next year which might also be part of the target population. Given the diversity of learners, it is impossible to sample all relevant cases from the target population. As such, the prediction model is inherently associated with an empirical risk of being built on features that might not account for all possible cases and thus misclassify some students. As an inductive form of inference, ML is implicitly guided by the following assumption, called the inductive learning hypothesis:

> Any hypothesis found to approximate the target function well over a sufficiently large set of training examples will also approximate the target function well over other unobserved examples" (Mitchell, 1997, p. 23).

When applying ML to a research problem, this is an important assumption to keep in mind. However, while empirical risk in this sense was long associated exclusively with inductive inference, similar concerns were outlined for deductive inference. Researchers noted that deductive inference is associated with risk, because "His argument [Popper's argument against induction, i.e., that not all cases onto which should be generalized can be known; authors] can also be used to make deduction useless for it, too, is based on an incomplete set of known facts. Even if the identified fact resembles the members of the set, how can we be sure that every possible feature of either the unknown or the members of the set itself has been considered?" (Rothchild, 2006, p. 3). Say, you extracted the features and rules on how to predict exam scores from theory and prior literature, this will not exempt you from the risk of missing important features and misclassifying students that might be particularly consequential in educational (high-stakes) settings. Ideally, both approaches should be utilized, and ML excels in the inductive extraction of patterns from data. As such, it features well to extend the methodological toolkit in any data-intensive scientific research.

Growth of ML and data-driven discovery

Inductive extraction of patterns from data is a particularly important means for researchers to gain insights into their objects (phenomena, processes, etc.) under scientific investigation. It was particularly important for the growth of ML that limitations with regards to human and/or computational processing power as well as limited general-purpose algorithms to find patterns in complex data were overcome. Technological advances in computing power, storage capabilities, reducing cost of training AI systems, massive capital investment, and algorithmic advances have boosted AI and ML, and data-driven discovery processes have become more feasible. Coupled with advances in cloud computing, benchmark data sets, and big data, ML has tremendously expanded the possibilities for modeling, predicting, and understanding complex systems. A key advantage of ML is that considerably unassuming (e.g., regarding specific content areas) algorithms perform well across many different classes of rather complex problems. For example, dimensionality reduction such as principal components analysis (PCA) can be used to find a common factor,

g, for intelligence-related tasks, explore phase transitions in physical systems (see Chap. 5), or probe measurement instruments for unidimensionality.

ML models have been found to generalize well in daunting problems where the number of model parameters (especially for ANNs) far outweighs the number of training examples. For example, with only a couple million images in the ImageNet competition artificial neural networks with 60 million parameters are trained successfully to recognize objects on some images. In addition to empirical success, theoretical investigations on the capabilities of ML algorithms have buttressed their capabilities to model and predict the behaviors of complex systems and thus facilitate understanding them. For example, with appropriate size, ANNs can already approximate functional relationships of interest which is stated by the universal approximation theorem. The extent to where data-driven discovery and knowledge extraction can ultimately go remains an interesting question. We will also encounter the no-free-lunch theorem, which restricts the possibility of extracting knowledge from mere data without assumptions. It was shown that any ML algorithm is as effective as any other when the generalization performance is evaluated over all possible data-generating distributions and tasks. Fortunately, in real-world settings, reasonable assumptions on algorithms and features already exist, allowing us to circumvent this sobering conclusion.

ML algorithm versus ML model

Differentiating ML from other forms of statistical modeling is not always straightforward (for a discussion of the two cultures in statistics, see Box). We chose to adopt the following rule of thumb: Whenever parameters, as in the case of PCA a lower dimensional transformation of the data or weights in a neural network, are learned from data, we refer to this as ML. We further differentiate ML algorithms (linear regression, ANN, decision tree, or k-nearest neighbor) from ML models. Algorithms are procedures implemented through computer code that are run on data. Models, then, are outputs of algorithms that contain model data and a prediction algorithm. The algorithm is used as a template and the model is then stored (e.g., the matrices of weights in the ANN) and used for prediction.

Algorithmic modeling versus data modeling

In an influential paper (Breiman, 2001), "two cultures" in statistics are differentiated, namely algorithmic modeling and data modeling. Algorithmic modeling refers to a culture in statistics where the data-generating mechanism is considered unknown, and algorithms (decision trees, neural nets) are fit to map predictor (features) and response variables. However, data modeling assumes a stochastic data model and seeks to estimate parameters from the data, by additionally including noise in the model. Although this differentiation is contested and many argue that both approaches are necessary (see Mitra (2021)),

ML can be categorized as an algorithmic modeling approach with the key goal of estimating the predictive accuracy of the trained model, whereas more established statistical approaches such as comparing competing data models by goodness-of-fit are grouped into the data modeling culture. We concur that both approaches are essential to understand complex phenomena and processes, however, algorithmic modeling offers a novel perspective that can complement and enrich traditional data-driven stochastic modeling. Adding prediction as a key component of modeling is what makes ML particularly powerful and also incited critique to move science from its traditional domain, namely explanation. In particular, data-driven modeling was beset with intricate problems of model misspecification that eventually yielded "irrelevant theory and questionable scientific conclusions" (Breiman, 2001). In the algorithmic modeling paradigm there is less concern about the mechanism producing the data, and rather gaining information "about how nature is associating the response variables to the input variables" (Breiman, 2001, p. 199). To what extent ML can yield more relevant theory is, to our estimation, still an open question, and considerate and critical application of ML is required to answer this question.

2.3 Forms of Machine Learning

ML algorithms are commonly differentiated into different forms, i.e., learning paradigms/categories/approaches that accomplish specific goals such as inferring patterns or predicting behavior. Popular forms of learning are called supervised (including: semi-supervised, active learning, and reinforcement learning), and unsupervised. In the following, we will briefly describe how these forms of learning work and the important features of each form of learning.

2.3.1 Supervised ML

Goals, tasks, and definition

The most common type of ML used in scientific research is supervised ML. The goal of supervised ML is to train an ML algorithm that picks up on patterns as presented by features (input variables), and then learns to imitate a researcher (teacher/expert, i.e., a ground truth) in classifying or scoring training data. Supervised ML typically involves performance improvement (P) in a desired task (T) by means of utilizing incoming data (E). P is assessed by comparing predictions with the ground truth provided by the researcher. For an ANN, Goodfellow et al. (2016, p. 274) define the training of an ML model as the process of "finding the parameters [..] of a neural network that significantly reduce a cost function [..], which typically includes a performance measure evaluated on the entire training set as well as additional regularization terms." To achieve this, the ML algorithm is provided training samples with correct labels or scores (ground truth/gold standard). In this sense, it is sometimes referred to as imitating a researcher or a teacher. Given a reasonable loss function that determines the discrepancy in prediction and ground truth, and an appropriate approach to optimize the parameters, the ML algorithm ideally finds the best model to predict unseen instances in the data.

Model validation

To what extent the trained ML model correctly classifies unseen samples and selects meaningful features is a crucial question when applying supervised ML in research. The capability of the trained ML model to generalize is one of the most central questions when applying supervised ML. It is commonly assessed in a procedure called (hold-out) cross-validation, where, in its simplest form, the data set is split into training and test data to evaluate the performance, P, of the trained ML model to predict unseen examples in the test data (as applied in Chap. 10). The test data need to be a representative sample of the target population about which scientific claims are made for otherwise the claims may be unjustified. There are more sophisticated forms of cross-validation, e.g., by performing multiple testing and aggregating accuracies. To prevent data leakage (see challenges below), another data set besides the training and test sets is extracted, namely the validation (or development) data set in order to fine-tune your model. Here, fine-tuning refers to setting up hyper-parameters, which are training parameters that are set before training the model and which control the ML algorithm in some way, e.g., by how much parameters should be updated (learning rate). Finding optimal hyper-parameters and weights means finding a balance between fitting the seen training data with keeping the ability to incorporate unseen data. Merely learning to imitate the training data (called overfitting) would translate into poor performance for unseen cases in most real-world scenarios, given the complexity and uncertainty of the real world (see above).

Model classes

Typical tasks in supervised ML are classification, regression, time-series modeling, and sequence-to-sequence models. To accomplish these tasks, researchers developed different model classes (covering different hypothesis spaces) for supervised ML. Model classes can be broadly subdivided into linear models, kernel-based methods, decision-trees, and artificial neural networks (ANNs):

- Linear models (e.g., regression models) transform a feature vector into an output vector by applying linear transformations. To cope with specific constraints in the feature vector (e.g., nominal data) or outcomes (e.g., binary decision), generalized linear models provide extensions to linear models. In linear models, the weights are learned to predict the data optimally. (Generalized) linear models are widely used in (science) education research.

- Kernel-based methods transform the feature space into a higher-dimensional space, and, in the case of support vector machines, classify data in this new space.
- Tree-based methods seek to split the feature space by applying binary splits with the goal to find clusters of data with minimal variance. Heuristics are used to find an optimal initial split of the training data, and methods such as pruning the trees are used to minimize overfitting the training data (we will elaborate on overfitting shortly). A tree with fewer leaves and the same accuracy on the test set is preferred over a tree with more leaves and similar accuracy on the test set. Random forests are an advancement of decision trees (see Box).
- Finally, ANNs apply linear- and non-linear transformations on the data at hand to reach the outputs. Many variations on ANNs were invented once deep learning has been shown to be a promising means for image data processing and processing of natural language data (such as LLMs).

Applications in science education

Supervised ML has been widely employed in science education to score and label responses. For example, Zhu et al. (2017) utilized c-rater-ML to automatically score science students' arguments according to quality of explanations about climate change, and uncertainty attribution. They found that the supervised ML model could reliably code the students' responses (quadratic weighted kappa values, as a measure for interrater agreement ranging from 0 (random agreement) to 1 (perfect agreement), above .70, see Williamson et al. (2012)). Uhl et al. (2021) built a constructed-response classifier to classify students' responses (ideas) on cellular respiration. They found that the tools based on ML could accurately identify ideas on cellular respiration in the constructed responses (see also: Sripathi et al. (2023)). Sripathi et al. (2023) utilized supervised ML to classify students' responses according to ideas on the question of weight loss (where does the mass go?). They achieved substantial human-machine agreement on unseen data (Sripathi et al., 2023). Putting these ML models into practice, science education researchers established automated feedback applications such as web interfaces (Donnelly et al., 2015; Gerard et al., 2019).

2.3.2   Unsupervised ML

Goals, tasks, and definition

In contrast to supervised ML, unsupervised ML builds upon complex input data without any labels with the goal "to infer the underlying structure or patterns of a system from the observed data" (Patriarcha et al., 2020, p. 10), or "to directly infer the properties of [a] probability density without the help of a supervisor [...] providing correct answers or degree-of-error for each observation" (Hastie et al., 2008, p.

486). Tasks then range from retrieving patterns and structures from data, and eventually reducing the input data (dimensionality reduction) for further analyses (feature engineering) or visualization. Another task is to cluster instances (i.e., grouping similar examples), de-noise data, model time-series (e.g., auto-regressive modeling) or estimate the probability distribution of the input data. Unsupervised ML can also provide generative models. Generative data models seek to model the data density distribution in a way that it coincides with the observed density distribution of the data, thereby reflecting or explaining the data-generation process. By sampling from generative models, it is possible to generate artificial training data, or explore the data generating process with different parameters. Others also considered the pre-training of language models as an instance of unsupervised ML. The pre-trained language models in a domain then provide meaningful learned representations that can be utilized in further tasks.

For machines to learn in an unsupervised form, an internal criterion that is independent of the task and data is used. Often this internal criterion is generated using vectors. The data as represented through features is transformed to a set of vectors in feature space. For example, students' constructed responses could be transformed into a term-document matrix (see Chaps. 3 and 7), and each document could then be represented as a vector in the space of all possible words that were used by the students. These vectors can then be grouped based on how close they are (i.e., to what extent similar words were used) to one another within that feature space. More sophisticated forms could then further decompose the term-document matrix and model it with more mathematically involved means such as latent semantic analysis.

Model validation

Validating an unsupervised ML model is typically more involved than validating supervised ML models, given that no gold standard labels are available that would indicate the accuracy of a model to predict unseen data. Unsupervised ML algorithms have many hyper-parameters that must be tuned. For example, in a clustering algorithm you might need to adjust the minimal distance between data points or the threshold value when clusters are formed. Moreover, algorithms such as latent Dirichlet allocation have multiple parameters such as the number of topics, and a parameter for how mixed these topics should be. Researchers typically have to make sense of unsupervised ML models by interpreting outcomes based on their substantive domain knowledge, and by systematically varying the parameters of the models and inspecting differences in outputs. Moreover, the generative capabilities of these unsupervised ML models can be probed and cross-validated using unseen samples. For example, the coherence of extracted topics can be evaluated by inserting intruder words into these topics and observing whether human raters can identify these intruder words. Ascertaining the model validity of unsupervised ML models requires multiple procedures and human expertise. In supervised ML, this expertise typically enters through sensible and meaningful labeling of the data in the first place, as this determines what the supervised ML model can learn.

Model classes

Similar to supervised ML, a vast variety of unsupervised ML algorithms have been proposed over the years. The respective Wikipedia article distinguishes quite sensibly between ANN-based techniques and probabilistic methods for unsupervised ML. ANNs can be used to attempt to mimic the data at hand, where the ANN seeks to generate sample data and receives an error signal of how far off it is. Such techniques are used in auto-encoders that then can represent and approximate the seen data. Such encoders can even be utilized to capture the behavior of physical systems. For probabilistic methods, several approaches such as clustering through hierarchical clustering, k-means clustering, t-SNE or (H)DBSCAN (see Chap. 5) are differentiated. Furthermore, approaches for learning discrete or continuous latent variable models such as a mixture of Gaussian, principal component analysis, or singular value decomposition can count as unsupervised ML (see Chap. 5). Latent variables relate to sub-manifolds which encapsulate meaningful patterns in the data set and can be utilized to model it (we will see an example in Chap. 5, see also Bishop (2006)).

Applications in science education

Unsupervised learning has been used in science education research to compress the data and exploratively differentiate (i.e., cluster) it (Sherin, 2013; Odden et al., 2020, 2021; Rosenberg and Krist, 2020; Wulff et al., 2022a). For example, Sherin (2013) utilized unsupervised ML to explore conceptions about the seasons in middle school students' interview transcripts. He showed that transforming the language responses into a vector space enabled clustering and identification of different explanations related to seasons. Odden et al. (2021) utilized a model called latent Dirichlet allocation to find thematic clusters in the articles retrieved from the journal Science Education over the last 100 years. Not only could they single out distinct themes over the last 100 years, but they could also quantify the rise and fall of these themes over time and track intellectual cross-pollination (Odden et al., 2021). Moreover, Wulff et al. (2022a) applied topic modeling with the help of LLMs and found that interpretable topics could be extracted from a rather small data set of pre-service physics teachers' written reflections.

2.3.3 Further Forms of ML

Self-supervised learning, semi-supervised learning, reinforcement learning, metalearning and transfer learning are further, specific learning approaches in ML research that become increasingly important where pre-trained ML models should be further developed to perform well in specific contexts. Even though these forms of ML are not the focus of this introductory textbook, we will briefly outline the main ideas behind some of these approaches:

Self-supervised learning

Self-supervised learning uses inherent relationships in the input data. The ML algorithm is then expected to predict a subset of the input based on another subset of the input. An illustrative example is natural language and the training of LLMs (see Chap. 7). A typical objective in these LLMs is to predict next words–a task that resembles the hypothesized brain mechanisms of predictive coding and active inference. Training models under this objective yielded efficient representations for language data, however, the extent to which meaningful world models have been developed is rather contested. While researchers showed that meaningful internal representations ("Emergent World Representations") could be learned by an LLM ("OthelloGPT") for a rather simple board game, that linguistic information is stored in LLMs, or that LLMs based on textual learning could ground (map) the concepts in a conceptual space, other researchers argued that there is a chasm between these world models on "ultrasimple" (Mitchell et al., 2023) worlds and humans more actionable world models.

Semi-supervised learning

Semi-supervised learning makes use of labeled and (typically much more) unlabeled data to enhance the predictive performance of the ML model. The availability of supervised (labeled) and unsupervised (unlabeled) ML was compared with the icing of a cake and the cake itself, indicating that unsupervised ML makes much more use of the available data, and much more data is available. For example, in semi-supervised classification, both unlabeled and labeled data can be used to train a classifier that is more capable than one trained on labeled data alone. Semi-supervised learning, among others, is important in contexts where labeling data is expensive; however, large amounts of unlabeled data are also available. For example, speech transcription (labeled) takes hours, whereas recording speech (unlabeled) is effortless. In addition, in protein 3D structure prediction, tremendous experimental efforts are required for crystallographers to determine the structure (label), whereas DNA sequences (unlabeled) are readily available (unlabeled).

Reinforcement learning

Reinforcement learning (the cherry on the cake, in the iced cake image from above) is concerned with finding the actions in an environment that maximizes specific rewards. In reinforcement learning there is typically a trade-off between exploration of the problem space, and exploitation of well-performing strategies to act in the environment. In games, a reinforcement learner can then play against each other in order to learn the environment and circumvent the problem that information on a good strategy is sparse. However, in science education research, mostly plain supervised or unsupervised ML approaches are currently being applied (Zhai et al., 2020). This might be explained by the fact that many researchers explored the capabilities of ML with existing data sets that were either labeled or unlabeled.

Metalearning

Metalearning is concerned with automating ML to address the challenge of finding the best ML algorithms for specific tasks and data sets. Metalearning typically encompasses multiple approaches such as meta-modeling, learning to learn, continuous learning, ensemble learning (combining the decisions by multiple base-learners), and transfer learning. And in all these cases it has been shown that metalearning can make ML more "efficient, easier, and more trustworthy" (Brazdil et al., 2022, p. 4). Metalearning makes use of meta-knowledge, such as knowledge of an algorithm's performance in a specific task. Typical tasks in metalearning include algorithm selection, hyper-parameter optimization, workflow synthesis, architecture search and/or synthesis, and few-shot learning (also referred to as transfer learning).

Transfer learning

Transfer learning is concerned with how ML models can transfer knowledge across tasks, given that "a learning algorithm should exhibit the ability to adapt through a mechanism dedicated to the transfer of knowledge gathered from previous experience" (as cited in Vilalta and Meshki (2022, p. 219)). Representational transfer refers to the case in which the source model has been trained in a certain context and the target model is trained later by using explicit knowledge from the source model. Functional transfer refers to the case in which source and target models are trained simultaneously, where the models can share some internal structure. Homogenous and non-homogenous transfer within representational transfer refer to either the same or a different feature space for source and target models. A common way to transfer knowledge is through parameter-based transfer learning as a form of representational learning. An initial set of parameters can be trained in a general setting and used in more specific tasks where less data is available, often called few-shot learning. Transfer learning has been used in science education research to enhance classification tasks and provide robust structures for clustering approaches even for comparably small sample sizes.

2.3.4 Sequencing Different Types of ML

The different ways of ML discussed so far are already powerful tools by themselves. However, they can become even more powerful when used in conjunction. Consider for example first using unsupervised learning to discover meaningful patterns in your data and then using supervised machine learning to automatically identify the discovered patterns in new data. An example of this can be found in the study of Tschisgale et al. (2023) who used a combination of supervised and unsupervised learning to first discover the different types of steps that students went through when solving a physics problem and then used supervised learning to create an algorithmic representation of the identified problem solving steps and provide further evidence for the validity of the found steps. In Chaps. 8 and 15 you will see this idea of sequencing supervised and unsupervised learning coming up again. For now the key message is that to answer a research question it can sometimes be helpful or even required to not just use one type of machine learning but combine multiple in productive ways.

2.4 Why ML in Your Research Project?

You saw that ML is quite a diverse field that offers a host of learning approaches that can help you answer research questions that rely on finding patterns in complex data sets. Applying ML is complex and requires careful consideration of the assumptions and constraints of your analyses. After all, ML is mostly a statistical modeling approach based on induction, and researchers in these fields cautioned that "the map is not the territory" (Korzybski, 1933) or quipped that "[a]ll models are wrong[,] but some are useful" (Box, 1979, p. 2). ML is certainly not a silver bullet that can produce meaningful outputs by itself based on your research data.

The reasons ML is used in your project should be carefully considered. Malik (2020) singles out hierarchical decisions for why using ML in the first place:

1. To use quantitative analysis over qualitative analysis;
2. To use probabilistic modeling over other mathematical modeling or simulation;
3. To use predictive modeling over explanatory modeling;
4. To rely on cross-validation to evaluate model performance.

Applying ML will in some ways limit you to these four dimensions, mostly to quantitative, probabilistic, and predictive modeling with the use of cross-validation. We consider these dimensions as very important to reflect on when deciding for or against the use of ML in your research, however, we also emphasize that ML can be used to bridge qualitative (more "narrative understandings of meaning-making" (Malik, 2020, p. 2)) and quantitative analysis (see Chap. 8) (Wulff et al., 2022b). Regarding point 2, there are interesting cross-pollinations where physics laws (mathematical modeling) are combined with ML to acquire informed and more accurate models. However, for educational research, no such laws are available to ground the models and further exploration in this area is necessary. Moreover, besides predictive modeling, researchers who utilize ML should also strive to explain model decisions. We will provide some methods and apply them in the following sections and chapters of this textbook. Moreover, Malik (2020) points out that "the results of cross-validation will only be as meaningful as the setup of a machine learning model, which will only be as meaningful as the way the world is put into observations and properties, which will only be as meaningful as how a phenomenon is quantified and measurement is done" (Malik, 2020, p. 2). As such, it is crucial for researchers who apply ML to recognize shortcomings, pitfalls, and limitations that are introduced either from specific procedures such as cross-validation, or more generally from the employed ML modeling pipeline.

2.5 Limitations and Challenges of ML

In various stages when applying ML in your research, different choices can limit and threaten the validity and generalizability of your findings. In fact, in many different disciplines, studies were found to lack rigor and standards of reporting and presenting ML-based findings. A valued goal in scientific research is reproducibility, which in the context of ML was defined as follows: Reproducible research is granted "if the code and data used to obtain the finding are available and the data are correctly analyzed" (Kapoor and Narayanan, 2023, p. 2). However, in practice, reproducibility was often threatened by the misapplication of ML, which is related to several intricate issues. Let us outline some important issues that might arise in your research, and ways to circumvent and report these issues. We grouped these issues into separate, yet highly interrelated, categories, namely data-related issues, procedural issues, and fundamental issues.

2.5.1 Data-related Issues

Insufficient quantity of training data

ML models such as deep ANNs or foundation models are trained on large data sets, given that many parameters need to be tuned for a representative sample of the target population (inductive learning). If meaningful natural language is to be generated, the ML algorithm should observe a representative and sufficient proportion of the target language. LLMs such as GPT are trained on trillions of tokens that enable these models to generate natural language (see Chap. 7), although they are not always accurate (knowledge hallucinations). Even more traditional deep learning models in computer vision are estimated to require about 1,000 examples per category (e.g., 1000 cats) to achieve a good performance. Science education researchers found that in science education research contexts sometimes 500 samples are sufficient to accurately classify constructed responses to a specific assessment task. However, researchers seldom acquire these big data sets in science education research. Moreover, learning paradigms such as transfer learning, fine-tuning or prompting with LLMs (see Chap. 7) can be utilized to augment educational data, or fine-tune large language models to your specific data set with fewer examples, which then leverages generic capabilities of LLMs and requires less training data. However, utilizing pre-trained models might introduce novel challenges such as a lack of representativeness in your research context.

Nonrepresentative training and test data

In your research you should explicitly specify for which population you want to build a model, and accordingly sample your training and test data. In other words, your training and test data should be representative of your target population (which also depends on your research goals and questions). Procedures to ensure representative sampling are then required and estimation of representativeness should be reported. For example, if you want to build an automated assessment tool for exam success in your class, you should be assured that you sample exams from students in your class. If, for some reason, a subgroup of students (e.g., those who are not interested in science) does not participate in the course and you have no feature scores for them, you will build a biased estimator, regardless of how good your ML algorithm is and whether you utilized LLMs to enhance classification performance. Amazon notoriously trained an ML algorithm to browse job applications and find quality candidates. They trained the ML algorithm on prior hiring data, which favored males for software developer and other technical jobs. The engineers neglected the fact that times were changing with more females entering these kinds of jobs, however, the trained ML models preferentially chose males because they were overrepresented in the training data, which then biased the hiring procedure and put women at a disadvantage. In your methods section, you might want to include information on your target population and build an argument for why, how, and to what extent you consider and assure that you were able to collect a representative sample. Especially when building applications in the education sector, which is considered a particularly sensitive area, for example, in the European Union's AI Act, scrutiny of the representativeness of your data set is important.

Poor-quality data

In computer science and meta-analysis research the "garbage in, garbage out" (short: GIGO) concept was introduced to emphasize problems with false, invalid inputs (computer science), or mixing high- and low-quality studies into the meta-analysis. This counts for ML research similarly: If your data is noisy or invalid for some reason, then the ML model will not learn anything meaningful. Typically, errors, outliers, and noise occur in empirical data. You then need to critically examine your data set for errors. For example, some students might have refused to specify their age. Handling missing data is an important procedure, to which we will return in Chap. 10. Moreover, it might make sense to delete outliers, as well as train multiple ML models and compare them to assess the impact of outliers or imputed values. A review study revealed that few researchers using ML currently address issues of missing data or data quality.

Irrelevant or unexpected features

The features that enter your ML algorithm should be chosen carefully. ML algorithms were found to pick up on any relationship or pattern in the data, and utilize it to maximize predictive accuracy, regardless of whether the researchers intended the ML algorithm to pick up these particular features. Early adopters of ANNs were surprised when the ML algorithm picked up on unexpected features:

> In the early days of the perceptron, the army decided to train an artificial neural network to recognize tanks partly hidden behind trees in the woods. They took a number of pictures of a woods without tanks, and then pictures of the same woods with tanks clearly sticking out from behind trees. They then trained a net to discriminate the two classes of pictures. The results were impressive, and the army was even more impressed when it turned out that the net could generalize its knowledge to pictures from each set that had not been used in training the net. Just to make sure that the net had indeed learned to recognize partially hidden tanks, however, the researchers took some more pictures in the same woods and showed them to the trained net. They were shocked and depressed to find that with the new pictures the net totally failed to discriminate between pictures of trees with partially concealed tanks behind them and just plain trees. The mystery was finally solved when someone noticed that the training pictures of the woods without tanks were taken on a cloudy day, whereas those with tanks were taken on a sunny day. The net had learned to recognize and generalize the difference between a woods with and without shadows! (Dreyfus and Dreyfus, 1992, p. 21)

Other vision models picked up on water to distinguish water birds from other birds, or, as mentioned above, identify cancer without seeing the relevant organs. You should therefore carefully craft and design the input features and be aware of their utmost relevance to any classification problem or similar problem amenable to ML.

Unbalanced data

A notoriously common characteristic of labeled data is class imbalance. This typically refers to the distribution of positive and negative examples (samples) in a classification problem. Say, you want to train a model that detects absenteeism with students. Hopefully this data set is largely imbalanced, with fewer students exhibiting absenteeism. Being absent from school can then be the positive class in a classification problem. I.e., you want to train a classifier that detects if a student will be absent in, say, the next week or so. A first problem with unbalanced data is that your model will have less training data for the positive class in this example. This might be traded-off with the fact that there might be less variability in students who become absent, however, this must certainly not be true. Another problem is that if you train your model on unbalanced data, it might simply always say no and achieve a high accuracy. If 99% of your students do not show absenteeism, then the accuracy of the model could easily be 99%. You might want to check this with a baseline classifier that always predicts the overpopulated class. Metrics such as confusion matrix, receiver-operator characteristic, and F1 score will help you identify such problems in classification problems and we will elaborate on these performance metrics in Chap. 10. In fact, utilizing inappropriate metrics (e.g., accuracy) for evaluating model performance on imbalanced data has been identified as a problem in several studies in ML research.

2.5.2 Procedural Issues

Bias-Variance Trade-Off

When deciding on a suitable ML algorithm for a given problem and during its training and testing, researchers face what is known as the bias-variance trade-off. In general, you can choose from a vast variety of ML algorithms (see Chaps. 4 and 5, among others). These algorithms differ by the degrees of freedom (e.g., parameters to vary) they have to fit your data. More degrees of freedom will also enable the model to more accurately capture patterns in more complex data sets. However, these patterns might be unique to this particular data set and may not generalize well to unseen data, making the trained ML model less useful in practice. This is the essence of the bias-variance trade-off:

Variance: One the one hand, complex, non-linear models such as ANNs can capture arbitrarily complex patterns (that might include spurious relationships) in the training data samples. This then often translates into poor generalizability, because the model is overly attuned to the training samples, i.e., it has a high variance. Consequently, unseen data samples that are slightly different from the training samples typically cannot be accurately classified.

Bias: On the other hand, utilizing a very simple, linear ML model might not be appropriate for the problem at hand. Bias refers to overly simplistic assumptions that go into modeling a phenomenon. The error in the training and test sets may be very high because the model is simply unable to capture any relevant variation in the data.

Marsland (2015, p. 35) captured the bias-variance trade-off with reference to the nature of quantum physics: "Just like the Heisenberg Uncertainty Principle in quantum physics, there is a fundamental law at work behind the scenes that says that we can't have everything at once." Researchers need to evaluate this, e.g., through comparing different ML algorithms, through cross-validation, and through preventing data leakage in their ML project (see next issue of data leakage). Moreover, regularization procedures are appropriate means to partly control for such issues. A widely shared rule-of-thump in scientific research is Ockham's razor: You should prefer a model that is simpler but has a similar power to explain and/or predict the phenomenon or problem. In ML, explainability may be replaced with predictability by choosing a simpler model with equal predictive accuracy. However, there might be another trade-off introduced here: balancing predictability and explainability. While a complex ANN model might be able to perfectly predict, say, students' performance in an assessment, it might be difficult to understand why the model makes these decisions. This lack of interpretability makes it unsuitable in high-stakes environments where accountability for decisions is of utmost importance.

Data leakage

Data leakage was among the most prevalent factors affecting the reproducibility of ML research. Data leakage can lead to an inflated performance estimation, and thus false scientific claims. Leakage refers to the "spurious relationship between the independent variables and the target variable that arises as an artifact of the data collection, sampling, or pre-processing strategy" (Kapoor and Narayanan, 2023, p. 2). "Data leakage occurs when one or more features used to train the algorithm has hidden within itself the result of the outcome, and is considered one of the most frequent mistakes in machine learning" (Filho et al., 2021, p. 1). Data leakage has been found even in high-stakes, consequential settings such as child maltreatment prediction, a field where "human judgments, [..] themselves are biased and imperfect" (Chouldechova et al., 2018, p. 1). Filho et al. (2021, p. 1) described a telling example of learning spurious correlations:

> A classic example from machine learning textbooks is the inclusion of the ID number of the patient as a predictor. While this should not have predictive importance if randomly assigned, it is common that patients coming from the same hospital have similar ID numbers in multicenter data sets. In the case of cancer prediction, for example, machine learning algorithms will learn that similar ID numbers that come from oncology hospitals have a higher probability of cancer.

One study in medicine sought to predict hypertension in patients one year prior. They utilized an ML algorithm called XGBoost and reached a high area under the curve (AUC), i.e., a good predictive accuracy (see Marsland (2015, pp. 24), and Chap. 10) of .870. However, researchers critically examined these findings and found that popular anti-hypertensive drugs were among the most important predictors. As such, the ML algorithms could have picked up on "predicting those [patients] already with hypertension but did not have this information on their medical record at baseline" (Filho et al., 2021, p. 2). Kapoor and Narayanan (2023) add that "the model would not have access to this information when predicting the health outcome for a new patient. Further, if the fact that a patient uses anti-hypertensive drugs is already known at prediction time, the prediction of hypertension becomes a trivial task" (p. 4). This stresses the importance for researchers to have a "conceptual pathway" of how predictors affect outcomes longitudinally.

To remind researchers of the intricacies related to data leakage, Kapoor and Narayanan (2023) present a taxonomy of data leakage, which we will outline below and illustrate with examples:

- **Lack of clean separation of training and test data set:** As we emphasized before, to adequately evaluate your model's performance on unseen data (generalizability), it is important to train and test the model in a setting where no information from the test set could already be used to train the model in the first place: "the model learns relationships between the predictors and the outcome that would not be available in additional data drawn from the distribution of interest" (Kapoor and Narayanan, 2023, p. 4). Several issues arise in this context:

  - **No test set:** This example refers to cases in which the same data set is used for training and testing. Of course, this would not provide you useful information about model performance in novel contexts, i.e., for unseen data. An example would be the scenario where only a few samples are available such that splitting them would prevent you from training the ML model.

– **Pre-processing on training and test sets:** This case is more subtle. Say, you build a term-document matrix (see Chap. 7) to train a classifier for your text data. To do this, you utilize the *entire* data set available to build the term-document matrix and then split the data set into training and test sets. Building the term-document matrix on the entire data set would cause data leakage, because you might inform the trained ML model of words that only occur in the test data. You should have separated the training and test data beforehand, because the test data distribution is considered unknown and cannot be used to inform anything for the trained ML model. In addition, when imputing values or correcting for under-/oversampling, the test data should always be separated. Separating the test data from the training data should be the first step, once the data are collected.

– **Feature selection on training and test sets:** Moreover, if feature engineering or feature selection is performed and the entire data set is used, data leakage is also caused, as the trained ML model will pick up on features in the test data. For example, if you perform dimensionality reduction based on the entire data set, the test data influence this procedure and likely inflate the performance.

– **Duplicates in data sets:** Duplicates are data samples that occur multiple times. This could lead to having the same samples in both the training and test data sets, which would then also inflate model performance estimates.

- **Model uses features that are not legitimate:** This problem can occur when a feature is a proxy for the outcome variable, as in the example above, with anti-hypertensive drugs being a proxy for hypertension. In educational research, you might want to predict social status of students and include information on whether some financial benefits are already granted to the students by the school. Kapoor and Narayanan (2023) emphasize that "the judgment of whether the use of a given feature is legitimate for a modeling task requires domain knowledge and can be highly problem specific." (p. 4). In fact, data set contamination seems to be a problem in LLMs.

- **Test set is not drawn from the distribution of scientific interest:** Differences in target distribution and test distribution will likely cause the model to be invalid in the context of scientific interest.

– **Temporal leakage:** Regarding longitudinal data, researchers should ensure that the training set does not contain information from the future that might constitute leakage and invalid conclusions. For example, if you want to predict students' end of year performance in a concept inventory based on their engagement in class and include their final exam score as a predictor, this might constitute temporal leakage. After all, in the real world you would not have the final exam score to predict conceptual understanding.

– **Nonindependence between training and test samples:** This is an often encountered threat to validity in many ML studies in different disciplines. Typically, you want to make predictions about unseen test data with a distribution that is similar to that of the training data. However, if the training and test data are dependent, e.g., if responses from one student appear in training and test data, this likely inflates the performance estimation of the model, given that the model has information that it would not have when applied in the field.

– **Sampling bias in test distribution:** This occurs when you test your model on data from a class, say, in your local region (spatial bias), however, the distribution of scientific interest refers to the entire country, or vice versa. As in a study on predicting autism, borderline cases were omitted from the test data, which constituted data leakage (and inflated performance estimation). Moreover, predicting borderline cases is considered the most difficult, and if cases are omitted from the test set this does not represent the target population anymore.

Kapoor and Narayanan (2023) outline three arguments that researchers engaging in ML should address to prevent data leakage: 1) clean train-test split, 2) each feature in the model is legitimate, and 3) the test set is drawn from the distribution of scientific interest. We will introduce the procedures in Python and R to ensure a clean train-test split. As indicated, ensuring that features are meaningful and sampling is representative of the distribution of scientific interest requires domain knowledge and knowledge of designing and conducting rigorous empirical research in education, which is equally important in ML-based science education research as in other empirical science education research.

2.5.3 Fundamental Issues

Explainability

Even though ML algorithms can be accurate in prediction tasks, many successful ML models, especially deep learning applications, remain largely black-boxes. Although prediction is a key goal for any scientific discipline, it is equally important to understand the influencing (causal) factors of phenomena and processes. An important branch of research in ML deals with interpreting and explaining model decisions, which is called explainable AI. While the best explanation of a simple model can be considered the model itself, this is not true for complex models such as ensemble models or deep ANNs. Billions of parameters in an ANN are not human-interpretable. Additional means to make sense of model decisions are required.

Methods for explaining ML models can be differentiated along several criterion dimensions:

- Intrinsic or extrinsic: The former refers to the case where an ML model is in itself interpretable such as a decision tree, Naive Bayes, or rule-based classifiers; the latter refers to the case where another interpretation method to make model decisions interpretable.
- Model specific or model agnostic: The former case refers to means of interpretation that only account for certain ML models, such as weights in logistic regression. Model agnostic methods can be used on any ML model to interpret outputs, such as the systematic analysis of false positive classification in classification models.

- Local or global: Local refers to methods that explain a specific sample, whereas global methods seek to interpret an entire model.

Different outcomes for interpretation procedures can be differentiated which helps researchers interpret the ML models. For example, (1) a feature summary statistic can highlight the importance of features. Some common methods use additive feature attributions, in which the contribution of each feature to the output is displayed. (2) Model internals such as the learned weights in a logistic regression classifier can be interpreted. (3) Moreover, researchers can create new data such as adversarial or counterfactual examples to make the model change the output class and interpret the created data. (4) Intrinsically interpretable models can be used as explanation models for black box models such as ANNs.

We will deal with various methods to illuminate ML black-boxes (i.e., white-boxing) and better understand model decisions. In Chap. 10 we introduce SHAP (SHapley Additive exPlanations) values and in Chap. 12 we introduce integrated gradients. Both concepts help to understand the contributions of features in the model for a certain prediction, which is an important quality to understand a model's output.

Explanation versus prediction

Researchers also highlighted the potentials that a shift towards prediction incited by ML can offer for scientific understanding of phenomena and processes in psychology. Yarkoni and Westfall (2017, p. 1) argued that historically the goal of psychological research is to understand (explain and predict) human behavior, with a focus on "explaining the causal mechanisms that give rise to behavior," by means of randomized controlled experiments, with "investigations of the various mediating and moderating variables that govern various behaviors," giving rise to "intricate theories of psychological mechanism, but that have little (or unknown) ability to predict future behaviors with any appreciable accuracy." We appreciate this call for improving the predictability of models in science education as well, and contend that both explanation and prediction are important guideposts for any scientific enterprise with the goal of developing an understanding of its relevant phenomena and processes.

Statistical causal inference

Focusing on predictive modeling as in ML also has its downsides. Even though an ML model might very well predict an outcome variable (called a dependent variable in the context of causal modeling) by merely accounting for all sorts of input features, nothing might be learned about the causal structure of the phenomena or processes under study. Even worse, controlling for certain input features might conflate effects. In causal modeling, mediators (e.g., smoke in the fire $\rightarrow$ smoke $\rightarrow$ alarm-chain; $X \rightarrow Y \rightarrow Z$), colliders ($X \rightarrow Y \leftarrow Z$), and confounders ($X \leftarrow Y \rightarrow Z$) are differentiated. Controlling for mediators and colliders might reverse these effects (introducing backdoor paths) and distort the assumed causal structure. Not controlling for confounding variables introduces further problems. ML researchers who simply include all available predictors (independent variables) in a model may be at risk of producing spurious relationships. Researchers must carefully consider the assumed underlying causal structure of the problem at hand and design their studies in order to be capable of causal modeling. For example, performing randomized controlled studies is still considered the gold standard, because it allows the estimation of a direct causal effect from a predictor without worrying about uncontrolled confounding variables. Even though we do not focus on it exclusively in this book, causal modeling can be performed with ML.

Generalizability

Inductive learning from data poses intricate risks of overfitting and propagating bias that are inherently related to ML algorithms and the studied research problem. Examples from ML research illustrate that overfitting is a serious concern in ML-based research. For example, some argued that radiologists could be replaced by computers in the foreseeable future because of ML models' superior performance on detecting cancer on X-ray images. However, it turned out that the performance of the developed ML models largely depended on the technical specifications of the X-ray scanner such that the test set performance declined when different scanners were used. Here, the ML models selected irrelevant features from the training data and overgeneralized the importance of these features. Images of military tanks were also observed to be separated from non-tank images with 100 percent accuracy. However, tank images were lighter (see above) than non-tank images, that is, there was a subtle (i.e., construct-irrelevant) bias in the data. In vision models for water-bird classification, for example, the ML model uses water to distinguish waterbirds from other birds, jeopardizing the generalizability of the model. Moreover, object identification algorithms can be tricked using adversarial examples to identify them as bears or whatever they are trained for. Researchers showed that merely random looking images (adversarial samples) could be constructed in a way to trick classification ML models into "seeing" a bear in these random-looking images. Thus, the ML model selects features that are not human-interpretable in any straightforward manner. Hence, the extent to which these models function reliably in real-world settings remains questionable. Similar tricks have been reported for LLMs, e.g., through clever questions (called prompts). Entire lists of failures have then been collected. If researchers do not understand why ML models make certain decisions, it is difficult to clearly evaluate the generalizability capabilities of the models, besides from field-testing them with out-of-sample cases, and documenting successes and failures. This becomes more intricate with increasing model sizes and data sets.

Bias and stereotypes

Bias, stereotypes, and more generally fairness of decisions by ML models become increasingly important as ML applications penetrates high-stakes decisions related to loans, hiring, trial and detention, and diagnostics. Biased decisions have been found in all types of ML applications, and may be caused by unbalanced representation in the training data. Unfairness can arise from data, ML algorithms, or user interactions (see Fig. 2.1). Data biases are manifold and can arise from measurement bias, such as when minority groups are arrested more often, and it is falsely concluded that they are more prone to criminal behavior. This might be due to a systemic bias in controlling one group more than the other. Data biases might also be related to representation biases which we will consider more deeply below. Algorithmic biases are related to the outcomes of algorithms that can affect user behavior. Presentation and ranking biases, for example, relate to the fact that presented and ranked outcomes can be manipulated by fake data (e.g., movie reviews). Moreover, evaluation biases relate to model evaluation with biased data sets, such as hand sanitizers that do not recognize certain skin colors. Furthermore, biased outcomes might interact with users that can then amplify and perpetuate biases, such as when a search engine's future recommendations are influenced early on by user behavior that tends to neglect search items lower in the list. Biases with optimization techniques and performance metrics can also have disadvantage, e.g., minority groups, without problems in the input data.

Representation bias may be alleviated by emerging LLMs. LLMs are currently trained on the Common Crawl of the Internet, Wikipedia, etc. (where all sorts of biases prevail). Special care must be taken to ensure that these models do not perpetuate, or even magnify, bias. For example, LLMs capture gender biases. When asked to predict the next word in a sentence like "She was a ..." more likely "nurse" is output by many LLMs, while males are more likely to be "doctors." Given that text on the Internet rather represents the world as it is, this is not surprising. In 2008, 90 percent of registered nurses were female. Similar distortions can be expected in natural sciences and engineering, where prestigious positions are still dominated by men. ML models that were designed to promote gender-neutral advertisements in science, technology, engineering and mathematics (STEM) were devised, however, gender-biases cropped in because fewer women saw the adverts and the ML algorithms "optimized" it in a way that women were at a disadvantage. However, societal

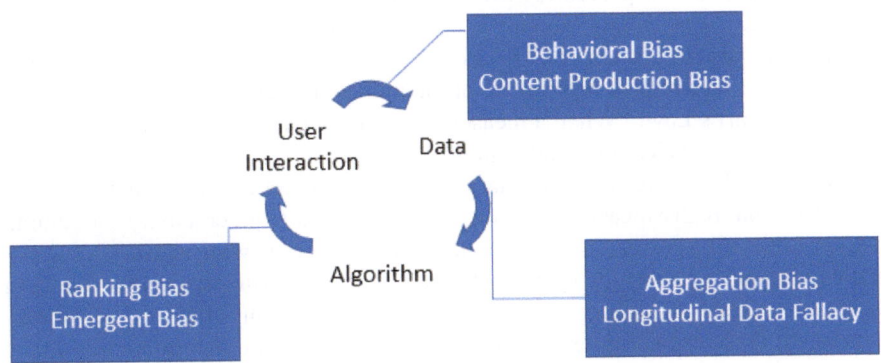

Fig. 2.1 Biases in data, algorithms, and use interactions, taken from Mehrabi et al. (2022)

efforts are in place to equilibrate gender imbalances, as laid out, among others, in the sustainable development goals by the UN. LLMs and other ML models are then at risk of perpetuating entrenched stereotypes and biases, and counteracting these goals.

Given the potential of ML, it also needs to be ensured that all humans will have equal access to these systems and non are instrumentalized to improve them. Otherwise, novel forms of modern colonialism are created. Certain people, languages, and other characteristics dominate most data sets and ML models. This leads to unequal distributions of wealth, agency, and empowerment. In many large companies, people-analytics are used to track work engagement and foster learning of future skills. While these systems can certainly enhance firms' productivity, it is crucial to closely monitor the individual agency and prevent that only the upper management oversees the data, thus controlling employers. Moreover, instrumentalizing humans to refine large generative LLMs in lower-income countries can be considered a modern form of abusing workforce. In such contexts, these systems may have adverse effects and increase a culture of control, surveillance, disengagement, and exploitation. The same holds true for learning analytics systems in educational contexts.

Human alignment

The more versatile ML programs and AI chatbots become, the more urgent is the discussion about regulating them (see: "AI Act" by the European Union), and aligning them with goals that humanity has. Aligning AI to human values has become a controversial issue. An entire research branch is dedicated to AI alignment, where it is sought to implement human values in AI systems and programs. Companies strive to regulate their models, however, rather heuristically. For example, generative LLMs and generative image models can be prevented from generating text or images when customers search for, say, hazardous information. While these early attempts work in certain circumstances, many agree that much more effort is required to define human values (some doubt that there even exist universal human values, or at least that they are easily discernable), enable machines to learn and adhere to them, and reconcile issues such as preventing ML from utilizing humans as a means to reach a certain end, e.g., eradicating humans to solve the climate crisis–given that they are the ultimate cause (emission of CO_2 into the earth's atmosphere) of the problem in the first place. This relates to the issue of incomplete representations of one's goals and to Goodhart's Law: "When a measure becomes a target, it ceases to be a good measure" (as cited in Thomas and Uminsky, 2022).

There are other issues related to tackling the alignment problem, such as defining metrics that really measure what one is interested in (e.g., creativity) or gaming metrics. Educational researchers, especially those concerned with measuring competencies, are well aware of the intricate difficulties in measuring innocuous-sounding constructs such as motivation, intelligence, emotions, and many others. As previously mentioned, these are complex constructs insofar as they are determined by many influencing factors both external and internal to the individual. Choosing unsuitable metrics for one's goals may result in unintended consequences. For example, using easy and short-term metrics, such as time spent watching YouTube was found to indirectly and unintentionally incentivize conspiracy theories (cited in: Thomas and Uminsky, 2022). Moreover, using students' test scores to evaluate teachers' effectiveness incentivized teachers to cheat test scores to earn higher rankings. This is a crucial aspect when applying ML as well.

Artificial General Intelligence and Artificial Superintelligence

Conversational AI programs such as ChatGPT are well-versed to converse with non-experts in almost any conceivable topic (given that the training data, i.e., the Common Crawl of the Internet, already cover much ground, and companies track user inputs and adjust model responses). While ChatGPT cannot solve domain-specific problems that reach a certain degree of sophistication, experts almost unanimously agree that the performance of ChatGPT is "thoroughly remarkable," even surprising experts in the field of AI. With these language- and even image-generative capabilities some researchers wondered to what extent an artificial general intelligence (AGI) or even artificial superintelligence (ASI) can be created. After all, ML models such as ANNs are in *some* structural and computational aspects *similar* to the human brain architecture, which is provably capable of inventing science and discovering insights about nature. From mere hardware considerations, computers even today seem to be superior to particulars of the (otherwise remarkable) human brain, e.g., in speed of signal passing, or operation speed of biological neurons versus processors. (However, they are also more inefficient, merely by observing that silicon computers generate a large amount of heat while the human brain remains at equilibrium). Physicist David Deutsch furthermore argues that AGI must be possible, "because of a deep property of the laws of physics, namely the universality of computation."[6]–though it might be a long way to accomplishing it.

While some researchers argued that some advanced LLMs already show "sparks of artificial general intelligence" (Bubeck et al., 2023), others remain more skeptical and remind the research community on the brittleness and limitations of LLMs. Among others, they consider the architectures to be fundamentally flawed to acquire whatever counts as general intelligence, or even sensible language processing. Others argued that the terms "intelligence" are misplaced, because intelligence is a multi-faceted construct that, among others, requires embodiment and exploration in the real world. As of this writing, it is reasonable to assume that we are nowhere near science fiction scenarios (Terminator) where AI-based systems (Skynet) become self-aware and nearly wipe out the human race with nuclear weapons. Current LLMs merely interpolate textual and image data, and are to some extent "stochastic parrots," indeed.

AGI inspired researchers to project futures with runaway systems. Specification of a certain goal (e.g., safeguarding the planet from global warming) for the AGI

---

[6] https://aeon.co/essays/how-close-are-we-to-creating-artificial-intelligence, last access: Dec 2023.

system could cause adverse effects (e.g., eradicating humanity as the most efficient solution). Alternatively, by failing to value humans into the value system of an AGI, and the task of producing a maximum number of paper clips, the AGI system with excess power over its environment could seek to convert all matter into paper clips, including human beings. A more subtle concept is "counterfeit people" by LLMs (Dennett, 2023). It is important for human societies to operate based on shared (though implicit) knowledge. If LLMs and AGIs overtake large amounts of tasks (e.g., consultation or therapy), AGIs could subvert this societal contract and consequently spoil civilization, given that one cannot necessarily know the knowledge base of one's interlocutor. Given these challenges, it is necessary to boost programs on AI safety, human alignment, and other ethical issues related to the implementation of AI and AGI systems.

Even today, AI technologies were shown to be powerful enough to invent thousands of highly toxic chemical and biological weapons (molecules). Alongside this prospect comes the concern about open source AI models that are capable of such inventions, given that eventually unauthorized people could use the technology for nefarious purposes.

Privacy and data sovereignty

Intricate issues with data privacy, privacy leakage of personally identifiable information, and data sovereignty occur with black-box models such as ANNs and LLMs. Once your private data is used for training an LLM (update weights), it is notoriously difficult to undo it and act upon your right for data withdrawal. Moreover, once your personally identifiable information is in the model, it is hard to undo this information, and subtle prompting strategies exist to recover forgotten personally identifiable information. Companies that use AI for decision-making processes should also be held accountable (e.g., by the EU AI Act) for transparently explaining (or at least: describing) how a certain decision was made, e.g., what influencing factors were. This is, of course, most pertinent in high-stakes decision making processes. With increasingly large models this is currently not impossible.

Ecological concerns

In addition, AI applications should be considered with increasing awareness of sustainable development. It was estimated that simulating the human brain would cost ten terawatts of power, while the remarkable human brain only uses (give or take) 20 watts to do the things it does. Moreover, training foundational models significantly taxes the environment in terms of CO_2 emission, and every single request to LLMs such as GPT-4 (or it's successors) costs a significant amount of energy. Only private companies can spearhead this development by investing billions of (mostly) dollars in data centers and computing resources. Moreover, the electricity use of the global tech sector is predicted to grow substantially by 2030 (approximately 60%), and much of the production of chips is fossil fuelled, given that supply chains are often not renewable. Given such concerns, it is worthwhile to consider simpler ML algorithms for ecological reasons compared with more complex ML models that disproportionally tax the environment.

2.6 Applying ML in Science Education Research

With all that said, it hopefully becomes clear that ML provides valuable opportunities for science education research to enhance data processing and data analysis, i.e., data-driven scientific discovery, yet it requires careful reflection of goals, data sets, algorithms, biases, ethics, privacy, and ecology. ML broadens the scope of research questions that can be addressed and the hypotheses that can be empirically tested. On the other hand, ML also introduces new challenges that science education researchers who apply ML have to deal with.

In the context of science education research, ML is probably best situated in a (post-)positivist research paradigm in science education (Treagust et al., 2014), where "meaningful knowledge claim[s] [...] should be supported by logical reasoning and empirical data that are self-evident and verifiable." (cited in Treagust et al. (2014, p. 4)). Positivists emphasize the possibility of drawing inferences (inductively) to underlying general relationships and theories. ML, then, can be of great value to advance this research paradigm and strengthen the inference step by capturing patterns in complex educational data sets, and reliably automating tasks that enhance science education research and instruction.

Throughout this book, we will provide glimpses of how other paradigms can (and should!) inform how we use, evaluate, and critique ML in science education. For example, in Chaps. 4 and 5 we seek to provide accessible examples of how supervised and unsupervised ML can be used, which is then applied to science education-specific problems in Chaps. 10 and 11. We emphasize that while a computer may treat the data we input in a (post)-positivist way, humans are still an essential part of the analytic system. Thus, critical evaluation of what and who are represented in data sets— and what and who are not represented—is essential for (at least) preventing uses of ML that cause harm, and (at best) generating uses of ML that work toward equity in science education, and improving analysis and implementation of teaching and learning processes. In addition, we describe in Chap. 8 how ML can be used within an interpretivist paradigm, and argue that such an approach is especially important in that it actively prevents the ceding of analytic agency from a human researcher.

References

Bishop, C. M. (2006). *Pattern recognition and machine learning. Information science and statistics.* New York, NY: Springer Science+Business Media LLC.

Box, G. E. P. (1979). *Robustness in the strategy of scientific model building: Technical Report #1954.*

Brazdil, P. B., van Rijn, J. N., Soares, C., & Vanschoren, J. (Eds.). (2022). *Metalearning: Applications to automated machine learning and data mining* (2nd ed.). Springer eBook Collection: Springer, Cham

Breiman, L. (2001). Statistical modeling: The two cultures. *Statistical Science, 16*(3), 199–231.

Bubeck, S., Chandrasekaran, V., Eldan, R., Gehrke, J., Horvitz, E., Kamar, E., Lee, P., Lee, Y. T., Li, Y., Lundberg, S., Nori, H., Palangi, H., Ribeiro, M. T., & Zhang, Y. (2023). Sparks of artificial general intelligence: Early experiments with gpt-4. *arXiv*.

Chouldechova, A., Benavides-Prado, D., Fialko, O., & Vaithianathan, R. (2018). A case study of algorithm-assisted decision making in child maltreatment hotline screening decisions. In Friedler, S. A. and Wilson, C., editors, *Proceedings of the 1st Conference on Fairness, Accountability and Transparency*, volume 81 of *Proceedings of Machine Learning Research*, pp. 134–148. PMLR.

Dennett, D. C. (2023). The problem with counterfeit people. *The Atlantic*.

Donnelly, D. F., Vitale, J. M., & Linn, M. C. (2015). Automated guidance for thermodynamics essays: Critiquing versus revisiting. *Journal of Science Education and Technology, 24*(6), 861–874.

Dreyfus, H. L., & Dreyfus, S. E. (1992). What artificial experts can and cannot do. *AI & Society, 6*, 18–26.

Filho, A. C., Batista, A. F. D. M., & Dos Santos, H. G. (2021). Data leakage in health outcomes prediction with machine learning. comment on 'prediction of incident hypertension within the next year: Prospective study using statewide electronic health records and machine learning'. *Journal of medical Internet research, 23*(2), e10969.

Gerard, L., Kidron, A., & Linn, M. C. (2019). Guiding collaborative revision of science explanations. *International Journal of Computer-Supported Collaborative Learning, 14*(3), 291–324.

Géron, A. (2017). *Hands-on machine learning with Scikit-Learn and TensorFlow: Concepts, tools, and techniques to build intelligent systems*. Beijing and Boston and Farnham and Sebastopol and Tokyo: O'Reilly.

Goodfellow, I., Bengio, Y., & Courville, A. (2016). *Deep learning*. Cambridge, Massachusetts and London, England: MIT Press.

Hastie, T., Tibshirani, R., & Friedman, J. (2008). *The elements of statistical learning: Data mining, inference, and prediction*. Springer.

Kapoor, S., & Narayanan, A. (2023). Leakage and the reproducibility crisis in machine-learning-based science. *Patterns (New York, N.Y.), 4*(9), 100804.

Korzybski, A. (1933). *Science and sanity: An introduction to non-Aristotelian systems and general semantics: International Non-Aristotelian Library*.

Krenn, M., Pollice, R., Guo, S. Y., Aldeghi, M., Cervera-Lierta, A., Friederich, P., Dos Passos Gomes, G., Häse, F., Jinich, A., Nigam, A., Yao, Z., & Aspuru-Guzik, A. (2022). On scientific understanding with artificial intelligence. *Nature Reviews Physics, 4*(12), 761–769.

Malik, M. M. (2020). A hierarchy of limitations in machine learning. *arXiv*.

Marsland, S. (2015). *Machine Learning: An Algorithmic Perspective* (2nd ed.). Chapman & Hall/CRC machine learning & pattern recognition series. Boca Raton, FL: CRC Press

Mehrabi, N., Morstatter, F., Saxena, N., Lerman, K., & Galstyan, A. (2022). A survey on bias and fairness in machine learning. *arXiv*.

Mitchell, M., Palmarini, A. B., & Moskvichev, A. (2023). Comparing humans, gpt-4, and gpt-4v on abstraction and reasoning tasks. *arXiv*.

Mitchell, T. (1997). *Machine learning*. New York, NY: McGraw-Hill Education.

Mitra, N. (2021). Introduction. *Observational Studies, 7*(1), 1–2.

Odden, T. O. B., Marin, A., & Caballero, M. D. (2020). Thematic analysis of 18 years of physics education research conference proceedings using natural language processing. *Physical Review Physics Education Research, 16*(1), 1–25.

Odden, T. O. B., Marin, A., & Rudolph, J. L. (2021). How has science education changed over the last 100 years? an analysis using natural language processing. *Science Education, 105*(4), 653–680.

Patriarcha, M., Heinsalu, E., & Léonard, J. L. (2020). *Languages in space and time: Models and methods from complex systems theory*. Cambridge University Press.

Rosenberg, J. M., & Krist, C. (2020). Combining machine learning and qualitative methods to elaborate students' ideas about the generality of their model-based explanations. *Journal of Science Education and Technology*.

Rothchild, I. (2006). Induction, deduction, and the scientific method: An eclectic overview of the practice of science. *SSR*.

Sherin, B. (2013). A computational study of commonsense science: An exploration in the automated analysis of clinical interview data. *Journal of the Learning Sciences, 22*(4), 600–638.

Sripathi, K. N., Moscarella, R. A., Steele, M., Yoho, R., You, H., Prevost, L. B., Urban-Lurain, M., Merrill, J., & Haudek, K. C. (2023). Machine learning mixed methods text analysis: An illustration from automated scoring models of student writing in biology education. *Journal of Mixed Methods Research*, 155868982311539.

Thomas, R. L., & Uminsky, D. (2022). Reliance on metrics is a fundamental challenge for ai. *arXiv*.

Treagust, D. F., Won, M., & Duit, R. (2014). Paradigms in science education research. In N. G. Lederman & S. Abell (Eds.), *Handbook of research on science education* (Vol. II, pp. 3–17). New York: Routledge.

Tschisgale, P., Wulff, P., & Kubsch, M. (2023). Integrating artificial intelligence-based methods into qualitative research in physics education research: A case for computational grounded theory. *Physical Review Physics Education Research, 19*(020123), 1–24.

Uhl, J. D., Sripathi, K. N., Meir, E., Merrill, J., Urban-Lurain, M., & Haudek, K. C. (2021). Automated writing assessments measure undergraduate learning after completion of a computer-based cellular respiration tutorial. *CBE Life Sciences Education, 20*(3), ar33.

Valiant, L. (2024). *The importance of being educable: A new theory of human uniqueness* (1st ed.). Princeton: Princeton University Press.

Vilalta, R., & Meskhi, M. M. (2022). Transfer of knowledge across tasks. In P. B. Brazdil, J. N. van Rijn, C. Soares, & J. Vanschoren (Eds.), *Metalearning, Springer eBook Collection, page 219*. Cham: Springer.

Vincent, J. (2019). Deepmind's ai agents conquer human pros at starcraft ii.

Williamson, D. M., Xi, X., & Breyer, F. J. (2012). A framework for evaluation and use of automated scoring. *Educational Measurement: Issues and Practice, 31*(1), 2–13.

Wulff, P., Mientus, L., Nowak, A., & Borowski, A. (2022a). Utilizing a pretrained language model (bert) to classify preservice physics teachers' written reflections. *International Journal of Artificial Intelligence in Education*.

Wulff, P., Buschhüter, D., Westphal, A., Mientus, L., Nowak, A., & Borowski, A. (2022b). Bridging the gap between qualitative and quantitative assessment in science education research with machine learning—A case for pretrained language models-based clustering. *Journal of Science Education and Technology, 31*, 490–513. https://doi.org/10.1007/s10956-022-09969-w

Yarkoni, T., & Westfall, J. (2017). Choosing prediction over explanation in psychology: Lessons from machine learning. *Perspectives on Psychological ScienceâŁ¯: A Journal of the Association for Psychological Science, 12*(6), 1100–1122.

Zhai, X., Haudek, K., Shi, L., Nehm, R., & Urban-Lurain, M. (2020). From substitution to redefinition: A framework of machine learning-based science assessment. *Journal of Research in Science Teaching, 57*(9), 1430–1459.

Zhu, M., Lee, H.-S., Wang, T., Liu, O. L., Belur, V., & Pallant, A. (2017). Investigating the impact of automated feedback on students' scientific argumentation. *International Journal of Science Education, 39*(12), 1648–1668.

Chapter 3
Data in Science Education Research

Peter Wulff, Marcus Kubsch, and Christina Krist

Abstract In this chapter, we explain why ML can be valuable for analyzing complex data in science (education) and show what types of data you might encounter in your research project. We single out important characteristics of these types of data that can become particularly important for the performance of ML algorithms.

3.1 The Importance of Data in Science

Several scientific disciplines hinge on (real-world) data to answer empirical research questions. Let us first briefly reflect on why data-driven discovery and ML might be important in many scientific disciplines such as science education research.

Data-driven discovery in science

Data have become a key "currency" in the 21st century, sometimes called the "information age" (Castells, 2010). Many commercial companies crave for customer data, and oftentimes you "pay" with your data to access certain services. In science, research approaches that are grounded in data were argued to increasingly complement (not replace) more traditional approaches that work with models and strong assumptions (see also Box): "The most pressing scientific and engineering problems of the modern era are not amenable to empirical models or derivations based on first-principles. Increasingly, researchers are turning to data-driven approaches for a diverse range of complex systems, such as turbulence, the brain, climate, epidemiology, finance, robotics, and autonomy. These systems are typically nonlinear, dynamic, multi-scale in space and time, high-dimensional, with dominant under-

P. Wulff (✉)
Heidelberg University of Education, Heidelberg, Baden-Württemberg, Germany
e-mail: peter.wulff@ph-heidelberg.de

M. Kubsch
Freie Universität Berlin, Berlin, Germany

C. Krist
Graduate School of Education, Stanford University, Stanford, CA, USA

© The Author(s) 2025
P. Wulff et al. (eds.), *Applying Machine Learning in Science Education Research*,
Springer Texts in Education, https://doi.org/10.1007/978-3-031-74227-9_3

lying patterns that should be characterized and modeled for the eventual goal of sensing, prediction, estimation, and control." (Brunton and Kutz, 2019, p. ix). While this might sound like a strong statement, we are reminded of many problems even in scientific disciplines that utilize analytics and reductionism, and thus often constrain their analyses to highly idealized systems. Think of the classical three-body problem in physics where no general (closed-form) analytical solution is attainable. When science deals with complex phenomena, such as the Earth's climate, folding of proteins, or the social behavior of humans, the importance of numerical approaches and simulations has long been recognized. In particular, data-driven discovery is a valuable means of scientific inquiry, especially when no theoretical expectations can guide the inquiry.

Data-driven discovery has been a feature of scientific investigation and part of the scientific method from early on, as empirical data enabled scientists to develop and refine theories. For example, in the early 17th century the laws of planetary motion were inferred by examining planetary motion data points (based on meticulous observations). Similarly, a modern understanding of quantum mechanics was derived, among others, from the inspection of atomic spectra. Moreover, Darwin's "Origin of Species," the theory of punctuated equilibrium in paleontology, and blood circulation stand out as inductive, data-driven discoveries, founded in observation and pattern recognition. As such, data-driven discovery is a crucial part of the scientific method: "We start a research project with observations made either in the field, the library, or the laboratory. How these observations are collected, classified, interpreted, and used as the basis of theorizing (from a hunch to a eureka) is, more or less, what science is about" (Rothchild, 2006, p. 4).

In addition, for science education research, data is key to answering research questions related to real-world phenomena. Below, we outline that science education researchers deal with particularly complex systems; thus, empirical, data-driven approaches are important to probe theories or develop them in the first place.

"Big Data"

In modern times, the availability of data increases rapidly in size: "Global data has been doubling approximately every two years and is expected to reach 175 zettabytes (i.e., 175 billion-million-megabytes) in 2025[1]. The unprecedented availability and size of data motivated the metaphor "Big Data." Big Data can be characterized by attributes such as velocity (of access and growth), volume (of size), and variety (of sources and types). However, other V's have been subsequently added, such as veracity, valence, variability, and value. Velocity refers to the speed at which data are created. Volume is attributed to the sheer size of the data (e.g., terabytes or petabytes), which requires specialized environments to access and store it. Variety refers to the different forms of data, such as structured (e.g., annotated Wikipedia articles). It is important to note that the vast majority of data is unlabeled, such as videos, emails,

---

[1] https://www.diamandis.com/blog/scaling-abundance-series-26, last access: Nov 2024.

etc. This means that no information on how to categorize these data into different buckets is presented alongside the raw data.

In the education sector, researchers have increasingly utilized Big Data to model, understand, and eventually improve individual learning processes, or parts of the educational system (e.g., institutions). The education sector is among the active sectors that produce vast amounts of data in high-stakes environments: educational data is most sensitive especially when related to the personal information of students. For example, the European Union passed the General Data Protection Regulation (GDPR) to ensure that data in all sectors are carefully handled in accordance with protective rights of individual citizens.

Complexity and complex systems

Complexity looms in the social world, and researchers are increasingly embracing this complexity in their research questions as mathematical tools to handle have advanced. Complex systems can be found in the natural and social sciences. For example, matter can be modeled as a complex system in which a collection of tiny particles can exhibit large-scale behaviors such as vortices in fluid flows or ferromagnetism (see Chap. 5). Ecosystems are complex systems involving multiple feedback loops and distal causes of events, such as how the application of fertilizer in a housing development contributes to fish killing in a pond several miles away. Diverse systems such as the economics of societies, the human immune systems, ant colonies, market behavior, earth and even cognitive processes and learning were modeled based on a complex systems perspective. For example, the human brain comprises many individual neurons that in their combined behavior give rise to cognition. Equally, (classical) computers can be thought of as complex systems with many interacting transistors. The brain is eventually better equipped to understand complex systems (e.g., seeing large-scale patterns and symmetries), whereas the computer is better at simulating them, as it can systematically keep track of a large number of arbitrary interacting objects.

Complex systems are not always precisely defined, but rather characterized by more or less typical attributes such as being comprised of many (interconnected) parts (e.g., molecules, cells, words, people, or human organizations), existing on different scales (such as biological organisms), and having a sense of purpose or function. Understanding complex systems requires understanding parts in relation to other parts of the system. Almost by design complex systems are closely related to complex data that might be linked to "Big Data" as well. Given the many parts of a complex system, they are characterized by high-dimensional state-spaces (i.e., possible configurations of the system). Importantly, small interactions can have large effects so that simple (e.g., linear) models are ill-equipped to capture the behavior of complex systems.

Considering the many interacting parts and non-linearities, adequate modeling techniques capable of capturing rich a hypothesis space[2] are necessary. Linear models that are widely used in research mostly assume that small causes lead to small effects. Especially early parametric models in statistics leveraged linear models. While these models are convenient to fit to data, they are by design incapable of appropriately capturing non-linear relationships appropriately. To model complex systems, it has been argued that first-principles approaches are rather ill-suited to capture relevant system dynamics and predicting behavior. Instead, data-intensive discovery tools such as ML provide promising tools for modeling such systems, and we seek to explore opportunities to do so in the context of science education research in this textbook.

3.2   Complexity in Science Education Research

Modeling teaching and learning as complex processes

Arguably, education sciences and disciplines that seek to understand and explain human behavior, human cognitive processes, learning and teaching, and social processes more generally are required to embrace complexity without expecting to reduce it to a few underlying laws (the reductionist approach). This is a central tenet in the study of emergentism. It is questionable to what extent it can be expected that a few underlying laws would be capable to capture all the variety and heterogeneity that is characteristic of teaching and learning situations. Science education researchers and educational psychologists have embraced complex systems perspectives in their research, recognizing that such perspectives can adequately approximate intra- (cognition, motivation) and inter- (teacher-student) person phenomena. As such, processes of learning and teaching, cognition, and language itself have been modeled with complex systems approaches. For example, it was shown that complex systems-based framing is adequate for research on scientific problem solving which is a complex cognitive process where equilibria and stasis play important roles (Stamovlasis, 2006, p. ix):

> [The] [p]roblem solving process through regressive steps and up to the fixed-point attractor (solution) could be seen also as following a punctuated equilibrium model analogous to the evolutionary dynamics seen in genetic systems, where long periods of stasis or slow changes are alternated by short periods of rapid changes. These changes are phase transitions and could be modeled as catastrophes.

Moreover, many teaching and learning processes are intricately related to language as a medium for knowledge generation, communication, and sense making. However, language and language use are complex (e.g., multi-dimensional, hierarchical, and compositional) such that no few underlying laws would be capable

---

[2] Hypothesis space in ML refers to the set of all possible models that an ML algorithm can utilize to predict target phenomena, given a set of features.

of fully explaining the progression of communication situations: "Perhaps when it comes to natural language processing and related fields [that model human behavior], we're doomed to complex theories that will never have the elegance of physics equations" (Halevy et al., 2009, p. 8). In a nutshell: "we can't reduce what we want to say to the free combination of a few abstract primitives" (Halevy et al., 2009, p. 9).

Reducing complexity

While no few abstract principles might be expected, it is nonetheless important to reduce complex data, e.g., by extracting patterns that exist in this data. Reducing complex data has been an important approach for educational research. Take the measurement of intelligence or personality factors as examples. The g-factor in intelligence research was derived by extracting the common variability in many different, but related tasks. These researchers demonstrated that complexity could be meaningfully reduced. Similarly, for extracting personality traits researchers hypothesized that these traits would manifest in language. Reducing complex language data then yielded a five factor model for personality traits that are widely used in psychological research. Complexity science research posits that such meaningful reductions can be expected for natural phenomena: "In many naturally occurring systems, it is observed that data exhibit dominant patterns, which may be characterized by a low-dimensional attractor or manifold" (Brunton and Kutz, 2019, p. 4). It might be considered among the great contributions of complexity science to show that in fact few underlying laws help to understand the systems.

Example: Problem solving research

Let us now provide an example that is more pertinent to science education research. An area of great interest in science education research, particularly in physics, is problem solving. Problem solving research engages, among others, in questions related to how problem solving proceeds, what cognitive and reasoning-related processes are involved, and what other factors determine successful problem solving. Researchers postulated models on how problem solving proceeds, both general and discipline-specific; however, most of them concentrate on isolated, well-defined problems. Even though most well-defined problems are studied, there still looms complexity. Constructed-responses for problem solving assessments exhibit a vast variety of problem solving strategies, fragmented knowledge, claims, statements, idealizations, meta-comments, etc. Validly and reliably assessing complex constructs such as problem solving can be guided by the assessment triangle (NRC, 2001). The assessment triangle differentiates a theory of cognition, observations, and a framework for interpretation of observations. Within the theory of cognition, researchers should detail what is known about the cognitive processes under study, such as problem solving as outlined in process models. Observations of student performance on tasks that elicit respective cognitive processes then form the data and empirical evidence. It was argued that constructed response formats (e.g., think-aloud interviews) are beneficial compared to closed-form questions (e.g., multiple-choice) to elicit relevant cognitive processes and document them via language. However, the complexity of the elicited data was considered a challenge. Finally, the framework for interpretation allows to draw inferences about students' problem solving processes. It should link performance (observations) with cognitive processes, based on substantive theory.

Data-driven discovery approaches can facilitate the use of constructed response item formats and pattern extraction. Thus, linking observations, interpretation thereof, and the theory of cognition can be enhanced. Ultimately, data-driven modeling approaches might enhance our understanding of some of the cognitive and meta-cognitive processes that undergird problem solving in greater detail, and are more grounded in empirical evidence.

3.3 Engaging with Your Research Data

We should expect our educational data to be complex. When engaging in an ML project to process and analyze this complex data, it is important to critically examine any step of data collection and processing. Therefore, we now engage in greater detail with important steps that help you understand your data and make it amenable for ML and NLP applications. Three broad steps for data handling in an ML research project can be differentiated: Obtaining the data, Exploring the data, and Preparing the data. We now focus on each of these three steps.

3.3.1 Get the Data

Suppose that your research questions require the collection of data, then opportunities for obtaining data and obtaining different types of data (see below) are vast and even expand with advanced in technology. For example, with the digitization of learning environments and technological advances novel measurement tools and novel opportunities for obtaining data open up. Researchers engaged to utilize eye-tracking, log-data, or to dynamically record classroom communications and transcribe it automatically with the help of AI-based technologies (see Chap. 15). While due care for privacy rights and data protection are necessary, evidence-based improvement of learning and teaching processes will also rely on the collection and analysis of such data.

Methodological advances such as ML and NLP allow you to expand data collection. Depending on your research questions, you may administer in an assessment project closed-form questionnaires or open-ended questions (constructed-response items). While constructed-response items were mostly used in small-scale projects, closed-form questions have been also used in large-scale projects. This is unfortunate, given that constructed-response items seem beneficial in terms of expressiveness and thus the anticipated diagnostic value is generally higher than that of closed-form items. Learners are encouraged to outline their reasoning in constructed-response items, as compared to closed-form questions. With the help of ML, large-scale use of constructed-response items may become possible.

Either for sensor data, constructed-response data, recordings, or Big Data such as journal editorials, once data is collected, further processing is required. Language data collected from constructed-response items have characteristics that are distinct from numerical data collected from Likert-scale[3] items. For the sake of this textbook, we focus on the common data types that you likely encounter in your research project: numerical data, categorical data, and language data. We will see that any type of data can be represented as vectors in numerical form.

Numerical data

Numerical data is a common form of data and comprise real-valued or integer-valued numbers that can be arranged into an array. In alignment with ML lingo, we refer to rows in the array as samples (also: examples, instances, or cases), and columns as features (also: inputs). For example, if a survey has a number of Likert-scale items, answered by a number of students, these data can be represented in a 2D array with as many columns as items, and as many rows as student answers. Each response will receive the respectively chosen answer option, say 1, which then refers to a lower agreement with this item compared to a response of 5 (given that the item is not inversely coded, e.g., through negation). Oftentimes in quantitative science education, numerical data refers to discrete, bounded measures such as Likert-scale items. If these scales are represented by numerical data (1,2,...), it is assumed that the distances between all subsequent response options are equal, which might not be true.

Not only forced-choice, Likert-scale questionnaire data can be represented as numerical arrays. In addition, images can be represented as numerical arrays (see Chap. 5), as well as language data (see below). Moreover, audio recordings or eye-tracking data can be represented as numerical arrays.

An important characteristic of numerical data is their distributional properties. The so-called normal distribution played a key role in many empirical scientific disciplines such as education and psychology. It is recognized that many naturally occurring processes give rise to normal distributions, which results (given large samples) from merely adding independently and identically distributed random variables. However, expectations of normal and normal error distributions are not always justified. Normal distributions tend to underestimate rarely-occurring events, however, many types of data such as language data are characterized by rare events that might be important for the meaning of the data. In addition, multinominal distributions are prevalent in educational data and would give rise to a vast variety of different cumulative probability distributions, only one among them being the normal distribution. It was noted that correlations among test items of .40 produce flatter distributions than normal. Moreover, all sorts of issues such as unidentified subgroups in a sample,

---

[3] Likert-scale refers to an ordered categorical rating scale where users can express their degree of (dis-)agreement with statements, etc.

ceiling and floor effects, etc. would threaten the assumption of normality of errors in educational measurements. Consequently, modeling tools are required that can deal with non-normal data and arbitrarily-shaped distributions in general, which presents a powerful argument for utilizing also ML in your research. After all, the extent to which your numerical data follow certain distributions (e.g., normal) is an empirical question, and can be tested, among others, with statistical tests (e.g., Shapiro-Wilk) and by visually inspecting the data distribution (e.g., histogram). It should also be noted that violation of normality does not always constitute problems with your analyses.

Categorical data

Categorical data are also prevalent in ML and discipline-based educational research. For example, classification problems (is this image a dog versus a cat; you identify as diverse, female, or male) produce categorical data where no inherent ordering is present, which is then called nominal data. Nominal variables can also be transformed into a numerical data array. This is known as dummy coding. For example, the self-reported gender of a person can be transformed into an array with columns for any gender that you define. A respective cell value of 1 then identifies a person with a certain gender. Multiple-choice questions produce such numerical arrays, where answer choice can be an integer-valued number representing the selected response option by the students. This can then be transformed into a binary (0 and 1) array, where correct responses are scored as 1, and 0 otherwise (missing data can be typically represented through nan/NA values).

Language data

Language data are increasingly used in quantitative science education research. Advances in NLP will eventually boost the popularity of constructed-response items. In addition, science education researchers utilized language data bacause language is an important medium for communication and sense-making. However, language, as data, can be difficult to handle. Language is comprised of units (e.g., phonemes and morphemes) that interact at different levels. Consequently, complex system perspectives have been applied to language as well. Language in general and in science disciplines can be characterized by underlying laws (e.g., subject-verb-object structures) that govern the behavior of language use. Unprocessed language data (spoken or written) represents a 1D sequence of symbols that are structured into words, sentences, paragraphs, or texts. For research purposes, mostly the complexity of language data needs to be reduced to be meaningfully processed. Language or language use can be characterized as unsegmented and ambiguous. Unsegmented, because spoken language comes as a stream of noise without boundaries. Researchers then have to segment it into meaningful chunks such as sentences. However, sentences are also connected to other sentences, because words in them might refer to words in other sentences, etc. Hence, segmentation of language data plays an important role in processing and analyzing it. Language is ambiguous, because meaning is context-dependent. Moreover, words in sequences are structured hierarchically: In a sentence the subject and object depend on the root verb, and texts more generally are hierarchically structured, where the headline predisposes the overall meaning, and paragraphs structure messages. In addition, pragmatics, i.e., the communication situation plays an important role for meaning making.

To process and analyze language data, more compressed representations have to be found. One method is to transform language data into numerical arrays, i.e., vectors and matrices. Take as an example the following sentence (Koponen and Huttunen, 2013, from: p. 2238):

The current creates the voltage between these points.

This example comprises a fairly simple English sentence (uttered by a hypothetical human who got it the wrong way around regarding current and voltage). A simple approach to analysis is to transform this sentence (we may call the sentence a document, given that this is the unit of interest) into a numerical representation. To do so, a so-called one-hot encoding is the method of choice:

Imagine that we form a vector out of this document by forming a column for each unique word in the vocabulary (of this one sentence) and indicating in each cell how often this word from the vocabulary occurs in the sentence:

```
v = array([[1, 1, 1, 1, 2, 1, 1]], dtype=int64)
```

For example, the value in the first cell indicates that the first word from the vocabulary occurs one time in our document, i.e., the sentence. Now, imagine a high-dimensional vector space with 7 dimensions, because we have 7 unique words and each word receives its own dimension (0 if not present, 1 if present). If we have a second document, say:

The voltage between these points creates a current.

We can use the vocabulary from above to represent it. The one-hot vector representation (similar encoding as above) for this new document would be:

```
w = array([[1, 1, 1, 1, 1, 1, 1]], dtype=int64)
```

Vectors can be—to some extent—represented in a diagram. Each element in the vectors represents a dimension in a respective vector space (also called feature space). We exemplify this with the terms 'current' and 'the' in the two documents. The vectors indicate that the documents are indeed different—if only for a single word. The dimensionality of 7 cannot be represented visually, however, but we can imagine that each term reserves another dimension in this vector space: They are independent of each other. Then, the documents can be compared by means of the similarity between two vectors in this space, e.g., by calculating the similarity of vectors with

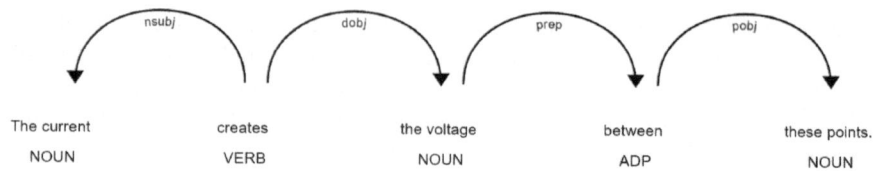

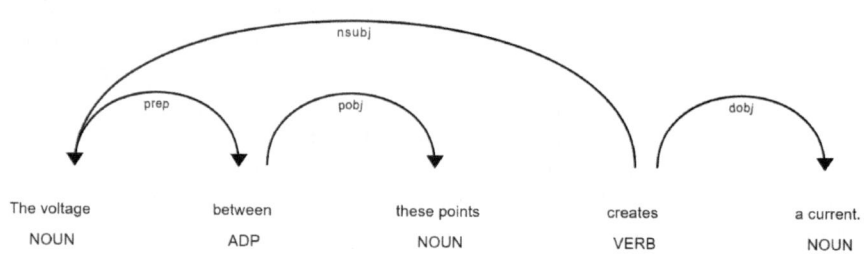

Fig. 3.1 Dependency tree representation of two sentences as generated through spaCy: Nsubj ... nominal subject, dobj ... direct object, pobj ... object of preposition, prep ... preposition

the scalar product. In our case, the scalar product[4] of these two documents would be 8., indicating positive alignment.

The documents differ only by the word 'the.' We might want to apply a technique called stopwords removal, to remove words that often bear little meaning. Further techniques could be employed to weight the terms by their importance. A frequently used technique is term-frequency inverse-document-frequency (tf-idf). As the name implies, terms are weighted by their frequency in a document and their overall occurrence. This technique gives greater weight to terms that appear in a document (and are potentially unique for this document), as compared to terms that occur across documents. This automatically filters stopwords to a certain extent.

Any form of representation such as these encodings has implicit assumptions. For the one-hot encoded language data, it is assumed that word ordering is irrelevant. This is a rather strong assumption that is (hands down) untrue in most circumstances. Moreover, the hierarchical structure of language as represented in Fig. 3.1 is destroyed through one-hot encoding. To find a representation with fewer assumptions it would be important to recognize the hierarchical structure of language.

---

[4] As a reminder: in a scalar product, each component (dimension) of one vector is summed with the respective component of a second vector. Visually, the scalar product is the projection of one vector onto the other. Hence, it ranges from 0 to a value that depends on the magnitudes of the vectors. If the scalar product is zero, the vectors are perpendicular; otherwise, they have components that are (anti-)parallel.

Dependency parsing is an important means for NLP researchers to determine dependency relations between words. For example, today, with the capabilities of LLMs, dependency parsing can be reliably automated. Patterns in language are complex: the dependency tree shows a hierarchical organization of sentences (the root verb rules noun phrases in the sentence). Moreover, noun phrases can be nested within each other ("the current that causes the wire to heat creates the voltage"). Thus, ordering in sentences matters for meaning making. In fact, sentence 2, where the voltage is the subject comes closer to the actual physical situation, where a voltage (in conducting materials) causes current. Sentence 1 represents a common preconception by students about current and voltage. Even more so, sentences are nested in paragraphs, paragraphs in chapters, etc. Patterns appear in language in a variety of forms. Dependency parsing could be utilized to capture false statements such as "current creates voltage." E.g., to capture sentences like "current creates voltage" we parse the dependency tree, extract subject, verb, and object and compare them to the misconception template.

Even with most sophisticated data structures, at some point, a numerical array is mostly the desired outcome, because it needs to be forwarded as features into the ML algorithm. Either for numerical data and textual data, the researcher has to conduct steps in a preprocessing phase to ensure the data is suited for the used ML algorithm. More concretely, in part II we engage with real-world numerical and language data.

3.3.2 Explore the Data

Given that you collected your data, now it is time to engage in exploring your data. In Chap. 2 we encountered many requirements that need to be checked for your data, such as missing values, or outliers (anomaly detection). Science education researchers often employ descriptive statistical analyses to get to know their data sets. One typically reports mean, median, standard deviation, and variance, i.e., distributional properties of the numerical data. In categorical data summary statistics such as samples per category can be reported, and in language data response length is an informative quantity to be reported. Some researchers go further to display other distributional properties for numerical data such as quantiles. Quantiles are scores from a given variable that output the values for which a certain percentage (e.g., 25%) of the values are below. Another valuable resource for getting to know your data is through a pairwise correlation matrix (they can also be tricky, see: Géron, 2017). It is also advisable to display the histograms of each numerical/categorical input feature, to visually inspect the distributions.

In ML, we typically assume that training and test data sets are generated by a probability distribution based on a certain (unknown) data-generating process. For example, if people are sampled randomly in some locations and their heights are measured, there is an (unknown) underlying distribution of all heights from which these people are sampled, and approximating this underlying distribution can be an important research goal. Typical further assumptions are then that your data is i.i.d., that is: independent, identically distributed. The former refers to the point that each sample in your data set is unrelated to another sample. This would not be true if you sampled multiple responses from one person in time-series data (see below). Sampling heights in one family that is above average will also not approximate the underlying distribution well. Moreover, one typically assumes that training and test data sets are drawn from the same unknown distribution. Critically reflecting upon your data and displaying descriptive statistics such as mean and standard deviation will already provide you valuable information about these assumptions (e.g., largely different standard deviations would hint at different underlying distributions), which can be buttressed with statistical tests (e.g., Levene's test).

While descriptive statistics are important for critical inspection, there are some particular challenges with data sampling and inherent relationships in data that should be inspected as well as they might introduce bias, affect your choice of algorithms or your conclusions. First, the representativeness of your data is an important dimension to consider, given that answering your research questions that are related to certain groups of people intricately depends on it. Moreover, distributional properties and sparsity should be critically inspected. Also, the linear separability of your data can affect the choice of algorithms that you might use, as well as considerations of global and local structures present in your data.

Representativeness of your data

One important quality of your research data is that they should be representative of your target population. Say, you want to assess the preconceptions about electrical current in grade 8 in country X. Your target population would then be the students in grade 8 from country X. For example, if boys and girls represent 49/51 percent, respectively, then these percentages should be similarly represented in your sample, if you really want to generalize to the population of grade 8 students in country X. Of course, a similar share would not prevent you from sampling bias. For example, if your study requires students to respond to questions online, you might find out that students who have no internet access at home might be disproportionally missing from your sample. Consequently, your conclusions and generalizations would be unwarranted. It is even more difficult. While there are known background characteristics that might be important (gender, ethnicity, race, prior knowledge, interests, ...), there might also be "unknown unknowns," i.e., variables that you might not even think of to be relevant to your research questions. For example, students with a certain religion might be missing on the day when your survey was administered. Alternatively, some religious students in your classroom refused to respond to a certain question, and this goes unnoticed and these response sets are then deleted because of missing data. Conclusions to the target population might then be unwarranted. If your research goals are related to such population characteristics, it is important to draw an unbiased sample. Prior literature might help you identify important background variables and procedures that might help you to anticipate sources of bias. It is then important to critically reflect upon such sources of bias and how they might have affected your conclusions. It is particularly important in the context of ML where ML models might be used in practice, because otherwise your trained ML model might make biased predictions. Checking for representativeness of your data involves checking for missing values and potentially for correlations of missing values with covariates such as religiosity, gender, age, etc. Once you notice that there is missing data and it relates to some important attributes, then you would address this as a limitation for your trained ML model.

Outliers in your data

Another important analysis can be a check for outliers. Outliers can screw prediction in many ML algorithms. For example, principal components analysis (a linear method) seeks to find a reduced-dimensional space to represent your data, and outliers will affect the representation. In k means clustering that is based on finding centroids of clusters, outliers might interfere with the determination of the centroids, and in k-nearest neighbors that is based on clustering data based on closeness to other samples, which might be spoiled. And least squares in regression analysis is highly sensitive to outliers such that model predictions for unseen data might be spoiled. You might be able to detect outliers by calculating z-scores and excluding samples with very high z-values or by inspecting specific visualizations of your data, such as a boxplot where outliers are typically points in the plot. You might even use ML techniques such as k-nearest neighbors to exclude samples with high average distances to other data points. You will likely use multiple methods and triangulate evidence to exclude data points. Note that detecting and potentially eliminating outliers is always a trade-off. You do not want to exclude any values just because they are extreme. This might be an important feature in your population under study, and eliminating this information certainly needs qualification and substantial arguments.

Sparsity of data

Another point of concern is sparsity of data, i.e., having few or no examples for relevant categories. For example, languages allow for an infinite number of sentences to be produced, and any reasonable sample (say of constructed responses) will only display a tiny subset of possible responses. Imagine you collected sample responses on the following item (see Box). For this item, $N = 448$ written responses by students were collected with a mean number of words per response of 15.3, an overall vocabulary of 942 different words was calculated. In fact, redundant words (so-called stopwords) were already deleted from the responses. Given that any response is encoded with the respective words that occur (one-hot encoding), this encoding would already have a dimensionality of 942.

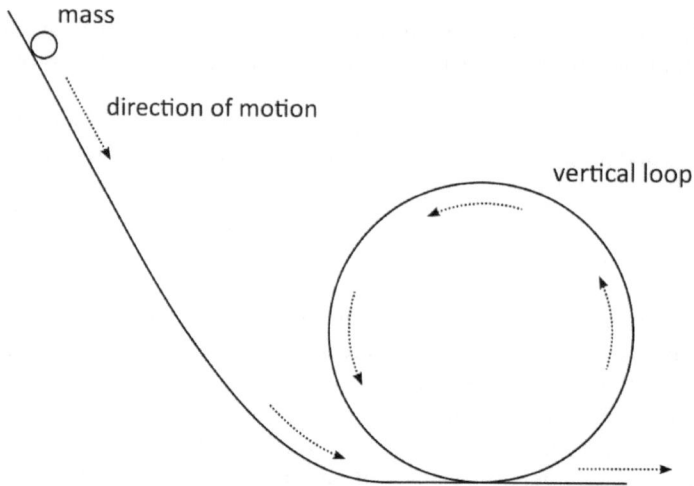

mass

direction of motion

vertical loop

Fig. 3.2 Vertical loop where a mass starts to move downwards through the loop

Conceptual physics problem: The vertical loop

Prompt: A very small mass slides along a track with a vertical loop (see figure). The mass starts from a height above the highest point of the loop. Assume the motion to be frictionless.

Determine the minimum starting height above the lowest point of the loop necessary for the mass to run through the loop without falling down.

Describe clearly and in full sentences how you would solve this problem and what physics ideas you would use (Fig. 3.2).

We can further calculate the co-occurrence of two word sequences (bigrams) in the text. An ML model could use bigrams for predicting text. For example, the word following "centripetal ..." is oftentimes "force". A model that encapsulates this pattern would be more predictive. To be capable to predict this, it would need to see representative text where bigrams co-occur in their natural frequencies for the representative target population, which are written responses on this particular item. However, even our data set for the vertical loop (which is quite large, given that 448 students were sampled) has no single bigrams that occurs even twice. This relates to sparsity of language data: One has to sample large amounts of text to get a representative language sample of the target distribution. This also introduces challenges for language prediction tasks, where architectural choices have led to significant progress in addressing the sparse-data challenge.

Linear separability of data

Test scores are often a product of the complex interactions between many different constructs. Since there is no clear path from one construct to the test score, the result can be non-linearity of the data and noise, because not all influences can be tracked and modeled. This might result in the data being non-linearly separable. Linear separable data is characterized by the fact that a lower-dimensional hyper-plane can separate all instances in one cluster from those in another cluster. If you don't know the group membership in the first place (unlabeled data) this cannot be calculated, however, if your clusters are known, linear separability of your data can be calculated. To calculate linear separability, you can use ML classification algorithms as tools such as support vector machines. Support vector machines with a linear kernel seek to find a (the best) separating hyper-plane for data points. Support vector machines have a hyperparameter, C, that can be set to very high values, which forces the machine to minimize errors when classifying points. If the machine is trained this way, it is likely to overfit the data, as it adjusts its parameters to perfectly predict every data point, including noise and outliers. If your data is truly linearly separable, a linear SVM with an appropriate margin will classify the data perfectly without overfitting. If your data is linearly separable, more simple (shallow) ML algorithms might perform well, e.g., in clustering the data. The following code is an implementation with `sklearn` in Python (see Code 3.3.2, in the online code repository, this code is applied to a real data set). If in this code a high accuracy results, your data can be linearly separated well.

Python code: Determine if your data is linearly separable

```python
from sklearn import svm
from sklearn.metrics import accuracy_score
import numpy as np

# Define the SVM with a high C value
clf = svm.SVC(C=1e6, kernel='linear')

# Train the model for your data set X1_ and y
clf.fit(X1_, y)

# Predict the labels for the training set
y_pred = clf.predict(X1_)

# Calculate the accuracy
accuracy = accuracy_score(y, y_pred)

print(f"Accuracy: {accuracy * 100}%")
```

Local and global structure of data

As you collect complex data such as language data, you might have long-range and short-range correlations. Interestingly, DNA behaves somewhat similarly to language in this regard. After all, as with language, DNA is comprised of an alphabet of the "characters" A, T, C, and G, that are arranged in 1D linear sequence to store information. As with language, the genome has long-range and short-range correlations, and understanding and predicting them is an important goal for science. Local and global structures also arise in semantic networks. In natural language, bigrams can account for short-range correlations, whereas themes/topics in a text account for long-range correlations. Once you simplify your data through models, it would be preferable to preserve both local and global correlations to accurately describe your data. Of course, there might also be settings where you are rather interested in one than the other, but in general, models are preferred to be capable of preserving both to avoid distorting the information.

Local and global structure are particularly important in ML approaches where distances between data points are utilized to cluster them or reduce dimensionality. In Chap. 5 we will introduce how unsupervised ML algorithms also seek to capture local and global structure. See the swiss roll data set colored in the spectrum colors (see Fig. 3.3, left). All points have specific distances towards each other in 3D (swiss roll), which could be calculated and depicted in a distance matrix (where each point's distance to each other point is stored as a number). Now, if you perform a projection technique such as PCA or, as done here, local linear embedding, you will distort the local and/or global structure (see Fig. 3.3, middle). Local linear embedding has the goal to preserve local structure, and proceeds by first choosing nearest neighbors of data points, then reconstructing the points from their respective neighbors, and finally finding a suitable low dimensional space. While the local structure (i.e., distances towards closer neighbors) is reasonably preserved for most samples (see Fig. 3.3, middle), global structure is distorted to a substantial degree. Multidimensional scaling, on the other hand, has the goal to preserve all distances (i.e., overall structure) of the data set. This projection can be seen in Fig. 3.3, right. The global whirl structure of the swiss roll is much better preserved in it.

While local and global correlations/structure are not specifically defined and might vary from data set to data set, it might help you as an idea to keep in mind when processing your data. In the data exploration stage, it is probably advisable to recognize in what ways your data might have local and global structures. In your project team you might ask questions like: Do we expect that our time-bound data exhibits correlations from time X to time Y? This is obviously a serious issues in most time-bound data, such as time-series data, where intra-individual correlations are present.

Depending on the specific assumptions that your algorithm makes towards data (we will go into more detail in Chap. 4), such as distributional assumptions, you will have to perform tailored analyses to check that these assumptions are met, or

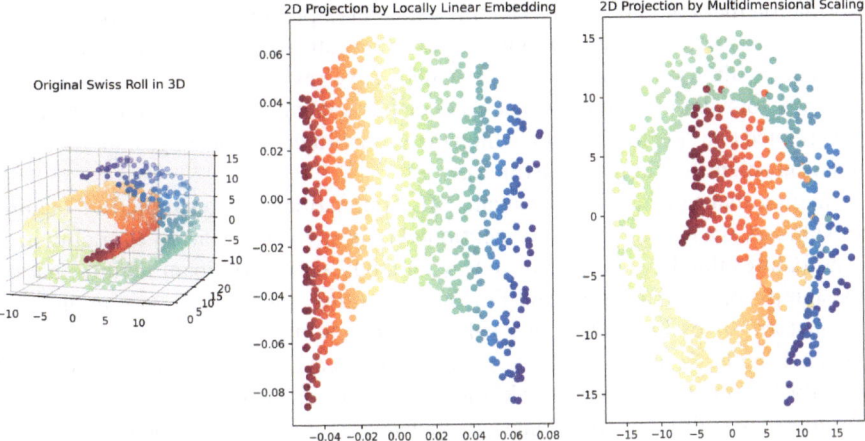

Fig. 3.3 The swiss roll data set colored with the spectrum colors (left) that is projected onto two dimensions with the goal to preserve local structure (middle), and with the goal to preserve overall structure (right). Find code for creating this figure in the accompanying notebook for this chapter

otherwise choose different ML algorithms. Having explored your data, you are now in a position to further process your data so that it can be fed into the ML algorithms.

3.3.3   Prepare the Data

Appropriately preparing your data for further input into the ML algorithms is crucial. First, all transformations to your data directly determine what assumptions you put into your analyses. For example, if you choose to represent your language data in a one-hot encoding, you eliminate the possibility that word order will be accounted by the ML algorithm. Second, preparing your data can speed up calculations or make them feasible in the first place. For example, color depth of images might increase the size of the data to handle and thus unreasonably tax your computer's working memory. Oftentimes, degrading color depth will not much affect the important features in the images related to your research questions. It will, however, boost performance of the ML algorithm and eventually speed up the training process. In any case, there will be some preparation steps necessary to make your data amenable for the ML algorithm to be trained on or explore patterns. Let us explore some important steps.

Data cleaning

Almost certainly, you will not directly utilize your collected raw data for analysis. It can be useful to clean your data to make it more manageable. Moreover, you might even be required to delete some information from the data set (e.g., sensitive information that could be used to identify students), pursuing the data sparsity principle. Cleaning data involves a range of different transformations that need to be in alignment with your research goals. In Chap. 2 we reviewed many ways that could screw up your data (remember that code or participant ID information was in one study related to a specific hospital which then constituted data leakage and eventually rendering scientific claims unjustified).

Imagine that your data, as in many ML applications, is constituted of features and values. Considering the abovementioned issues, you can start by dropping irrelevant features that have no relevance to your study, e.g., participant ID information, timestamps, or similar meta-data. For other features, such as age, you might want to transform dates into actual age values, or create dummy codes for gender (0–female, 1–male, 2–unspecified, etc.). This can be easily performed in Python with the `LabelBinarizer` from the `sklearn.preprocessing` module. For language data, there are additional important steps such as stopwords removal or lemmatization (see Chap. 7; find implementation of stopwords removal in the accompanying notebook for the LLM chapter). Next, you have to consider normalizing your features, and checking for missing values and outliers. For many ML algorithms, normalized features are important to enable sensible learning and that is not dominated by some features. The same argument counts for outliers. Moreover, some ML algorithms are designed to deal well with missing values, however, others cannot handle them and you have to find ways to account for this. We will engage in greater depth with handling missing values and outliers in Chap. 10.

Feature engineering

The features that represent the data are crucial for the performance of the ML models (we also stressed this and potential pitfalls in Chap. 2). Two ways of feature engineering are differentiated: feature selection, and feature extraction. In the former method you select the most relevant features from a given set of existing features (e.g., emotions among a set of emotions to predict learning). In the latter approach you combine different features to form new features. A famous method for feature extraction would be dimensionality reduction with the goal of reducing the number of input features by retrieving as much information as possible (see Chap. 5). As we discussed earlier, data gathered from complex systems can often be decomposed into few dimensions (attractor states) of variability that capture important properties or dynamics of the system. It is then possible to reduce the data to these states without losing (relevant) information. A major difference between shallow (sometimes referred to as traditional) ML and deep learning ML algorithms is that the latter also excels at representing the raw features and finding effective representations by themselves, e.g., in the first layers of an ANN, sparing you this to be worked out manually. However, to understand model decisions, the human researcher then would have to understand what features the deep learning ML model picked up on, which might be difficult.

Feature scaling

Feature scaling, i.e., bringing features on a common scale, is crucial for ML algorithms (especially ANNs or regression) to work well. And it is worth mentioning that you need to perform scaling only on the training data, for otherwise sensible information would leak into the training process and generalizability of your trained ML model cannot be tested anymore (some speak of data leakage, see Chap. 2). Feature normalization and feature standardization are differentiated. The former refers to fitting values in the range between 0 and 1 (subtracting min and dividing by max), and the latter refers to z-standardization (subtracting by mean and dividing by standard deviation).

3.4 Summary

The 21st century will offer science education researchers a vast variety of data sets and different types of data to make sense of teaching and learning processes, such as language, audio, video, numerical, or sensor data. Different types of data will be encountered, such as natural language data or numerical data, as well as categorical data, and even image and audio data. As explained above, a complex systems perspective can be a helpful analytical lens to understand relevant phenomena. Moreover, ML can be a valuable modeling tool, and in order to apply ML in your analyses you need to get, explore, and prepare this data. Thoughtful preprocessing and careful reflection on issues of representativeness and bias are important for drawing valid conclusions from the analyses.

References

Brunton, S. L., & Kutz, J. N. (2019). *Data-Driven science and engineering*. Cambridge University Press.

Castells, M. (2010). *The information age: Economy, society and culture* (2nd ed.). Chichester, West Sussex and Malden, MA: Wiley-Blackwell, with a new pref edition.

Géron, A. (2017). *Hands-on machine learning with Scikit-Learn and TensorFlow: Concepts, tools, and techniques to build intelligent systems*. Beijing and Boston and Farnham and Sebastopol and Tokyo: O'Reilly.

Halevy, A., Norvig, P., & Pereira, F. (2009). The unreasonable effectiveness of data. *IEEE Intelligent Systems*, 8–12.

Koponen, I. T., & Huttunen, L. (2013). Concept development in learning physics: The case of electric current and voltage revisited. *Science & Education, 22*(9), 2227–2254.

NRC (2001). *Knowing what students know: The science and design of educational*. National Academies Press.

Rothchild, I. (2006). Induction, deduction, and the scientific method: An eclectic overview of the practice of science. *SSR*.

Stamovlasis, D. (2006). The nonlinear dynamical hypothesis in science education problem solving: A catastrophe theory approach. *Nonlinear Dynamics, Psychology and Life Science, 10*, 37–70.

Chapter 4
Applying Supervised ML

Peter Wulff, Marcus Kubsch, and Christina Krist

Abstract This chapter introduces the basics of how supervised ML works. We present a pipeline which encapsulates the essential parts of an ML research project that utilizes supervised ML.

4.1 Basics of Supervised ML

Learning a mapping from inputs to outputs

Supervised ML is most commonly applied in science education research, because it allows the automation of tasks ranging from scoring students' written responses to making predictions about being at risk of failing a class (see Chap. 2). Automation and prediction are key goals for supervised ML, but how can we automate something using supervised ML? Supervised ML allows us to automate things because in supervised ML the goal is to learn a rule/mapping that relates inputs (features) to outputs (outcome variables), e.g., score on a test and give a grade. When this mapping from inputs to outputs has been learned, we have a ML model that encapsulates the relation between inputs and outputs. Then, the task of assigning outputs (e.g., grades) based on new inputs can be automated using this ML model.

We already emphasized that ML is an inductive learning approach. As such, in supervised ML the mapping is learned by providing a set of examples, called the training data. The training data needs to be split from test data that is later used to estimate the generalizability of your ML model. The training data needs to provide representative examples to the mapping between the inputs and outputs, e.g., a set of

P. Wulff (✉)
Heidelberg University of Education, Heidelberg, Baden-Württemberg, Germany
e-mail: peter.wulff@ph-heidelberg.de

M. Kubsch
Freie Universität Berlin, Berlin, Germany

C. Krist
Graduate School of Education, Stanford University, Stanford, CA, USA

© The Author(s) 2025
P. Wulff et al. (eds.), *Applying Machine Learning in Science Education Research*,
Springer Texts in Education, https://doi.org/10.1007/978-3-031-74227-9_4

test scores and the respective grades. You then need to specify which supervised ML algorithms should be used for the problem at hand (we outline multiple candidate algorithms below). There are already ways to automate this decision, for instructive purposes, however, we stick to the setup from scratch here. With the particular ML algorithm, you already constrain the hypothesis space of possible mappings that can be learned. For example, some supervised ML algorithms assume linear relationships between inputs and outputs, which might or might not be appropriate for your problem at hand.

Having split the data and chosen an ML algorithm, you then train the ML algorithm with your training data. This can be done instance based or batch based (i.e., providing multiple examples at a time). In either case, your ML algorithm provides predictions, which are rather random in the beginning of the training phase. These predictions are then compared to the actual (gold standard) outputs, e.g., grades. The difference (called: loss) between predictions and gold standards is then used to adjust the model with the goal of minimizing this difference. Large differences will typically lead to greater learning. The ML algorithm is trained for multiple epochs (runs through the training data), and the training loss is expected to decrease during training otherwise the ML algorithms gets no better at capturing the mapping from inputs to outputs. This was compared to a teacher pointing out the mistakes of a student, that is, supervising them. The training stops when the difference between predictions and gold standard either reaches a threshold or cannot be (reasonably) reduced any further. Now, the model trained using the training data is evaluated by applying it to another set of examples—called test data—and compared to how well the predictions match the gold standard values in the test data.

Supervised ML workflow

A workflow for implementing supervised ML that also recognizes the overall research problem that is to be addressed as well as the loops within the decision making related to supervised ML is depicted in Fig. 4.1.

We place particular emphasis on the following four major steps (after having posed your research question), as they are crucial for validation studies and assuring reproducibility:

1. Data splitting: splitting the data into a training and a test set.
2. Model setup: decide on a (or more) supervised ML model.
3. Model training: learning the mapping between inputs and outputs by minimizing the difference between the model predictions and the gold standard values in the training data.
4. Model evaluation: Using the test set to check to what extent the learned mapping between inputs and outputs generalizes beyond the training set.

This process is potentially iterative, e.g., when the evaluation step provides no satisfactory outcome, the model may be modified or a different type of model may be trained.

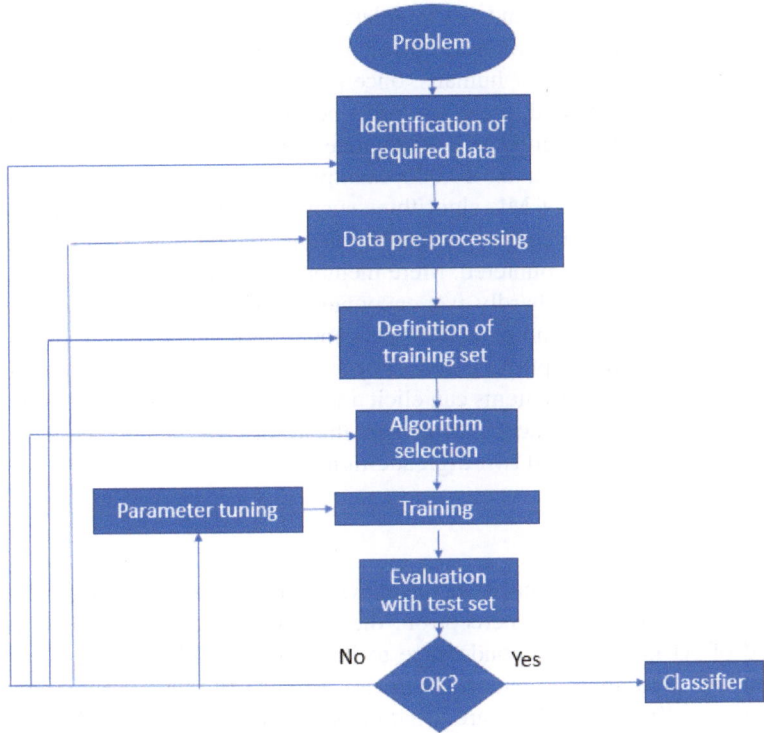

Fig. 4.1 Workflow of supervised ML, taken from Kotsiantis (2007, p. 250)

4.2   Example: Adding Two-Digit Numbers

Research problem

To see how supervised ML works and can be applied, we will start with a simple example that played quite an interesting role in the history of AI (McCloskey and Cohen, 1989): addition of natural, two-digit numbers. Let us first operationalize the problem of adding two natural numbers: Adding two-digit natural numbers is as simple as "9+5=?". Humans learn addition early on in school, however, it requires rather complex cognitive processing such as keeping track of carryover, generalizing what one knows about single-digit addition, etc.

Why using ML?

How can a machine learn to do this? The fundamental idea of supervised ML is to learn from data and involves the four steps as outlined above. The advantage of this problem is that we can easily utilize the computer's mathematics engine (calculator) to generate a wealth of train and test samples. Actually, there are overall 10,000 different possible two-digit additions. Equivalent additions (e.g., $4 + 3 = 3 + 4$) count as two distinct additions. For humans, once principles such as commutativity and associativity are understood, these problems become easier. However, we will not explicitly program the computer to recognize these underlying principles. If we were to show the computer all 10,000 examples and train it for sufficient time, then, given the capacity of ML algorithms such as the multilayer perceptron (see Fig. 4.2), we can expect it to memorize all additions and become capable of solving all of them if they are encountered. Mere memorization would give us little advantage, because in reality we hardly ever encounter problems were all data is available or all possible examples and outcomes are known. In practice, much less than the actual set of all response options is seen. For example, in language processing, even very simple, short-response items can elicit a wealth of possible responses of which only very few are seen in practice. Moreover, the trained ML model would likely not be able to generalize beyond two-digit addition.

Choosing ML algorithms

In this example, we will engage ourselves with two important ML algorithms, the single-layer and multi-layer perceptron, the sometimes called hydrogen atom of the field of AI given its preponderance to illustrate capabilities. There are many different ML algorithms to choose from (for a brief overview of some important algorithms see Table 4.1). They are sometimes differentiated into shallow and deep learning algorithms. Shallow algorithms rather depend on the given features. Hence, the researcher has to wisely choose how the input data is represented, i.e., what features are given to the ML algorithm. For example, imagine you want to predict class performance and insert self-efficacy as measured through validated Likert-

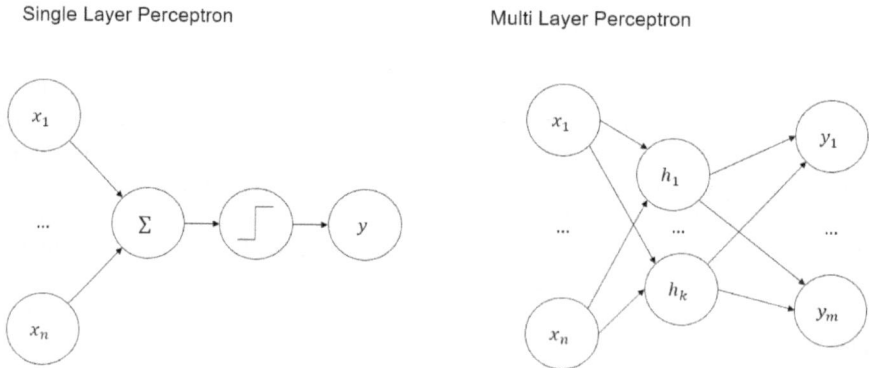

Fig. 4.2 Structure of single and multi-layer percepton. Left in both algorithms are inputs (features). The h_1 to h_k refer to hidden nodes. y are outputs. The step-function is called activation function and might also be used in the multi-layer perceptron

scale items. You could insert the average scores or the individual scores as features. Both might result in different performance of the ML algorithm. On the other hand, deep learning-based algorithms (based on ANNs) have the capacity to train their own representation of the input data. Here, you would most probably insert the individual scores and let the ANN decide which representations are formed. These representations are typically formed in the early layers of the ANNs.

Science education researchers are familiar with logistic regression models, as they also underlie the Rasch model used in item response theory. In AI research, logistic regression models are referred to as single layer perceptron (SLP, see Fig. 4.2). They might be called ANNs, because they consist of nodes and connections between the nodes. However, they are shallow, because they have no intermediate (hidden) nodes, which are quite important to make ANNs capable of representing the inputs in an efficient manner. It has been established that SLPs can classify linearly separable problems (it has been proven that they find a separating hyperplane for linearly separable data). Linearly separable means that the data points can be separated by a (hyper-)plane (in 2D a line). ML algorithms such as logistic regression and versions of support vector machines with linear kernels (see Table 4.1) are designed to find the optimal hyperplanes that separate the classes of data points. These algorithms are eventually less useful if the problem at hand is known to be non-linear.

The Multi Layer Perceptron (MLP) advances the SLP by introducing a hidden layer. It was established that MLPs are universal function approximators. Hence, they can fit almost arbitrary mappings, including non-linear mappings. This makes them ideal candidates for applying them to fit the addition example. Moreover, MLPs are the workhorse of ML researchers, actually forming the template for as advanced architectures as transformer models that power modern NLP applications such as LLMs (see Chap. 7).

Supervised ML—in a nutshell

In supervised ML a machine is shown a set of N training examples which can be complex vectors. In supervised ML each training example comes with a so-called "label," which can be complex as well, but is typically simpler compared to the input vectors. The learning goal in supervised ML is to find a function that approximates this mapping from inputs to outputs. Typical tasks are classification or regression. In classification, we search for a functional relationship from input vectors to a number of desired categories. In text-sequence classification, categories could refer to elements in argumentation, such as claim, evidence/data, warrants, reasoning (Lee et al., 2019), explanatory elements in students' evolutionary explanations (Nehm and Härtig, 2012), or elements in reflections on physics teaching enactments (Wulff et al., 2021). In regression, the mapping is from inputs to a target space that can be integer as well. This includes predicting a real-valued score, or similar. For example, education researchers predicted the utility-score for science given the students' essays

Table 4.1 Common supervised machine learning approaches. Generated with assistance by Chat-GPT (version GPT-4), conversation can be accessed here: https://chat.openai.com/share/a521048d-7993-4434-9b5d-acd7b99e9a25. For a more detailed ranking of different ML algorithms with respect to various criteria see Table 4 in Kotsiantis (2007)

Name	Main Idea	Benefits	Challenges	Application example
Logistic Regression	Probabilistic approach for binary classification	Simple, interpretable	Not suitable for complex relationships	Medical diagnosis
Decision trees	Iterative splitting of data by feature values to fine optimal grouping	Simple; well interpretable	Not suitable for complex relationships; simple trees can only partition data into hyper-rectangles (Kotsiantis, 2007)	Classifying written reflections
Support Vector Machine (SVM)	Finds the optimal hyperplane for classification	Effective in high dimensional spaces	Requires careful kernel selection	Image classification
Naive Bayes classifier	Uses Bayes' theorem for probabilistic inference	Handles uncertainty well; speed; tolerance to missing values; explainability	Computationally intensive; independence assumption (i.e., naive assumption) of features is most certainly wrong (semi-naive versions are available)	Spam filtering
k-Nearest Neighbours	Identify the k nearest (in feature space) instances to the sample and classify based on the most frequent class label	Accurate classification possible, stable compared to decision trees and some kinds of neural nets (Breiman, 1996)	Large storage requirements, sensitive to choice of similarity function, lack of principled way to choose k (Kotsiantis, 2007)	Recommendation systems
Feedforward Neural Network	Layers of artificial neurons (with weights as parameters) are trained by gradient descent and back-propagation to achieve optimal weights for classification or regression problems	Flexible, handles complex patterns	Requires a lot of data, prone to overfitting	Speech recognition

(Beigman Klebanov et al., 2017). A recently evolved strategy in the context of deep learning is sequence-to-sequence modeling (Carleo et al., 2019). Here, input and output dimensions are greater than 1. It is commonly used in NLP applications such as machine translation or textual summarization. It is possible, to also predict multiple categories (multi-class classification), or multiple scores and values.

The function hypothesis space (i.e., all possible mappings) has to be restricted. In recent times, a powerful ML model became ANNs that can cope with high-dimensional data which is desirable in many contexts. In this book, we will concern ourselves with ANNs, because they proved to be versatile tools for a multitude of problems oftentimes with better generalization performance compared to other ML models such as support vector machines (Bishop, 2006, p. 226). A task for ML researchers is then to choose wisely from among the many algorithms. This choice depends special structure of the problem at hand. For example, language-related tasks typically have to attribute for the dependency structure inherent in language, e.g., similar words in certain positions. In our case studies we will highlight widely used functions that can be utilized in science education research problems.

In order to learn the mapping from inputs to outputs it is important to provide the model feedback as to how far off it is from the desired labels. This is accomplished through a loss function (also: objective function) that expresses the deviance between predicted outputs and the targets. Designing the loss function is of utmost importance to building an accurate ANN. The loss function can be extended to include penalties for too complex models, or even physics laws that must not be violated by the outputs. We will discuss these issues later on. In fact, ANNs will shortcut any loops that you build into the loss function. Common choices are mean squared error loss for regression problems or cross-entropy loss for classification problems. Based on the loss function it is then calculated, a quantity called empirical risk. During training, the model parameters (weights) are adjusted so as to minimize the empirical risk and the training error tracks the success of this strategy.

In the learning process these parameters are constantly updated in a process called gradient descent. Then, the learning rate should be carefully chosen, because it typically impacts the training process substantially. Too low training rates will make convergence slow or impossible, i.e., the ANN does not learn anything; too high training rates will cause overshooting the optimum and eventually not converging either. In batch learning, the parameters are updated after a pass through the entire training data set. In online learning, on the other hand, the updates are performed after each data point. Typically the empirical risk is not calculated over the entire training data set, but rather over a mini-batch, i.e., a fraction of the training data set. The model parameters are then constantly adjusted.

Once a set of parameters has been trained, the model can be used for inference. To do so, an input vector is passed through the model. As discussed above, at this point performance should be assessed on the unseen test data to get some estimation for the models' generalizability capabilities. It is crucial that the entire data is randomly split and data leakage avoided at all costs, which can be intricate especially with time-series data sets or multiple responses by the same persons. Assessing generalizability for regression or simple binary classification is straightforward, as mean square error or accuracy measures such as Cohen's κ are well understood. However, for multi-class classification problems additional problems occur. For example, oftentimes certain categories are overrepresented (see Chap. 2). In this case the area under curve (AUC), precision, recall, and F1-score are used to assess the performance.

Feed-forward Neural Networks, artificial neurons, and deep learning

Artificial neural networks are widely used ML algorithms. In ANNs the number of basis functions is fixed and they are adaptive through tunable parameters that can be fixed during training. They are a directed, acyclic graph of layers. At the core of many ANNs are linear functions that operate in a paradigm called "integrate-and-fire." Again, there are input variables, weights as parameters, and biases (also parameters) that determine the baseline firing threshold. Similar to a biological neuron, the artificial neuron fires once the threshold is surpassed. Each of these functions is then wrapped into a nonlinear activation function to produce outputs. Layers can be arbitrarily stacked onto each other, and except for experience and intuition, there are few guideposts that help researchers to setup ANNs (e.g., determine width, or depth of ANNs). Each layer-wise application of weights onto the inputs represents a data transformation in the form of a tensor multiplication. Chollet (2018) compares the application of a feed-forward neural network with the unwrapping of a paper.

In a regression context, the final output is simply the output of the activation function. For multiple binary outputs, the final activations can be passed through a sigmoid function. For multiclass problems, a softmax function can be used. Sufficiently deep feed-forward neural networks can approximate arbitrarily well any functional relationship, hence they are universal function approximators. Even a two layer feed-forward neural network with arbitrary size of hidden units can approximate "any continuous function on a compact input domain to arbitrary accuracy" (Bishop, 2006, p. 319). Researchers found a class of functions that can be approximated much more efficiently with deep networks (i.e., more hidden units). Shallow networks would need exponentially more units—a phenomenon called *depth efficiency*. To what extent real-world problems fall into this class is another open problem in ML research.

It is also insightful to note the resemblance of ANNs (and artificial neurons) with human neurons that are made for information processing, i.e., passing a signal in a robust way. Inputs (in the form of activation potentials) arrive in dendrites. If these inputs are sufficiently large in sum (see summing function in SLP above), the neuron will "fire", i.e., pass the signal through the myelinated axon trunk to the terminals that are connected to other neurons. However, many researchers noted the degeneracy of analogizing the human brain with ML, noting that "there is scant evidence that brain computation works in the same way as neural networks" (Prince, 2023, p. 36).

Implementing ML algorithms in Python

The SLP and MLP can be implemented in Python as follows (see Code snippet 4.2). We will rely on the `torch` module to implement the ANNs. The SLP and MLP are then implemented as class objects in Python. They inherit some attributes from the `nn.Module` from the `torch` module. There are two functions for these classes. The former initialized the ANN, i.e., set up the linear layers, whereas the latter is used during inference time to process the input and provide predictions. All details regarding the updating of the model parameters (i.e., learning) are abstracted away by the `torch` module, which helps us to focus on the important aspects such as defining the architecture of the ANNs. The ReLU (rectified linear unit) is a so-called activation function which is important to introduce non-linearities into the model. It simply keeps positive values and sets negative values to zero. The sigmoid function maps any input value to the open interval $(0, 1)$.

Python code: Build ANNs from scratch in `pytorch`

```python
# define network:
class SLP(nn.Module):
    # Single layer perceptron
    def __init__(self,
                 dim_input=52,
                 dim_output=36 ):

        super(SLP, self).__init__()

        self.input_layer = nn.Linear(dim_input,dim_output)
        self.relu = nn.ReLU()
        self.sigmoid = nn.Sigmoid()

    def forward(self,input_):
```

```
            output = self.sigmoid( self.relu( self.input_layer(input_) ) )

         return output
class MLP(nn.Module):
    # multilayer perceptron
    def __init__(self,
                    dim_input=52,
                    dim_hidden=80,
                    dim_hidden2=80,
                    dim_output=36 ):

        super(MLP, self).__init__()

        self.input_layer = nn.Linear(dim_input,dim_hidden)
        self.hidden_layer = nn.Linear(dim_hidden,dim_hidden2)
        self.hidden_layer2 = nn.Linear(dim_hidden2,dim_output)
        self.relu = nn.ReLU()
        self.sigmoid = nn.Sigmoid()

    def forward(self,input_):
        x = self.relu( self.input_layer(input_) )
        x = self.relu( self.hidden_layer(x) )
        output = self.sigmoid( self.hidden_layer2(x) )

        return output
```

Given that ML models can learn arbitrary mapping from inputs to outputs, we should be able to train an ML model to learn this two-digit addition. Even more, it should be capable of generalizing to unseen output. For example, if the ML model has not seen the problem $74 + 23$ in the training data, we would expect it to be able to generalize from the other examples and output the correct response.

Data representation

Before diving into training the ML algorithms, we have to think about our data more carefully. Data representation is a crucial part of ML research, both in supervised and unsupervised ML. Most naively, we could simply create 100 input nodes for the first summand, and another 100 nodes for the second, leaving us with 200 input nodes that are mapped to the output nodes, which would be the natural numbers from 0 ($0 + 0$) to 198 ($99 + 99$), i.e., 199 output nodes. While this might work, it creates a large network. A more efficient representation is through a so called (distributed) one-hot encoding of the numbers 0 to 9. In this encoding, every number is represented as a vector in 11-dimensional vector space with three dimensions 1.0 and the other dimensions 0.0. We start with zero occupying the first three dimensions, and then subsequently move through the dimensions. The numbers 0 to 2 are then represented as a one-hot encoded distributed vector:

Python output: Distributed representations of numbers 0 to 2

```
        0    1    2    3    4    5    6    7    8    9   10   11
0     1.0  1.0  1.0  0.0  0.0  0.0  0.0  0.0  0.0  0.0  0.0  0.0
1     0.0  1.0  1.0  1.0  0.0  0.0  0.0  0.0  0.0  0.0  0.0  0.0
2     0.0  0.0  1.0  1.0  1.0  0.0  0.0  0.0  0.0  0.0  0.0  0.0
...
```

Each number in an addition is represented through its respective vector. To do so, we concatenate the vector for 1 with the vector for 0 (this is: 10), and so forth. For example, the addition example of $10 + 23$ will be represented as:

Python output: Distributed representation of $10 + 23$

```
tensor([0., 1., 1., 1., 0., 0., 0., 0., 0., 0., 0., 0.,
        1., 1., 1., 0., 0., 0., 0., 0., 0., 0., 0., 0.,
        0., 0., 1., 1., 1., 0., 0., 0., 0., 0., 0., 0.,
        0., 0., 0., 1., 1., 1., 0., 0., 0., 0., 0., 0.])
```

Testing for linear separability of your data set

Linear separability is an important property of data in ML research (see Chap. 2). Linear models are robust, and MLPs might be an overkill, if data is linearly separable. However, if the data is not linearly separably, linear models are not adequate for your data. Linear separability is also important in human object recognition, where non-linearly separable objects are harder to learn. ML algorithms such as SVMs actually find a linear hyperplane to separate the data, however, they also find linear hyperplanes for non-linear separable data (they first map the data into some other Euclidean space).

Testing for linear separability can be done with, among others, quadratic programming or ANNs, namely the SLP. In quadratic programming, a SVM is fit to the data. It is important to note that there are also ways to estimate the difficulty/complexity of a classification problem, which can help researchers determine which ML algorithms to use.

To generate correct responses, we rely on the internal calculator of Python to generate additions. We then write a wrapping function (see notebook in online supplement) to generate random samples of two-digit addition, specifying input, output, and the actual problem in plain notion.

We can now initialize the SLP and MLP models, and input a randomly generated value to see what the ANN returns:

Python code: Initialize ML models

```
SLP_model = SLP()
MLP_model = MLP()

x = dataset_generator.generate()
SLP_model( x['input'] )
```

An appropriate output is presented, i.e., another concatenated vector for the result of this addition. As expected with randomly initialized ML algorithms that were not trained so far, outputs are randomly scattered around 0.5, which yield zero when applied to the sigmoid function. This is good and well, given that we haven't trained the ANN.

Generate and split train and test data

Now, let's generate the train and test data. We simply randomly generate 10,000 train samples (with replacement). We then generate 1,000 test samples. Moreover, we exclude any case, where a test sample occurred in the train data. As such, we can genuinely test if the model acquired generalizability, given that we assure that it has not seen the test data during training. Note that in practice train and test data will overlap to some extent, given, for example, that we can assume that students in the future will write similar responses compared to students in the past. However, this is not necessarily the case and always should be critically reflected in ML projects.

Decide for a loss function

The ANN is meant to learn the mapping based on the train set. Therefore, the ANN is shown an input vector and is supposed to predict the correct result. The degree of difference between both is captured in the loss function (sometimes called objective function). Minimization of the loss relates to the empirical risk associated with ML. Researchers can come up with any kind of loss function, and in fact, the loss function to a large extent can determine the learning behavior. There are important differences between traditional ML versus deep learning, as well as for the specific tasks of regression, classification, and unsupervised ML (clustering, and dimension reduction).

In the original paper with the addition problem, McCloskey and Cohen (1989) used the following loss function: $d = (f - a)(a)(l - a)$, where d is the loss value, f is the target activation level (i.e., 0.0 or 1.0 of the gold standard output vector), a is the activation level of the prediction, and $(a)(l - a)$ is the derivative of the logistic function (sigmoid). A sensible first choice is to simply measure the mean absolute difference between model output (predicted response) and true response, which is called absolute loss or L1 loss. In our case of adding two-digit numbers this loss function was found to not lead to successful learning (i.e., reduce the training loss). We then tried binary cross-entropy loss, which is a widely used loss function. Cross-entropy loss generally captures the difference in the probability distributions between predicted and expected output. The smaller the value, the closer these distributions are. Cross-entropy loss noticeably improved the learning behavior.

Optimization procedure

Besides the particular loss function, a procedure needs to be decided on which updates the parameters of the ML algorithm optimally, called optimizer, so that the loss function becomes minimal in fact. In other terms, we need to find a parameter set that minimizes the loss function and then forms, alongside the algorithm, the final ML model. Derivatives of functions are important here, because they indicate what small changes in parameters might do with the loss, and thus how to minimize the loss function (find mathematical details in Goodfellow et al. (2016), p. 79. One important such method is gradient descent, a procedure that can be envisioned as hill-climbing a mountain range. One problem is that you might get stuck in a local valley and believe that you have found the global minimum. It is important for ML researchers to ensure that they have found a good solution, e.g., through probing the generalizability capacity of the trained ML model. In our case, the optimizer Adam is used to update the model parameters which is typically a very good optimizer in many ML applications. It is an efficient improvement of stochastic gradient descent.

To pass the loss signal through the model, in this case the network, back-propagation is a method of choice. Back-propagation played historically an important role in AI research, because it was found to enable ANNs to learn efficiently, making it the connectionists' master algorithm. In the original paper, they utilized the back-propagation algorithm to update the weights in the ANN. We also used back-propagation. This can be implemented with the `loss_value.backprop()` command. This command computes the gradients for the nodes, and the following command then updates the parameters: `optimizer.step()`.

Hyperparameters

As you can see from all the setups we have done so far, there is much for researchers to control and optimize. While the model parameters are updated during training, there are other parameters that are not updated during training and set in advance that control the learning process. These are called hyperparameters. There are many. For once, the model architecture (number of layers, i.e., depth of ANN; the width of the layers), the training (e.g., number of epochs, i.e., times to cycle through the training data), the learning rate (typically very important, too low values will result in essentially no learning, and too high values will prevent the algorithm from finding the optima), optimization function and loss function are all hyperparameters that can be controlled by the researcher. There are rules-of-thumb and empirical evaluations on how to set the hyperparameters. However, it is always also the task for the researcher to understand the impact of important hyperparameters. With procedures such as grid search you can systematically browse through hyperparameter configurations and find an optimal set of hyperparameters.

Training the ML model

We now train the SLP and MLP architecture. We actually tracked the training loss by writing out the actual loss for a sample in the training data set, and adding all training losses up with each other. We verified that this sum decreases over the epochs of training (see notebook in online supplement). We also fit the model after one epoch to the validation data set (not the test data), and found that this loss also decreased.

Evaluating the ML model

In the following plots, we track the loss value on predicting cases in the train and validation data plotted over different sizes of the training data (see Fig. 4.3). The MLP solves the problem for any sizes of the training data from 1000 samples to 7000 samples. It can be seen that the training loss increases slightly with more training samples (which can be expected, because it has to learn more), however, the loss on the validation data set decreases, given that probably the overlap with the training data increases. The trained ML model (MLP) then perfectly predicts the test data (see Python output). We also tested a simple logistic regression algorithm (SLP) for this problem and found that it did not succeed in solving the problem.

Python output: Classification report for MLP on test set for addition problem

	precision	recall	f1-score	support
004	1.00	1.00	1.00	97
005	1.00	1.00	1.00	109
006	1.00	1.00	1.00	98

007	1.00	1.00	1.00	95
008	1.00	1.00	1.00	114
009	1.00	1.00	1.00	93
010	1.00	1.00	1.00	111
011	1.00	1.00	1.00	90
012	1.00	1.00	1.00	101
013	1.00	1.00	1.00	92
accuracy			1.00	1000
macro avg	1.00	1.00	1.00	1000
weighted avg	1.00	1.00	1.00	1000

Conclusion

Two-digit addition is a well-defined task of specific operations. The utilized MLP was able to perfectly learn this task, based on incomplete information. Just like human learners, we would now move on to train the model to also excel at subtraction. We pose a novel problem: simple addition with ones $(0 + 1, 1 + 1, 2 + 1, \ldots, 1 + 9)$. After 150 epochs the model had the capacity to solve this problem with 100% accuracy. We then trained this model on addition with seven $(0 + 7, 1 + 7, \ldots, 7 + 9)$. The model also learns solving the addition with seven with 100% accuracy. While doing so, however, the model completely forgot how to solve addition with ones (0% accuracy). This behavior is arguably much different from human learning and became known as "catastrophic inference" or "catastrophic forgetting" (McCloskey and Cohen, 1989). This was a challenge for ML researchers for quite some time until

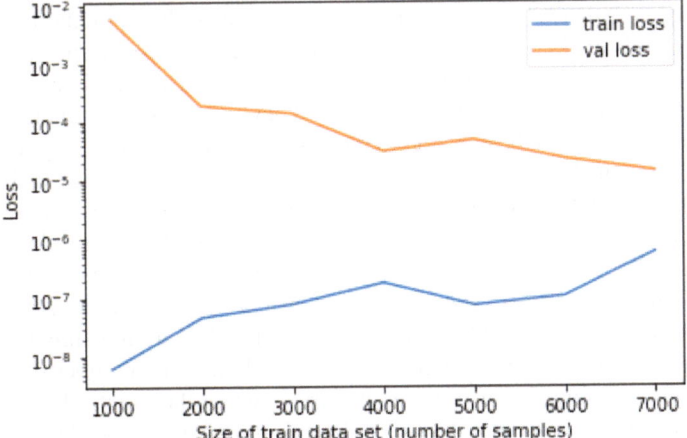

Fig. 4.3 Loss plotted over train set size for train loss (blue), and val loss (orange)

advanced ANN architectures were devised that could better incorporate their prior capabilities, e.g., by slowing down weight upgrading for important weights required for prior tasks. Moreover, different types of ANN architectures were advanced that could process information better and preserve task performance. Also, sleep-like procedures where found to partly alleviate catastrophic forgetting in ANNs. And architectures such as baseline or foundational models, e.g., transformer-based language models or memory-augmented ANNs where past events are remembered more efficiently, are quite robust to not forget capabilities when novel tasks are learned, i.e., the models upgraded.

Recent developments

The Perceptron was a milestone in ML research. A subsequent book on the perceptron by Minsky and Papert (1972) (first edition in 1969) showed limitations of the perceptron. In fact, the authors concluded that the SLP can only classify linearly separable problems. They erroneously (somewhat ironically) generalized that MLPs have similar limitations, allegedly affecting a so-called AI winter where funding plummeted. In the early days of AI (good-old-fashioned-AI, GOFAI, symbolic AI), researchers focused on expert systems, hence deductive systems based on formal logic. These were considered well versed in solving well-defined, logic problems, however, they suffered at more complex, fuzzy problems such as speech recognition or language translation. With the increasing power of hardware, it became feasible to train large ANNs. It was called the "bitter lesson of AI" (Sutton, 2019) that these statistical, inductive approaches in fact excelled at many relevant problems: "Seeking an improvement [for a relevant task] that makes a difference in the shorter term, researchers seek to leverage their human knowledge of the domain, but the only thing that matters in the long run is the leveraging of computation."

The networks then have a capacity for patterns that they can learn through back-propagation and specific training data. Specific ANN architectures of these networks helped to boost ML into a new AI spring. Alongside ANNs, a plethora of different algorithms has been developed over the decades that are tailored to specific tasks such as regression, classification, clustering, density estimation, dimensionality reduction, or representation learning. Modern deep ANNs have adjustable parameters in the billions. While humans can be considered an existence proof for a quite general learner where ANNs with billions of parameters and connections should be capable to learn in principle, the ANNs in deep learning are less general and more specific to certain classes of tasks (e.g., language processing), even though this increasingly changes with versatile NLP tools such as LLMs that learn problem solving in domains which it hasn't been explicitly trained for. We will consider LLMs later on in this book (see Chap. 7). If researchers are able to choose a suitable ANN architecture, it was shown extensively that deep ANNs can generalize even though the number of parameters far outweighs the number of training examples, up to the point where ANNs can learn even random mappings. Another robust finding is that more (e.g., parameters in the model, training samples, training time) is typically better in these kinds of architectures.

Implications

While we can be quite certain that we have a well functioning model for two-digit addition in our case, such evaluations are much more difficult in practice. For once, testing the generalizability of the trained ML model requires researchers to have a solid understanding of the expected responses. Oftentimes, generalizability can be estimated through random splitting of the given data in train and test sets. However, certain stratifications of the data might be important to consider when testing for generalizability. Notoriously, certain groups related to, among others, genders, ethnicities, or races might not be adequately represented in the training data, such that probing for generalizability is inherently flawed (see Chap. 2). More generally, the models oftentimes live in their own world, which is known as the open category problem (Christian, 2021, p. 281). A computer vision model would only know a world of 1000 categories that are provided by the researchers, and decide for a complete white noise picture that it looks more like a dog than a cat, given that it is forced to make a choice. These categories drastic simplifications of what would be encountered in real life to what would be encountered in real life. Uncertainty was often not well integrated into these models (Christian, 2021, p. 283ff).

4.3   Considering Limitations

Under- and overfitting

When training your model, two goals are important: making the training error small, and making the gap between training and test error small. Two major challenges are related to these goals: underfitting and overfitting, also related to the variance and bias of your model. Bias is often associated with simpler models (e.g., linear models) that approximate complex real world problems. A high bias relates to underfitting the training data. Variance refers to a model's sensitivity to variability in the training data set. Deep learning models are typically high in variance, i.e., they are capable of approximating arbitrarily complex training data. This is related to overfitting. In ML research problems, you typically face the bias-variance trade-off. You will likely find no optimal solution, but you have to balance bias and variance in order to build a practically meaningful ML model.

Underfitting occurs when your training error does not get sufficiently small. This could be caused by an inappropriate hypothesis space. Say, your data distribution is quadratic and you only allow linear models to be fit then the training error cannot get smaller than a certain value. Here, you could explore different hypothesis spaces and thus provide the model more flexibility to approximate the training data.

On the other hand, you might be in risk of overfitting your data. This refers to the case where your model will perform well on the training data, but does not perform well for unseen samples. Here, your model approximates the training distribution too closely which does not map well onto the test distribution. Remember: training and test distributions are ideally identical (i.i.d. assumption). However, in practice you will sample only a finite amount of examples and both distributions will necessarily differ to some extent. This can have multiple reasons such as data for training and test sets were collected differently and some background variables in the students differ. For ANNs it was found that dropping certain artificial neurons during training ("Dropout") could work as a regularizer and help prevent the network from overfitting the training data. Useful techniques to counteract overfitting are regularizations. These add a penalty to the cost function in your model training which effectively enables the model to eliminate or shrink some parameters and generate a sparse model, i.e., increase the bias.

In any case, this needs to be critically monitored during the training process of the model, and it is therefore crucially important to use cross-validation and other forms to assure that your model is neither under- nor overfitting, and generalizes well to unseen samples.

Noise and the gold standard

In supervised learning, researchers train models that seek to map inputs to outputs. However, if inputs (features) and outputs are empirical measurements, they are characterized by noise. For example, if features (called predictors in predictive modeling) are characterized by noise, then this can be propagated through the model and hamper predictive performance. Another source of noise is including uninformative predictors, i.e., those which have no relationship with the outcome. Finally, the outcome (response variable) might be noisy. Some outcomes may only be measured with some degree of unwanted, systematic noise. Say, you want to categorize sentences in students' written arguments according to some argumentation model. We noted that language is ambiguous, and if your training data is mislabeled some percentage of times, then you can only expect a certain performance from the model. This is: you will chase ghosts if you seek to tune hyper-parameters and improve model performance if already your training data classification is only 90% accurate. Therefore, it is crucial to evaluate human inter-rater performance for the classification of your training data.

Constraining hypotheses spaces and the no-free-lunch theorem

When applying ML it is important to recognize that "the designer of the learning algorithm implicitly defines the space of all hypotheses that the program can ever represent" (Mitchell, 1997, p. 23). This is called the hypothesis space, and an example would be that of a linear regression algorithm that has all linear functions in this set. This also relates to the No-free-Lunch theorem: As models can be thought of as simplifications of the observations/data, one has to make certain assumptions about the data in order to find a best performing model. However, these assumptions constrain the hypothesis space. The no-free-lunch theorem states that over all possible data-generating distributions and all possible tasks any ML algorithm performs as well as any other, say a classifier that always predicts the same class. While this is true in theory, in practical applications of ML this is of minor relevance. Baxter and Lederman (1999) outlined that the no-free-lunch theorem assumes that training and testing distributions are non-overlapping. This is not often true in real-world applications. Nevertheless, researchers who apply ML in their research should be explicit about the constraining assumptions on hypotheses that can be modeled with their ML algorithm. This is particularly true for shallow ML algorithms, while deep learning ML algorithms such as ANNs can approximate arbitrary data distributions.

References

Baxter, J. A., & Lederman, N. G. (1999). Assessment and measurement of pedagogical content knowledge. *Examining pedagogical content knowledge* (pp. 147–161). Dordrecht: Springer.

Beigman Klebanov, B., Burstein, J., Harackiewicz, J. M., Priniski, S. J., & Mulholland, M. (2017). Reflective writing about the utility value of science as a tool for increasing stem motivation and retention - can AI help scale up? *International Journal of Artificial Intelligence in Education, 27*(4), 791–818.

Bishop, C. M. (2006). *Pattern recognition and machine learning.* Information science and statistics. New York, NY: Springer Science+Business Media LLC.

Breiman, L. (1996). Bagging predictors. *Machine Learning, 24*, 123–140.

Carleo, G., Cirac, I., Cranmer, K., Daudet, L., Schuld, M., Tishby, N., Vogt-Maranto, L., & Zdeborová, L. (2019). Machine learning and the physical sciences. *Reviews of Modern Physics, 91*(4).

Chollet, F. (2018). *Deep learning with Python.* Safari Tech Books Online, Manning, Shelter Island, NY.

Christian, B. (2021). *The alignment problem: How can machines learn human values?* London: Atlantic Books.

Goodfellow, I., Bengio, Y., & Courville, A. (2016). *Deep learning.* Cambridge, Massachusetts and London, England: MIT Press.

Kortemeyer, G. (2023). Could an artificial-intelligence agent pass an introductory physics course? *Physical Review Physics Education Research, 19*(1), 15.

Kotsiantis, S. B. (2007). Supervised machine learning: A review of classification techniques. *Informatica, 31*, 249–268.

Lee, H.-S., Pallant, A., Pryputniewicz, S., Lord, T., Mulholland, M., & Liu, O. L. (2019). Automated text scoring and real-time adjustable feedback: Supporting revision of scientific arguments involving uncertainty. *Science Education, 103*(3), 590–622.

McCloskey, M., & Cohen, N. J. (1989). Catastrophic interference in connectionist networks: The sequential learning problem. *Psychology of learning and motivation* (vol. 24, pp. 109–165). Academic Press.

Minsky, M., & Papert, S. A. (1972). *Perceptrons: An introduction to computational geometry.* The MIT Press, Cambridge/Mass., 2. print. with corr edition.

Mitchell, T. (1997). *Machine learning.* New York, NY: McGraw-Hill Education.

Nehm, R. H., & Härtig, H. (2012). Human vs. computer diagnosis of students' natural selection knowledge: Testing the efficacy of text analytic software. *Journal of Science Education and Technology, 21*(1), 56–73.

Prince, S. J. D. (2023). *Understanding deep learning*. MIT Press.

Sutton, R. S. (2019). The bitter lesson.

Wulff, P., Mientus, L., Nowak, A., & Borowski, A. (2021). Stärkung praxisorientierter hochschullehre durch computerbasierte rückmeldung zu reflexionstexten. *die hochschullehre, 11.*

Chapter 5
Applying Unsupervised ML

Peter Wulff, Marcus Kubsch, and Christina Krist

Abstract This chapter provides a more in-depth treatment of unsupervised ML alongside a workflow for applying unsupervised ML in your research. We showcase the workflow with a toy example for unsupervised ML with numerical data.

5.1 Basics of Unsupervised ML

The availability of unlabelled data

While supervised ML is the prevailing method for utilizing ML in science education research and other scientific disciplines, the vast majority (more than 90%) of the estimated 1600 Exabytes of data that exist in the world today are unstructured, i.e., unlabelled, according to some estimates more than 95% of it. It is oftentimes too restrictive and resource consuming to label data, say billions of pictures. Moreover, it is not always known a priori what labels should be given. Labeling data, after all, is a theoretically involved procedure. Content analytical procedures such as inductive-deductive coding are almost unfeasible for big data. Principled and systematic procedures are required, and unsupervised ML is a promising technique that provides researchers with the means to explore patterns such as clusters in unstructured data. We reviewed that most complex dynamical systems can be reduced to a few dimensions (attractor states, or manifolds) that govern important aspects of the system's behavior (see Chap. 3). For example, in complex fluid dynamics, regular vortexes might introduce non-random patterns in the system's behavior. Unsupervised ML methods can be utilized to uncover relevant dimensions in high-dimensional data

P. Wulff (✉)
Heidelberg University of Education, Heidelberg, Baden-Württemberg, Germany
e-mail: peter.wulff@ph-heidelberg.de

M. Kubsch
Freie Universität Berlin, Berlin, Germany

C. Krist
Graduate School of Education, Stanford University, Stanford, CA, USA

© The Author(s) 2025
P. Wulff et al. (eds.), *Applying Machine Learning in Science Education Research*,
Springer Texts in Education, https://doi.org/10.1007/978-3-031-74227-9_5

sets. Even texts and images have this property. Dimensionality reduction can be used to compress data sets to make them manageable and visualize relationships in them.

Unsupervised ML workflow

The general principle of many unsupervised ML approaches is to come up with some mathematical description of what we mean by a pattern or structure. Typically, this comes down to comparing how similar or different instances in our data are and grouping them based on that similarity. The result is an assignment of the instances in our data to a set of groups. The challenge is that the groupings are not speaking for themselves: knowing that, based on answers to an interest survey, 20 students are in group 'A' and 20 other students are in group 'B' is not informative. Furthermore, these groupings can be meaningful or not–depending on the problem or question at hand. In contrast to supervised learning, we do not know what the *true* grouping (assuming a true grouping even exists) is. In consequence, the groups found using unsupervised ML need to be interpreted qualitatively, e.g., students in group 'A' might be primarily interested in biological and chemical phenomena whereas students in group 'B' might be primarily interested in physics and astronomy. With the qualitatively described groups at hand the question of validity comes up. This includes questions such as *Are the groups distinct?*, *Do the groups generalize?*, and *Can we craft a validity argument in light of our question or problem?*.

Overall, several major steps in unsupervised ML can be identified

1. Set your research goals

   - Determine your research questions according to unsupervised ML, such as dimensionality reduction, denoising, or clustering.

2. Gather unlabelled data

   - E.g., images/drawing, natural language data, questionnaire responses, network data
   - Choose representations for your data, e.g., embeddings for language data

3. Pattern recognition: using a mathematical model to find patterns in data

   - Choose appropriate ML algorithm (if problem is linear, some algorithms might be more appropriate than others; more complex algorithms also capture local and global structures that can be non-linear)

4. Fit algorithm to the data

   - E.g., find appropriate hyperparameter configuration

5. Qualitative pattern interpretation: describing the groups qualitatively

   - Visualize low-dimensional map points, visualize clusters, output representative samples for clusters, determine relation to covariates

- Compare different approaches for the problem

6. Pattern validation: Depending on the context, determining how the found pattern can be used

Tasks and goals for unsupervised ML

Among the primary goals for unsupervised learning are complexity reduction (e.g., dimensionality reduction, sometimes referred to as feature extraction), density estimation, denoising data, or clustering. It is then the target to retrieve important aspects of the original data. Unsupervised ML is often used as a preprocessing procedure (see Chap. 2) before training supervised ML or to support humans in interpreting complex data sets, given that it is capable of reducing the data to be processed to the essential dimensions. Early theoretical work dates back at least 150 years in the year 1873 (Brunton and Kutz, 2019). Science education researchers have employed unsupervised techniques for decades, namely principal components analysis (PCA) that is used, among others, to determine the underlying structures of a questionnaire. For example, Huffman and Heller (1995), and Eaton and Willoughby (2018) used exploratory and confirmatory approaches to find or confirm patterns in complex data as collected by administering the widely used force concept inventory in physics. Such approaches can advance our understanding of actual students' conception in these inventories, and how to improve the measurement instruments.

5.2 Examples: Dimensionality Reduction and Clustering

We will start introducing unsupervised ML with a most generic and widely employed technique called singular value decomposition (SVD, see Box 5.2.1) applied to a simple image, that can take us quite far in terms of preprocessing data. We then illustrate the capabilities of data-driven discovery of unsupervised ML. We will utilize complex data gleaned from the simulation of a real-world system in a physics context, namely identifying phase transitions in ferromagnetic materials (Wang, 2016). Classifying states of matter and phase transitions is a major goal for condensed matter physicists. However, identifying phases and phase transitions can be challenging when problems are complex, e.g., certain order parameters are elusive. Unsupervised ML can be used to "extract information of phases and phase transitions directly from many-body configurations" (Wang, 2016, p. 1). Finally, we will take an example from astrophysics, where researchers can use unsupervised ML to cluster astronomical objects such as stars and galaxies.

5.2.1   Dimensionality Reduction with Image Data

Why dimensionality reduction?

Reducing dimensionality of data is important for various reasons. First, when visualizing data, two or three dimensions cannot be surpassed to make them visually interpretable in our spatially 3D world. Also, the dimensionality of the data affects the computational cost for many ML algorithms, hence, lowering dimensionality also has a benefit there. Moreover, noise might be removed from the data. High-dimensional data of complex systems can be typically reduced to a set of essential dimensions that capture much of the dynamics and complexity. However, it is often important to preserve important structures in the data because it might reveal important properties about the problem of interest.

The problem: dimensionality reduction with image data

We consider the case of a single image. Analysis of images is used in many scientific disciplines. In particular, in science education researchers analyzed students' drawings with supervised ML algorithms (Zhai et al., 2022). Images have symmetries that can be exploited. For example, translational symmetry in space and time, i.e., close pixels tend to have the same color and, in a sequence of images (movie), a pixel tends to have the same color throughout some time. Consequently, an image of 10,000 pixels (i.e., dimensions) might be captured with three latent dimensions that relate to geometrical translation and rotation.

Gathering and representing the data

The following image can be retrieved from the Internet and represented in array-form (see Python output 5.2.1), where, in this case, the array is of dimension $(2736, 3648, 3)$ for height, width, and color-intensity, respectively (and, if color information is omitted, only height and width dimensionality remains as well as, for black and white images, one dimension for indicating whether a pixel is black or white). However, the pixels are highly correlated (in complex ways) with each other. For example, larger areas in the image share a similar color, hence, knowing the color of one pixel there almost determines the color of the other pixels.

Python output: picture as array

```
array([[172, 172, 172, ..., 164, 162, 164],
       [170, 172, 175, ..., 161, 162, 162],
       [171, 172, 174, ..., 159, 161, 161],
       ...,
```

```
[ 83,   81,   79, ...,  153, 151, 147],
[ 85,   82,   79, ...,  147, 144, 137],
[ 92,   88,   83, ...,  146, 139, 128]], dtype=uint8)
```

Choosing an unsupervised ML algorithm

We now might be satisfied with a smaller version of this image, which is of reduced size. This could be due to resource-related issues, such as storage capacity, working memory limitations, etc. If we are only interested in the object in the picture, we might not need minute details and resolution. Removing color would be a first step, given that colors are represented with three numbers for red, green, and blue, rather than one number for grayscale intensity. Let's apply SVD to our sample image. We perform SVD with Python as follows (see Box 5.2.1 and Python code below). For SVD, the mathematics library numpy in its linalg (linear algebra) subroutines has a function called svd that is used to perform the decomposition.

Python code: Perform the SVD

```
u, s, vh = np.linalg.svd(img, full_matrices=True)
fig,ax = plt.subplots(2,2)
sizes = [10,20,50,200]
for n,i in enumerate(sizes):
    a,b = np.where( np.ones(len(sizes)).reshape(2,2)==1. )
    r = i
    X_hat = u[:,:r]@np.diag( s[:r] )@vh[:r,:]
    s_ = u[:,:r].size + s[:r].size + vh[:r,:].size
    ax[a[n]][b[n]].set_title( f'{s_} /
                              {s_/org_size*100:.2f} \%',  )
    ax[a[n]][b[n]].imshow(X_hat, cmap='gray', vmin=0, vmax=255)
```

Find appropriate hyperparameter configuration

We can then access a submatrix which captures underlying dimensions. The larger the matrix we access, the more closely will the image resemble the final image. In Fig. 5.1 we represent 10, 20, 50, and 200 dimensions (originally the dimension was 1725). We can see that even 50 dimensions almost capture all the information that is required to generate the image. This is typically also true for language data, where entire texts might be summarized with a few number of topics.

5.2.1 Background: Singular Value Decomposition (SVD) and Principal Component Analysis (PCA)

Many techniques to perform dimensionality reduction transform the data to identify a lower-dimensional set of axes (dimensions) that accurately represent main characteristics (e.g., variation) of the data (Marsland, 2015). In essence, SVD is a matrix decomposition, where the original matrix such as the array of grayscale values for pixels is approximated with smaller matrices. The original high-dimensional data matrix is decomposed into the matrices where in the truncated form only a subset of the columns is considered. The subset is chosen so that the matrices can still reconstruct the original matrix well (Note: Details about the computations and characteristics of these matrices can be found in Brunton and Kutz (2019)).

Reference to PCA

The SVD is an optimal representation of the original matrix, given mean squared error loss (Brunton and Kutz, 2019). PCA makes use of SVD, however, first the data in the original matrix is centered, and variance is set to unity (Brunton and Kutz, 2019). Then SVD is computed to extract principal components (which are orthogonal to each other) that capture the maximal variance within the data. The eigenvalues then indicate the importance of the principal component (explained variance ratios can be calculated through dividing the eigenvalues by the sum of the eigenvalues). Researchers typically only keep the first few dimensions (i.e., principal components), hence, the dimensionality of the data is reduced. SVD is a widely used technique also in particle physics, geophysics, and many other fields.

Model validation

In this example, we primarily used "face validity" to confirm that the retrieved lower-dimensional image data meaningfully captures patterns in the data: The authors are well recognizable even with only a small fraction of the original dimensionality. For resource-related issues, we indicated the percentage of storage requirements for the respective images above them (see Fig. 5.1). An interesting parallel of SVD as a lossy compression of data is that generative LLMs also resemble lossy compressions of the Internet and other large data repositories. You might get an approximate reconstruction, but not an entirely precice reconstruction from this compression.

Fig. 5.1 SVD with increasingly more dimensions as applied to the picture of the book editors (Stina, Marcus, and Peter, from left to right) in Heidelberg (photograph taken as a "selfie"). The first number above the pictures refers to the dimensions (`u[:,:r].size + s[:r].size + vh[:r,:].size`), and the second number refers to the size of the image as compared to the original size

Dimensionality reduction with text data

Dimensionality reduction can also be valuable in working with text data. Text can be represented in a co-occurrence matrix, such as the term-document matrix (see Chap. 3). We have done this for some 5000 sentences of Jane Austen's masterpiece "Pride and Prejudice" (see code in online supplement). There are over 7000 unique words in these sentences, and the term-document (a document here refers to a sentence, as extracted with the spaCy library in Python) matrix has the dimensionality of (5129×7457) (i.e., sentences over unique words). We then applied PCA to decompose the term-document matrix (note that it is a sparse matrix, with only 0,3% of the cells occupied with actual co-occurrence values). While there are better techniques, we found that even with only 1 dimension, the reconstruction loss (mean squared error of the reconstructed matrix and the original term-document matrix) could be substantially lowered, compared to a matrix with any cell zero (see Fig. 5.2).

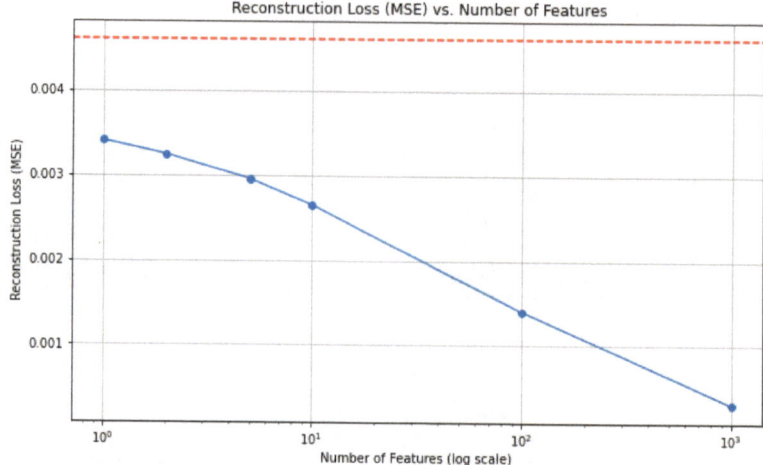

Fig. 5.2 PCA applied to text data. Displayed is the reconstruction loss (MSE) between the original term document matrix and the reconstructed matrix based on different numbers of features 1,2,5,10,100, and 1000, as well as for the matrix with all zeros (red dashed line)

5.2.2   Clustering Different States of Matter

What is clustering?

Another goal for unsupervised ML is grouping or clustering data. "Clustering is the problem of finding a set of groups of similar objects within a data set while keeping dissimilar objects separated in different groups or the group of noise" (Campello et al., 2020, p. 1). Science education researchers examined, among others, in what ways students' explanations on the seasons can be clustered with unsupervised ML (Sherin, 2013). Unsupervised ML offers different types of clustering procedures, such as density-based clustering, or hierarchical agglomerative clustering, or some mixture. Many clustering approaches in fact seek to minimize some within-cluster distance measure and increase between-cluster distance, essentially a parametric approach where the parameters of the underlying (Gaussian) distribution (a mixture of as many distributions as there are clusters). In contrast, density-based clustering approaches do not make parametric assumptions, but rather consider high-density volumes as clusters. Again, density-based clustering provides a less restrictive way to cluster complex data (e.g., within-distances in a cluster do not necessarily need to be low) that might be used to improve clustering results. Moreover, density-based clustering approaches do not need a pre-specification of numbers of clusters, which might be advantageous in explorative research contexts.

The problem: ferromagnetism and states of matter

To illustrate some of the underlying principles of clustering techniques and how to apply unsupervised ML to actual data, we will take data as can be gathered from physics systems. A classic example in condensed matter physics of a phase transition is the emergence of magnetism (ferromagnetism), given a collection (many bodies) of elementary spins. Those who experimented with ferromagnets probably noticed that above a certain temperature (Curie temperature) the magnetic field of the ferromagnet vanishes, and reappears once the ferromagnet is cooled below this temperature again (by analogy, think of evaporating water molecules (fluid) at 100°C [212°F]—a phase transition from fluid to gas). Whether a ferromagnet has magnetic properties or not then results from the interaction between minuscule elementary magnets that may or may not align in one direction. If they point in one direction, their magnetic fields add up and a magnet is formed on macroscopic scales. However, increasing movement of these elementary magnets due to rising temperature causes elementary magnets to randomly flip. In ferromagnetic materials (at ambient temperature iron, nickel, and cobalt are ferromagnetic) the elementary magnets can interact with each other. This results in areas of magnetization (Weiss domains), that add up to form a noticeable magnetic field, which can reinforce or diminish an external magnetic field.

Gathering and representing the data

In ferromagnets, the interaction of two elementary magnets is related to a certain amount of energy that this alignment costs or stores. In the simplest case (Ising model, see Box 5.2.2) we assume that the elementary magnets have only one spatial directionality, up or down, and they can align parallel (ferromagnetic) or anti-parallel (anti-ferromagnetic). Also we assume that only neighboring elementary magnets will interact with each other. The alignment of two elementary magnets will add/subtract a certain amount of energy to the system. When they align, the overall energy of the system is reduced, and increased if they do not align. Physical systems in general strive to minimize the energy in the system. Hence, at low temperatures, the ideal configuration would be for all elementary magnets to align with each other. A ferromagnet is the result.

5.2.2 Background: The Ising model of ferromagnetism

A well-studied model for the emergence of ferromagnetism is the Ising model. The Ising model assumes that only spins σ_i (up or down, i.e., $\sigma_i = +1, -1$) are important and that each body i in the problem can be represented by one spin. See Fig. 5.3 for a 5×5 square lattice with white and black spins (i.e., $+1, -1$) representing the spins. As indicated, each elementary magnet interacts with other elementary magnets. However, we assume interactions of spins only

Fig. 5.3 Very simple
2D-Ising model visualization
(see: https://de.wikipedia.
org/wiki/Ising-Modell)

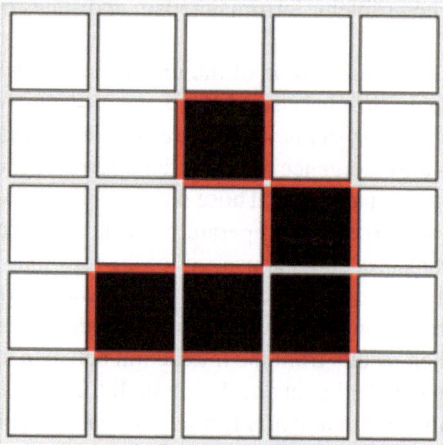

with direct neighbors, and we also assume there to be no external magnetic field. A specific configuration of white-black-fields on the lattice is referred to as a state of the system. The red lines indicate opposite elementary magnets, which store a certain amount of energy, that depends on the material, can be considered constant for all opposite elementary magnets and is stored in the system. In order to reach the energy minimum, red borders (especially at low temperatures) are to be minimized. The Hamilton function, i.e., the energy, (without external magnetic field) can then be calculated.

The minimum in energy can be achieved when all neighboring elementary magnets align parallel with each other (ferromagnetic state). This is what physical systems strive to achieve, however, things are complicated by thermal movement which introduces random flipping of the spins. So much so that above a critical temperature T_C ($T/J = 2.269$) no alignment will be achieved, and hence the material cannot become ferromagnetic anymore.

Note: The Ising model is not only capable of explaining ferromagnetism. Rather, physicists used it (the Hopfield net with energy function) to model associative memory in ANNs and calculated the memory capacity of these networks (see: https://de.wikipedia.org/wiki/Hopfield-Netz, last access Sept 2023). This work was recognized in the Nobel Prize in Physics 2024, which went, among others, to John Hopfield.

We now assume that we only have a measurement of the elementary magnets of the system (see notes). Actually, we then have an array of minus and plus ones quite similar to the data representation of the image above. We also know the temperature for the configurations. We use a Monte Carlo sampling to generate data points for a square lattice of binary spins at a given temperature using the 2-dimensional Ising model without an external magnetic field present. The temperature will be varied and each time we draw 100 samples on a 40×40 grid with the simulation. Examples for each temperature (y-axis) can be seen in Fig. 5.4.

Merely from inspecting these images visually, one notices that at very low temperatures we find spin configurations where all spins show in either one direction (i.e., all black or white). At times, however, at these low temperatures, there is a half-half split where certain areas point in one direction and other areas in another direction. In either case, the material will be magnetic to the outside, given that all spins align and the magnetic fields add up. At higher temperatures there is a much more fine-grained splitting of spins in either one direction. The threshold temperature where the phase transition happens is referred to as the critical temperature below which the spins will align. However, it would be barely possible with visual inspection to identify this critical temperature state. The system will be scale invariant at critical temperature.

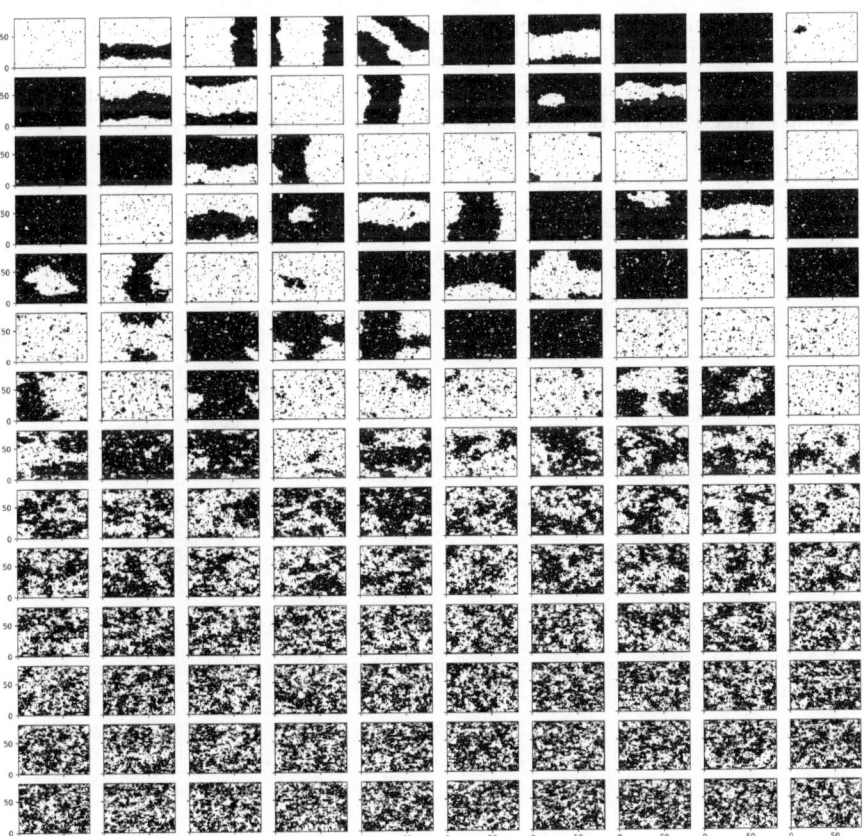

Fig. 5.4 Spin configurations for 10 MC samples for each temperature

Reduce dimensionality to process data more efficiently

Given these unlabelled arrays, it is now our goal, to use unsupervised ML to cluster
the images into groups and reconstruct the order parameter (magnetization) and the
critical temperature of the system. As we indicated when we talked about dimensionality reduction and unsupervised ML, dimensionality reduction can often be
a beneficial first step (feature engineering and extraction) in order to make your
data more amenable to further processing and clustering. Thus, we will first utilize dimensionality reduction here. To do so, we will use the t-SNE algorithm (see
Box 5.2.2) to reduce the samples in dimensionality. T-SNE can be implemented via
Python as indicated in Code 5.2.2. The TSNE method takes multiple arguments, such
as `n_components` which refers to the target dimensionality, here 2, `perplexity`,
which indicates to what extent the algorithm considers local versus global structure of the data, and some parameters on the learning process (`learning_rate`,
and `n_iter`). The argument `random_state` can be enabled to make the output of
this stochastic algorithm reproducible. We can hypothesize that either all spin (i.e.,
elementary magnet) up or all spin down should occur as separate clusters, given
that they are well recognizable configurations. We specify that t-SNE will map the
$40 \times 40 = 1600$ dimensional samples into 2D mapped samples (projections). The
resulting projections can be visualized as depicted in Fig. 5.5.

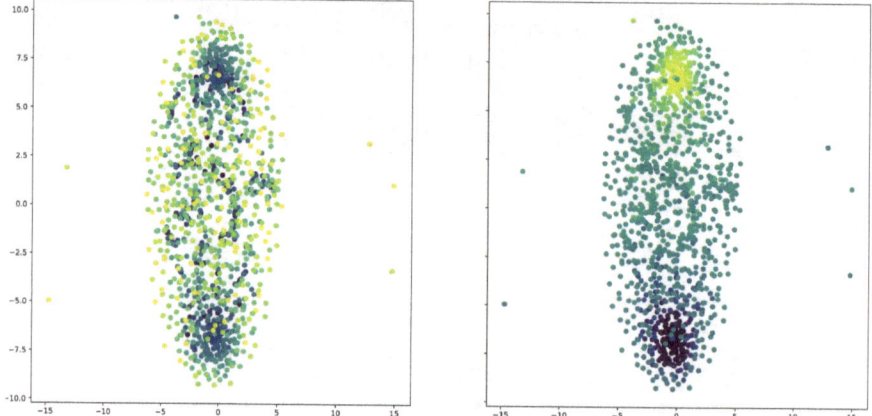

Fig. 5.5 Identical t-SNE projections of the configurations. Colors refer to temperature (left) and
magnetization (right)

Python code: Apply t-SNE to Ising samples

```python
from sklearn.manifold import TSNE
np.random.seed(42)  # Set the random seed

# Create t-SNE instance and perform the embedding
tsne = TSNE(n_components=2,
            perplexity=200,
            learning_rate=200.0,
            n_iter=10000,
            random_state=42)
tsne_embeddings = tsne.fit_transform( mc_samples_arr )
```

Note that each point in this 2D space (Fig. 5.5) refers to one configuration of the lattice of elementary magnets. The color-coding refers to the temperature (left) and magnetization (right) for this configuration. We see two discernible clusters at low temperatures (dark shaded data points) and high magnetizations. They are clusters of all elementary magnets up versus all elementary magnets down, which result at low temperatures. The high temperature samples arrange themselves in between the low temperature clusters.

Choosing an unsupervised ML algorithm for clustering

We now want to extract clusters from the t-SNE-reduced data representations. We apply a hierarchical, density-based clustering technique called HDBSCAN (see Box 5.2.2) do perform this task. HDBSCAN makes no parametric assumptions and can find clusters of different densities. Moreover, it suggests a number of clusters based on systematic processing of the data. We can see that a clustering algorithm extracts three clusters (Fig. 5.6). HDBSCAN can be implemented as indicated in Code 5.2.2.

Interpret the clusters

The three clusters in fact refer to the constellations of all spin up, all spin down, and random mixing, as we see when visualizing samples from each cluster in Fig. 5.6 (right). In fact, we hypothesize that the bottom-row cluster refers to the paramagnetic state (see also: Carrasquilla and Melko (2017), pp. 11). The sample spin configurations that belong to each cluster (see Fig. 5.6) confirm that one cluster (top row) attributes to only elementary magnets down, the other cluster (middle row) attributes to elementary magnets up, and the final cluster (down row) attributes to configurations of random mixing of elementary magnets.

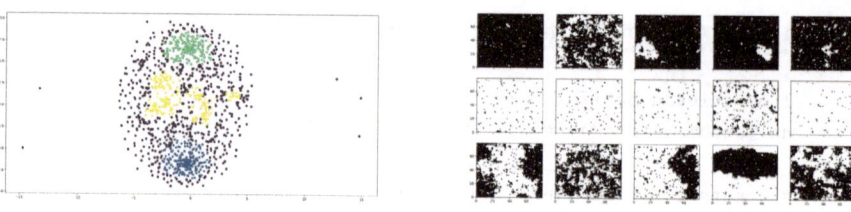

Fig. 5.6 HDBSCAN clustering algorithm applied to t-SNE embeddings (left), and 5 random samples (columns) from each cluster (rows) (right)

5.2.2 Background: HDBSCAN

Hierarchical density-based spatial clustering of applications with noise (HDB-SCAN) is an unsupervised ML algorithm to extract clusters from complex data. The basic idea of density-based clustering is to find dense volumes in the data that are then interpreted to be clusters. In contrast to clustering approaches where parameters of known (assumed) probability density distributions for the clusters are fit, density-based clustering makes no such parametric assumptions (Campello et al., 2020). In a simplified water analogy, one can imagine that the probability density distributions for the data points represents a mountainous landscape. Water is then added to this landscape. A certain water level corresponds to a threshold for cluster extraction. The extracted clusters can be thought of as the islands (i.e., regions above the water level) that remain (see Campello et al. (2020) for an illustrative explanation of this method). HDBSCAN then utilizes a data transformation into the space of pairwise dissimilarities, utilizing linkage trees and minimal spanning trees of data points and finds dense clusters that can even be of varying density for all possible epsilon values (a distance threshold) with the parameter min points that fixes the minimal count of points in a cluster, i.e., a density threshold (Campello et al., 2020). It was found that HDBSCAN is quite robust across certain parameter variations which can be an advantageous feature. Moreover, clusters of different densities can be extracted and the algorithms per design suggests stable clusters and noise points.

Python code snippet: Apply HDBSCAN to t-SNE embeddings

```
import hdbscan
cluster = hdbscan.HDBSCAN(min_cluster_size=50,
                          metric='euclidean',
                          cluster_selection_method='eom'
            ).fit(tsne_embeddings)
```

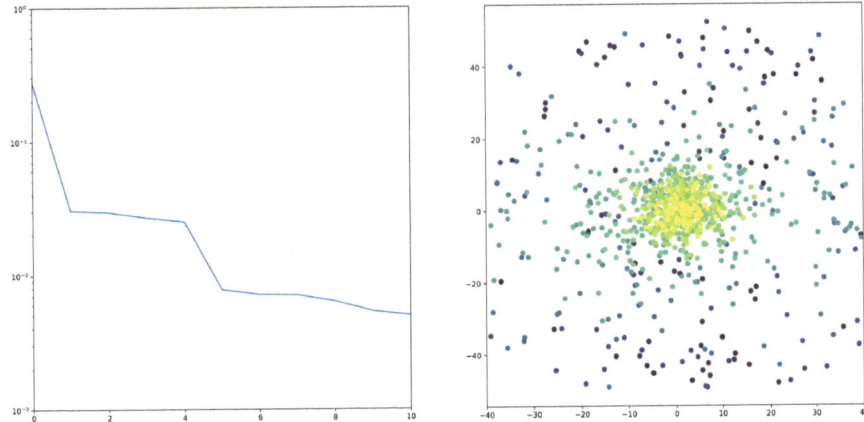

Fig. 5.7 PCA of the configuration points

Using PCA to extract patterns

Moreover, we can utilize a more familiar algorithm for educational researchers, such as PCA to get information on the underlying data. We fit a PCA on the data. The explained variance ratios indicate that really only the first principal component determines the variance in the data. As can be seen in Fig. 5.7 as projected into the first two principal components, the variance relates to the temperature. This suggests that the temperature is an important (order) parameter in this problem.

Data-driven discovery with unsupervised ML

Finally, we want to find the phase transition with the unsupervised ML approach: "Discovering a phase transition amounts to finding a hypersurface which divides the data points into several groups, each representing a phase." (Wang, 2016, p. 2). We employ HDBSCAN to group the data. In fact, the lower (third) cluster should capture the critical temperature, i.e., the temperature when the material is undergoing a phase transition (e.g., from magnetic to non-magnetic). Calculating the mean temperature for these points yields 2.36, which can be considered close to the actual critical temperature (see Box 5.2.2).

5.2.2 Background: t-SNE

Van der Maaten and Hinton van der Maaten and Hinton (2008) introduced t-SNE, which proved to be a powerful method for dimensionality reduction for high-dimensional data. The aim of dimensionality reduction methods is often to preserve important structures in the original data set when transformed into a low-dimensional representation. Science educators might be well familiar with principal components analysis (PCA) (Hotelling, 1933). PCA is a linear technique, where axes are found that capture most of the variation in the data set. However, these techniques are typically unable to adequately keep structure for high-dimensional data, especially with keeping non-linear local structure (find reasoning in van der Maaten and Hinton (2008)). Among the non-linear techniques, t-SNE proves particularly capable of retaining global and local structure of the data.

SNE seeks to align the probability distribution for data points in the high- and low-dimensional versions of the data set (van der Maaten and Hinton, 2008). The similarity between two data points is denoted by the conditional probability. The similarity, i.e., the conditional probability, will be modeled with a normalized Gaussian. In the low-dimensional space, an analogous distribution for the new, corresponding, datapoints will also be calculated. It is now the goal to make these probability density distributions in both spaces (high- and low-dimensional) approximately equal, which would indicate that the similarity is modeled similarly in the low-dimensional space. Statisticians use the Kullback-Leibler divergence to calculate the match between two probability distributions as a cost function.

However, given that the density may vary across the space, there might not be an optimal value for all data points (van der Maaten and Hinton, 2008). The value for the variance is determined through a search to produce a probability distribution with a certain perplexity, which is a hyperparameter that is meant to be set by the researcher. The perplexity is related to the effective number of neighbors (see kNN technique). Typical values of perplexity range from 5 to 50, and SNE is robust to changes in it.

Stochastic gradient descent can be used to find an optimal set of map points. First, map points are randomly sampled from an isotropic Gaussian. The gradient function resembles an energy function of a network of springs along the directions between each of the map points (can be attracting or repelling). The force exerted by a spring is proportional to the length and to the mismatch in probability distributions (i.e., similarities). Finding an optimal spring constellation will then result in a set of map points. A learning rate has to be specified by the researcher as well. This optimization problem is difficult and requires extra simulations to find appropriate parameter choices.

t-SNE uses a more suitable version of the cost function (symmetrized) and rather than Gaussian to compute similarity between the map points it uses the heavy-tailed student-t distribution. Furthermore, given that distances in high- and low-dimensional need not match on face value, in the low dimensional space a heavy-tailed distribution (student-t with one degree of freedom) is used. This allows distances to be better modeled in low-dimensional space (find reasoning in van der Maaten and Hinton (2008)). It is also important to note that t-SNE and UMAP depend critically on initialization of parameter values for preserving global data structure (Kobak and Linderman, 2021).

5.3 Considering Limitations

Algorithms, hyperparameters, and data representation

Unsupervised ML is seemingly innocuous: algorithms are utilized to reduce dimensionality or cluster complex data. However, as with supervised ML, many decisions go into setting up an unsupervised ML procedure (see workflow above). There is a vast variety of different algorithms to choose from for different purposes. Moreover, each algorithm is determined by a set of hyperparameters that have more or less impact on the quality of found dimensions or the extracted clusters. Moreover, assumptions that go into data representations are also of importance. In Chap. 3 we outlined in what ways your data representation impacts the potential structures and patterns that you can extract. We saw that bag-of-words language representations cancel out information on word ordering in sequences and hence no patterns related to word ordering can be found.

Model validation

Model validation is typically rather complex, given that we have no gold-standards that would suggest to what extent the model is capable of extracting meaningful clusters. We outlined in the workflow section that interpretation of clusters is a crucial phase. In this phase, it is important that you find representations and indicators that enable you to give meaning to the clusters. From our provided toy examples, this could relate to 2D representations or considering eigenvalues in PCA. With educational data, this will become even more complex, given that oftentimes complex language data is assessed or no sound theoretical expectations exist as to concrete parameter values as in the example of ferromagnetism (i.e., the Curie temperature). We will provide some means of validating model decisions when we consider a case study that utilizes unsupervised ML in Chaps. 11 and 13.

References

Brunton, S. L., & Kutz, J. N. (2019). *Data-Driven Science and Engineering.* Cambridge University Press.

Campello, R. J., Kröger, P., Sander, J., & Zimek, A. (2020). Density-based clustering. *Wiley Interdisciplinary Reviews: Data Mining and Knowledge Discovery, 10*(2).

Carrasquilla, J., & Melko, R. G. (2017). Machine learning phases of matter. *Nature Physics, 13*(5), 431–434.

Eaton, P., & Willoughby, S. D. (2018). Confirmatory factor analysis applied to the force concept inventory. *Phys Rev Spec Top Phys Edu Res.*

Hotelling, H. (1933). Analysis of a complex of statistical variables into principal components. *Journal of Educational Psychology, 24*(6), 417–441.

Huffman, D., & Heller, P. (1995). What does the force concept inventory actually measure? *The Physics Teacher, 33*(3), 138–143.

Kobak, D., & Linderman, G. C. (2021). Initialization is critical for preserving global data structure in both t-sne and umap. *Nature biotechnology, 39*(2), 156–157.

Marsland, S. (2015). *Machine Learning: An Algorithmic Perspective* (2nd ed.). Chapman & Hall / CRC machine learning & pattern recognition series. Boca Raton, FL: CRC Press.

Sherin, B. (2013). A computational study of commonsense science: An exploration in the automated analysis of clinical interview data. *Journal of the Learning Sciences, 22*(4), 600–638.

van der Maaten, L., & Hinton, G. (2008). Visualizing data using t-sne. *Journal of Machine Learning Research, 9*, 2579–2605.

Wang, L. (2016). Discovering phase transitions with unsupervised learning. *Physical Review B, 94*(19).

Zhai, X., Haudek, K. C., & Ma, W. (2022). Assessing argumentation using machine learning and cognitive diagnostic modeling. *Research in Science Education.*

Chapter 6
Sequencing Unsupervised and Supervised ML

Peter Wulff, Marcus Kubsch, and Christina Krist

Abstract In this chapter we provide an example application where unsupervised and supervised ML are sequenced. Unsupervised ML is first used to identify clusters in a complex data set (of galaxies, stars, and quasars). Then, supervised ML picks up on these clusters in order to be used as an automated classifier for unseen data.

Sequencing unsupervised and supervised ML is a sophisticated process of applying ML in science education, where human involvement becomes crucial (Nelson, 2020). For the purpose of demonstrating the process of sequencing unsupervised and supervised ML we will first engage with a data set from astronomy, which is arguably better suited to demonstrate the main rationale behind this approach. Afterwards, we will comment on a study that applied this approach in science education research and highlight crucial steps which are likely more pertinent to our readers.

6.1 Applying Unsupervised ML to Find Clusters

Classifying celestial objects is of great importance for astronomy to better map and understand our universe (Clarke et al., 2020). Future telescopes are expected to increase data availability on newly observed objects in ways humans will not be capable of classifying by hand in reasonable time (which is already true with current data). Moreover, gaining detailed data on all objects (spectroscopical and multi-wavelength observations) is also not feasible, and researchers would fare well with capable algorithms to classify them from less data, e.g., from photometry alone. Let

P. Wulff (✉)
Heidelberg University of Education, Heidelberg, Baden-Württemberg, Germany
e-mail: peter.wulff@ph-heidelberg.de

M. Kubsch
Freie Universität Berlin, Berlin, Germany

C. Krist
Graduate School of Education, Stanford University, Stanford, CA, USA

P. Wulff et al. (eds.), *Applying Machine Learning in Science Education Research*,
Springer Texts in Education, https://doi.org/10.1007/978-3-031-74227-9_6

us seek to sequence unsupervised and supervised ML in order to extract meaningful clusters from observational data that is provided by Clarke et al. (2020), and then utilize these clusters further to train a classifier that can classify unseen data samples automatically.

Clustering is generally an important goal in many different applications. We already saw that spin configurations of (simulated) matter can be grouped into different states. In other circumstances, one might have different properties of objects (or students) and may want to cluster them according to these attributes. For example, in astrophysics, researchers might want to cluster astronomical objects into clusters (galaxies, quasars, stars) (see Box 1). As for the purpose of computational grounded theory (CGT) we now pretend that we do not know the true underlying categories, and begin with an unsupervised clustering approach (see also Chap. 5).

Box: 6.1 the data set

The data set is publicly made available by Clarke et al. (2020). It contains photometric measurements on five optical bands (first five columns), and four infrared bands (next four columns). Finally, a measure for how extended the object is is provided (last column in Code 6.1). Moreover, the researchers provide predicted classes (STAR, GALAXY, QSO [quasar]) by their trained ML model. The first four samples in a data set comprising 10k samples are displayed in Box 1. A detailed description of the careful procedure to retrieve and clean the data can be found in Clarke et al. (2020).

Python output: Sample of galaxy data set, and classes

```
STAR
[[21.5 18.6 17.2 16.5 16.1 14.   14.   13.3  9.1  0. ]
 [19.2 18.   17.5 17.2 17.1 15.3 15.1 12.8  9.2  0. ]]
GALAXY
[[21.4 20.   19.5 19.1 18.9 14.6 14.4 11.1  8.9  2.4]
 [22.8 22.5 21.3 20.4 22.8 13.5 13.6 12.   9.   0.6]]
QSO
[[20.6 20.1 20.1 20.1 19.9 17.1 16.   13.1  8.9  0. ]
 [19.9 19.8 19.5 19.6 19.6 15.4 14.2 11.4  9.3  0. ]]
```

Setting up the data

To make the problem computationally more tractable, we sampled 10,000 instances from the original data set, and retrieved an unbalanced sample with GALAXY:

7141, STAR: 1622, and QSO: 1237. Any instance is represented through 10 features (see Box 1). Before performing any analyses, we split the data set into train, validation, and test set to prevent data leakage (see Chap. 2). To test our model on unseen data, we first split the data set into a train and test data set (note: if hyperparameters should be tuned, another validation data set would be necessary). This can be done with the `train_test_split`-function from the `sklearn` library in Python.

Choosing an unsupervised ML algorithm for dimensionality reduction

Now, based on these features, the goal is to find clusters for these astronomical objects (overall 10,000 objects) and combine them with a supervised ML algorithm to have a classification model for unseen data. If we inspect the mere correlations of the different variables (see Chap. 3) among each other as pairwise scatter plots (see Fig. 6.1), we hardly identify any discernible patterns such as major clusters. From the original paper, we then take the recommendation to use the non-linear dimensionality reduction technique called UMAP (see Box 1). UMAP is capable of

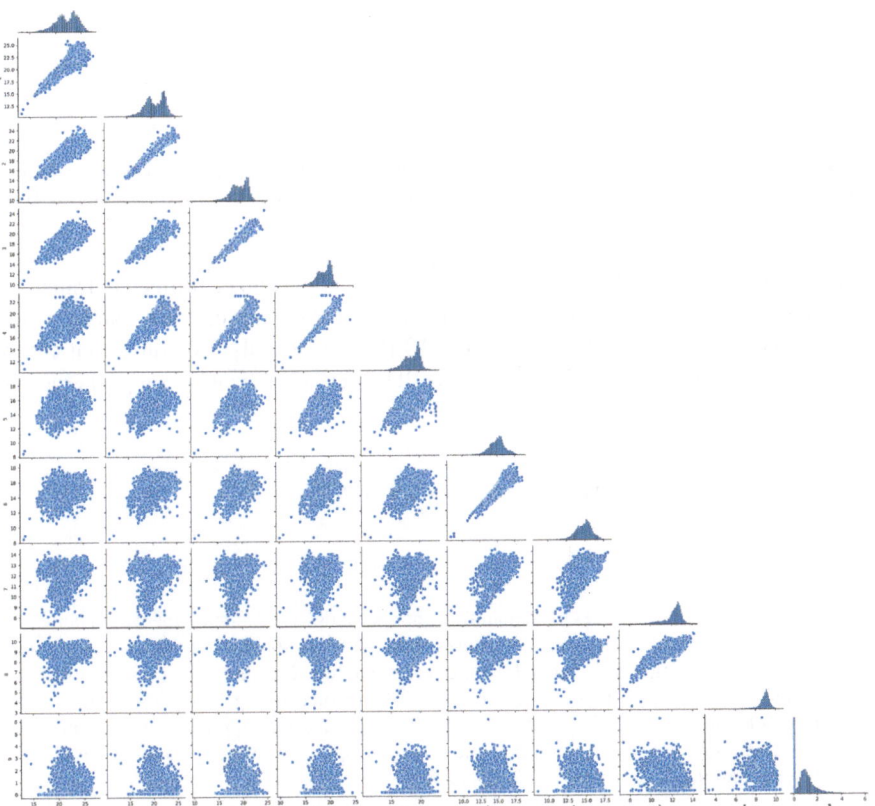

Fig. 6.1 Pairwise scatterplots of the variables in the training data set

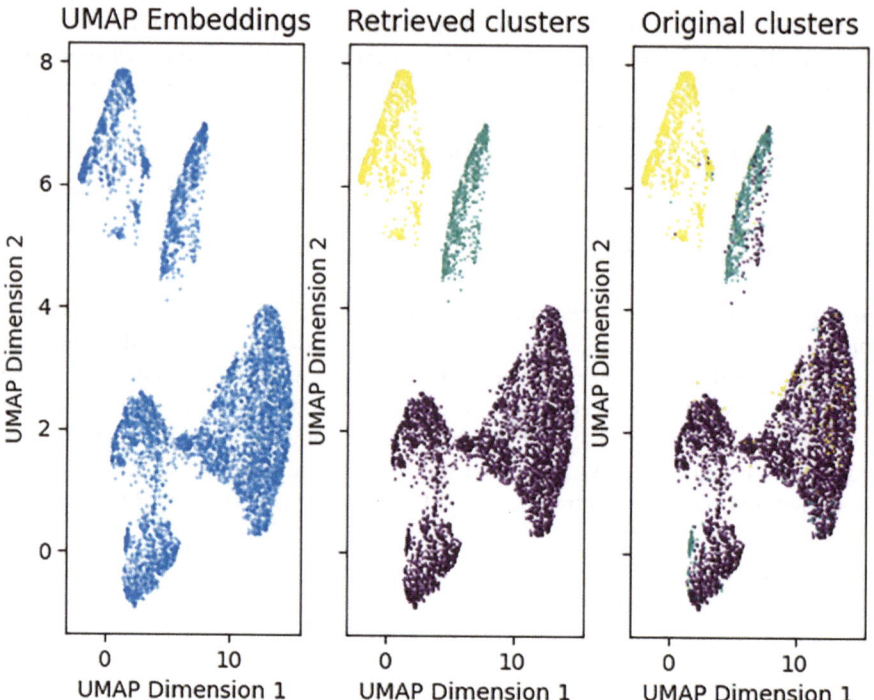

Fig. 6.2 UMAP reduced embedding space visualization for galaxies, stars, quasars data set (left). Cluster membership indicated with colors (middle). Original clusters in UMAP embedding space (right)

restoring global and local features in complex data, and it can be utilized to transform new data according to the learned model (this is, for example, not possible with t-SNE). First, UMAP is used to reduce the 10 features to only 2 features that can be displayed in a 2D space (see Fig. 6.2). The 2D space visualization suggests there to be three distinct clusters.

Box: 6.2 Dimensionality reduction with UMAP

McInnes et al. (2020) write:

At a high level, UMAP uses local manifold approximations and patches them together with their local fuzzy simplicial set representations to construct a topological representation of the high dimensional data. Given some low dimensional representation of the data, a similar process can be used to construct an equivalent topological representation. UMAP then optimizes the layout of the data representation in the low dimensional space, to minimize the cross-entropy between the two topological representations. (McInnes et al., 2020, p. 4)

As any ML algorithm, also UMAP requires the specification of hyperparameters. In particular, dimensionality of the target embeddings (low-dimensional data space), number of neighbors (lower values prefer more local structure to be preserved), and minimum allowed distance between points in embedding space (lower values preserve true manifold structure, but may lead to dense clouds) (Allaoui et al., 2020).

While t-SNE (see Box 2) is considered more focused on local structure, UMAP to some extent preserves both local and global structure. Moreover, UMAP seems to better scale with large data sets compared to t-SNE (Allaoui et al., 2020). Similar to t-SNE, it is capable of capturing non-linear relationships, as compared to PCA, which is a linear procedure (Allaoui et al., 2020). Actually, UMAP was found to improve clustering algorithms' (e.g., k-means, HDBSCAN, Gaussian mixture models, and Agglomerative Hierarchical Clustering) performance substantially as reported by Allaoui et al. (2020).

Choosing an unsupervised ML algorithm for clustering

For clustering purposes, we utilize HDBSCAN (see Box 2), since it suggests a number of clusters. It can be seen that some clusters are probably not correctly classified, which might be mended by tinkering with the UMAP hyperparameters to get a clearer separation of the clusters. Moreover, hyperparameters in HDBSCAN were chosen by us to retrieve a fairly reasonable looking distribution of clusters. In reality, cluster interpretation should be more theoretically involved. The feature values for the different classes could be compared and experts in astronomy will likely be able to discern galaxies, quasars, and stars from average feature values. However, we will accept these clusters for our present purposes.

Choosing a supervised ML algorithm for classification

We finally want to utilize this clustering to predict unseen data. We utilize a random forest classifier (see Box 1) is fit to the train data, which is an advanced version of decision trees. The random forest classifier can be implemented in Python as displayed in the Python code. Note that UMAP is applied as well to the unseen test data. We took care that no information from the validation or test data leaked into the UMAP or clustering algorithms. The Python output shows the random forest classifier could accurately classify instances in the unseen test data set with slight differences for the different categories. As we hinted at before proves true in this example problem as well: the category with the most support could be classified with the highest accuracy. The unbalanced nature of this data set can be seen by inspecting the F1 scores. The macro F1 score is noticeably lower compared to the weighted score. This indicates that some categories are underrepresented and have different (lower) class-based F1 scores.

Random forest classifier

We already encountered the decision tree algorithm in Chap. 4. Decision trees seek to separate the input data through binary decisions into classes that best approximate the gold-standards. Random forests utilize multiple trees: "The idea is largely that if one tree is good, then many trees (a forest) should be better, provided that there is enough variety between them" (Marsland, 2015, p. 275). Random forests generate random decision trees by either presenting each decision tree with a different training data set, and by restricting the features that each tree can use to make decisions (Marsland, 2015). The subset of features to consider is actually a new parameters, but random forests do not seem to be very sensitive to this parameter. Moreover, the number of trees to put into the forest is another hyperparameter to set up (Marsland, 2015). The introduction of randomness through restricting training data and features reduces the variance, but does not affect the bias (Marsland, 2015) (See Chap. 2 for more information on variance and bias). Decision making is then formed by a majority vote of decision trees, hence it is a committee/ensemble method.

Python code: Apply UMAP and Random Forest Classifier to the train/test data

```
rf = RandomForestClassifier(n_estimators=100, random_state=42)
rf.fit(X_train_umap, y_train_pred)

# Using UMAP to transform the test data
X_test_umap = reducer.transform(X_test)

# Using the Random Forest model to predict the labels
# on the UMAP-transformed test data
y_pred = rf.predict(X_test_umap)

print( classification_report(y_test, y_pred) )
```

Classification report: Random forest

	precision	recall	f1-score	support
GALAXY	0.94	0.98	0.96	1444
QSO	0.81	0.78	0.79	240
STAR	0.99	0.83	0.91	316

accuracy			0.93	2000
macro avg	0.91	0.86	0.89	2000
weighted avg	0.93	0.93	0.93	2000

6.1.1 Considering Limitations

In this chapter, we showed how sequencing unsupervised and supervised ML could be used to extract patterns in data and utilize them for classifying unseen samples. For this example, we found a quite high accuracy as measured, among others, through the F1 score. In our example, we utilized our knowledge about astronomy (actually about this data set) that three distinct structures are to be expected in the embedding spaces. In reality, more theory-driven arguments and systematic ablation studies (e.g., through systematic variation of hyperparameters, see Tschisgale et al. (2023)) are necessary in order to come up with robust and justifiable ways to extract meaningful patterns in these data sets. Moreover, researchers could utilize means to balance the unbalanced data set (see Chap. 2). Furthermore, explainable AI approaches could be reasonable ways to find important features for classification and clustering. In the original paper, Clark et al. (2020) retrieved the relative feature importance for the random forest classifier. They found that the measure of how extended the object is played the most important role in classification. This was not possible for our analysis, given that we utilized random forests on the UMAP embeddings which are inherently difficult to interpret. Researchers could therefore decide that it would be more meaningful to utilize all 10 features in a classifier and find out the feature with greatest importance in order to learn something about the problem at hand.

6.2 An Applied Example in Science Education Research

While the stars, galaxies, and quasars data set illustrates the main steps of sequencing unsupervised and supervised ML in a comparably straightforward context, applying this method in science education with real-world, complex data is somewhat more intricate. Tschisgale et al. (2023) used this method in physics education to extract clusters in students' problem textual solutions, and open sourced the data set and Python code.[1] You can implement the entire analysis on your own computer with these resources. Here, we will briefly highlight the main steps in the analysis with the accompanying code.

---

[1] Find the data and code here: https://osf.io/d68ch/, last access: May 2024.

The goal of this study was to identify clusters in students' textual descriptions of their problem solutions to the vertical loop problem already mentioned in Box 3.3.2 (Chap. 3), and to train a classifier that assigns a cluster to the respective sentences. It is important to note that this study did not attempt to develop a classifier that would perform well on unseen data, as this would entail splitting the data into train, validation, and test set beforehand (see Chap. 2).

In order to utilize unsupervised ML, in a first step LLMs were used with the `SentenceTransformer`-class from the `sentence_transformers` library in Python. These embeddings (high-dimensional vectors) could then be piped into a clustering algorithm. However, beforehand, dimensionality reduction was performed (UMAP), given that we already mentioned that complex data can typically be reduced to a few dimensions. Finally, a clustering algorithm (HDBSCAN) was utilized to calculate the resulting clusters. The influence of hyperparameters (e.g., the number of dimensions that the embeddings should be reduced to) was carefully evaluated against resulting clusters, and by considering the theory articles on these techniques and prior analyses with similar goals.

Python code snippet: embeddings, dimensionality reduction, and clustering

```
model = SentenceTransformer(
            modules = [word_embedding_model,
                        pooling_model])
embeddings_orig = model.encode(
                    sentence_corpus,
                    show_progress_bar = False)
...
UMAP_object = umap.UMAP(n_neighbors = 15,
                    n_components = 5,
                    metric = 'cosine',
                    random_state = 7276744,
                    min_dist = 0).fit(embeddings_orig)
embeddings = UMAP_object.transform(embeddings_orig)
...
cluster = hdbscan.HDBSCAN(min_cluster_size = 15,
                    metric = 'euclidean',
                    cluster_selection_method =
                    'eom').fit(embeddings)
df_step_1['Cluster'] = cluster.labels_
```

The resulting clusters (see Fig. 8 in Tschisgale et al. (2023)) were then read by human raters with complementary information such as a condensed tree (see Fig. 9 in Tschisgale et al. (2023)). A rubric of clusters was established (Table I), and based on theoretical arguments and the complementary information human raters considered merging several clusters with one another (see Python code). In the end,

5 distinct clusters were identified: assumptions and idealizations (AI), conceptual aspects (CA), quantitative aspects (QA), formulation of a solution (FS), and general descriptions (GD).

Python code: pattern refinement

```
dat_CGT_step_2['Cluster_global'] =
    ['' for k in range(np.shape(dat_CGT_step_2)[0])]
for k in range(np.shape(dat_CGT_step_2)[0]):
    if dat_CGT_step_2.loc[k,'Cluster_adapted'] in [0]:
        dat_CGT_step_2.loc[k,'Cluster_global'] = 0
    elif dat_CGT_step_2.loc[k,'Cluster_adapted'] in [5, 6, 9, 10]:
        dat_CGT_step_2.loc[k,'Cluster_global'] = 1
    ...
```

In the last step, a relevance vector machine (RVM) was trained (supervised ML) with the embeddings as input and the cluster label as output. Accuracies were 0.77 and 0.76 for females and males, such that it could be concluded that the classifier did not exhibit bias with regards to gender. Even though this process requires substantial work, the end product would be a classifier for this particular physics problem (we assume here that the test data would be equally distributed as the training data). Such a classifier might be used as a means to indicate to students which steps in the problem-solving process (e.g., making assumptions) are still missing.

Python code: pattern confirmation

```
model = EMRVC(kernel = "rbf")
model.fit(X_train, y_train)
y_predict = model.predict(X_test)
```

References

Allaoui, M., Kherfi, M. L., & Cheriet, A. (2020). Considerably improving clustering algorithms using umap dimensionality reduction technique: A comparative study. In A. El Moataz, D. Mammass, A. Mansouri, & F. Nouboud (Eds.), *Image and Signal Processing, Springer eBook Collection* (pp. 317–325). Cham: Springer International Publishing and Imprint Springer.

Clarke, A. O., Scaife, A. M. M., Greenhalgh, R., & Griguta, V. (2020). Identifying galaxies, quasars, and stars with machine learning: A new catalogue of classifications for 111 million sdss sources without spectra. *Astronomy & Astrophysics, 639*, A84.

Marsland, S. (2015). *Machine Learning: An Algorithmic Perspective* (2nd ed.). Chapman & Hall / CRC machine learning & pattern recognition series. Boca Raton, FL: CRC Press.

McInnes, L., Healy, J., & Melville, J. (2020). Umap: Uniform manifold approximation and projection for dimension reduction. *arXiv*.

Nelson, L. K. (2020). Computational grounded theory: A methodological framework. *Sociological Methods & Research, 49*(1), 3–42.

Tschisgale, P., Wulff, P., & Kubsch, M. (2023). Integrating artificial intelligence-based methods into qualitative research in physics education research: A case for computational grounded theory. *Physical Review Physics Education Research, 19*(020123), 1–24.

Chapter 7
Natural Language Processing and Large Language Models

Peter Wulff, Marcus Kubsch, and Christina Krist

Abstract In this chapter we introduce the basics of natural language processing techniques that are important to systematically analyze language data. In particular, we will utilize simple large language models and showcase examples of how to apply them in science education research contexts. We will also point to recently advanced large language models that are capable of solving problems without further training, which opens up novel potentials (and challenges) for science education research.

7.1 Natural Language Processing

The necessity to systematically process language data

Language data was singled out as a key resource for evidence on learning and teaching-related processes (see Chap. 3). The difficulty with language is that it is ambiguous, unsegmented, and noisy. For example, the meaning of a term is dependent on the context in which it is used (distributional semantics), i.e., "the semantic values associated with words are flexible, open-ended and highly dependent on the utterance context in which they are embedded" (Evans, 2006, p. 491). It is therefore difficult to specify "the" meaning of a term in a lexicon. For example, the term "work" has many different meanings, depending on whether you relate it to everyday life ("I go to work.") or in science contexts ("Work is force exerted to an object along a path."). However, even in science (here: physics) at least half a dozen different usages of the term "work" can be differentiated. Rather, flexible methods are required to make sense of a term in context, a task also called word sense disambiguation. Moreover, language typically comes as a 1D sequence of words, where only in written

P. Wulff (✉)
Heidelberg University of Education, Heidelberg, Baden-Württemberg, Germany
e-mail: peter.wulff@ph-heidelberg.de

M. Kubsch
Freie Universität Berlin, Berlin, Germany

C. Krist
Graduate School of Education, Stanford University, Stanford, CA, USA

P. Wulff et al. (eds.), *Applying Machine Learning in Science Education Research*,
Springer Texts in Education, https://doi.org/10.1007/978-3-031-74227-9_7

117

language speakers (often) use signs (e.g., punctuation such as period, semicolon, or comma) to signify segmentation boundaries in language. However, these boundaries might not be most relevant in your research, and other coding units might have to be defined. Thus, systematic means are required to segment language data into meaningful units in order to further analyze it. Finally, real world experiences and situations are high-dimensional and language use about such experiences and situations can only capture parts of it. Hence, noise and ambiguities are characteristic of language utterances, both spoken and written.

It is important to remember, however, that systematic tools for processing language cannot compensate for substantive domain knowledge as human experts have (see Chap. 8). While it is crucial to systematically segment language and retrieve specific word senses in context, this does not exempt researchers from assuring that the processing of language is meaningful given their research goals and interpret the results. For example, one study might be interested in calculating the frequency of specific words (types) used in science textbooks. The idea that human raters perform this task is quite absurd given the redundant nature of the tasks and the size of the textbooks. Anyways, computers excel at systematically browsing through large data sets such as science textbooks. Yet, a capable segmentation procedure is required to extract specific words. Simply segmenting text by spaces might not be sufficient for the research question. It might be important to treat terms such as "free fall" or "cell division" as one entity, given that they are intentionally used by the writer to designate a specific concept. Human researchers need to critically monitor such assumptions that go into the systematic processing of their language data. This can at times be daunting, and we seek to provide some guidance in this chapter on what assumptions might crop up and what NLP procedures can facilitate your research. Let's start with the historical development of language processing by computers, which can be considered as old as computers are themselves.

A historical brief of Natural Language Processing

Natural language processing (NLP) is an umbrella term for a vast variety of methods related to computer-based processing of spoken and written language data in a systematic, computer-based way. NLP researchers developed a rich tool set of methods to systematically process and analyze natural language utterances with boosts through advances in ML. The history of the field of NLP parallels the development of AI. NLP can be divided into four phases: (1) machine translation (1950–1969), (2) syntax and co-reference handling (1970–1992), (3) adoption of shallow ML algorithms (1993-2012), and (4) deep learning and ANN (2013-present) (Manning, 2022). In phase 1, NLP systems were mere word-level lookup tables to perform translations of "comically small" (Manning, 2022) data sets (compared to the present). In phase 2, some of the complexity of human language was embraced and put into rule-based NLP systems that separated declarative linguistic knowledge and the procedural processing of it. A prequel to the fame of ChatGPT in 2022/23 was already demonstrated by the program ELIZA. ELIZA enables through preset scripts human-machine interaction in natural language. Most famously, ELIZA could model a non-directive psychotherapist. In it's simplest form, ELIZA had access to a rule-based list of words. A request by a human is then searched for a word in this list. For most words, also hypernyms are linked to them. An interaction could then look like:

User: I have problems with my father.
ELIZA: Tell me more about your family.

Allegedly, this simple rule-based NLP system was used by many people. This so-called ELIZA-effect is nowadays observed in many NLP systems and we will come to it later when talking about LLMs. In any case, the ELIZA program was hard-coded and not able to learn from data unless the human designer improved the word list.

In phase 3, large data sets (some millions of words) became available and NLP researchers embraced statistical modeling, ML algorithms in particular, to extract information from these data sets (empirical orientation). It was noticed in 1996: "In the space of the last ten years, statistical methods have gone from being virtually unknown in computational linguistics to being a fundamental given" (see: https://norvig.com/chomsky.html). In particular, researchers sought to develop annotated data sets, so that supervised ML algorithms could be trained on specific NLP tasks, such as part-of-speech tagging (i.e., singling out a word as a noun, etc.). Finally, in phase 4, the empirical orientation was extended and ever larger data sets (now billions to trillions of words, i.e., Big Data) and powerful ML algorithms became available that also required NLP researchers to embrace novel strategies to efficiently and effectively train the ML models. Deep learning methods enabled more performant models. A hallmark moment came in 2017/2018 when it could be shown that self-supervised learning alongside a self-attention mechanism enabled ML models to develop linguistic capabilities which made them useful in many different tasks. The now omnipresent LLMs were introduced.

Large Language Models

Since November 2022, when the conversational AI ChatGPT was made available by OpenAI, the public as well as researchers and practitioners from various disciplines suddenly became acquainted with a novel technology: LLMs. While modeling language such as texts was an important objective for NLP researchers from early on, the introduction of the transformer architecture enabled new degrees in predictive accuracy and generative capabilities for LLMs. LLMs can be used in different ways, e.g., fine-tuning or prompting (see below), and are trained in a self-supervised form of ML (see Chap. 2). Given the self-supervised training, similarities to mechanisms of how human brains interact with the world (not only language) were noticed. The human brain was conceptualized as an experience machine that is constantly predicting its environment, incorporates received error messages (i.e., unfulfilled predictions), and updates expectations as a key mechanism (also known as predictive coding, or active inference). It is then probably not surprising that predicting language in context is a valuable training objective and enabled LLMs to be versatile tools for a broad range of (linguistic) tasks, such as textual summarization or translation between languages (we will inspect limitations later on), with the (debated) limitation to be restricted to (false) knowledge that has been seen during training (see Chap. 2). This training objective enables English LLMs to possess "basic capabilities in syntax, semantics, pragmatics, world knowledge, and reasoning, but these capabilities are sensitive to specific inputs and surface features" (Chang and Bergen, 2023, p. 1), which tricked some people into believing these LLMs are human writers (see Box 1).

Box 1: The imitation game and LLMs

In the mid-20th century, the imitation game was introduced. In this game an interlocutor who does not know if s/he is talking to a human or AI will have to figure this out through questioning. Conditions were specified under which this game could be won: "an average interrogator will not have more than 70 percent chance of making the right identification after five minutes of questioning" (Turing, 1950, p. 442). This is a quite famous experimental setup in AI research and has spurred much research as well as criticism. An early AI system was said to have won the imitation game. The Russian chatbot Eugene Goostman was able to convince 33 percent of its human interlocutors that it was human rather than a chatbot.[1] This program came second in the Loebner prize, which is awarded to a computer system that can interact for at least 25 minutes with the interlocutor (a strong Turing test). In fact, Eugene Goostman "pretends to be a thirteen-year-old foreigner and proceeds mainly by ducking questions and returning canned one-liners" (Marcus et al., 2016, p. 3). Many researchers considered this example to be flawed.

Since using natural language is a characteristic feature of humans (also argued to be related to science literacy in a fundamental sense (Norris and Phillips, 2003)), it is certainly understandable to utilize such a test as an approximation of intelligent behavior. However, the importance of natural language should not be overrated for intelligent behavior. Some of the raised concerns with the Turing test then relate to the fact that intelligence is multidimensional and no simple test can measure it. Others relate to the fact that intelligence requires embodiment and physical manipulation of things, etc. (Marcus et al., 2016). After all, it would be very easy for a human to probe whether s/he is interlocuting with a human or AI: simply ask a question that only an AI can quickly answer, such as "write 100 words with the letter 'a' included," or, more advanced in modern AI systems, "calculate 980 divided by 35." If no human biases (as in the testing of AlphaStar system, see Chap. 1) are implemented in the AI system, that should definitively do the job.

With recent advances in LLMs, researchers also tested if these systems (such as GPT-4) could pass the Turing test. Jones and Bergen (2023) utilized an online environment where "average" humans from social media were invited to interlocute with several different AI systems. The ranking was as follows, where percentages in parenthesis indicate how many games were passed: GPT-4 with prompting (41%), ELIZA (27%), and GPT-3.5 (14%). The authors also investigated participants' decisions and summarize that "[p]articipants' decisions were based mainly on linguistic style (35%) and socio-emotional traits (27%), supporting the idea that intelligence is not sufficient to pass the Turing Test" (Jones and Bergen, 2023). This is also interesting in the sense that ELIZA as a simple rule-based system (see above) outperformed GPT-3.5, and indicated that even in the year 2023 fairly easy programs can fool people. To be fair, however, it is also not clear what can be excepted from such conversations, and either ELIZA and GPT-4 show that generation of well-sounding text can be performed with such systems.

NLP in science education

Science education researchers utilized NLP methods to answer their research questions. For example, Nehm and Härtig (2012) devised a rule-based NLP system to extract concepts in students' short-form constructed responses on evolutionary biology. Sherin (2013) then utilized a transformation of documents (here: interview excerpts) into a high-dimensional vector space that was then forwarded into an unsupervised ML algorithm to identify clusters in the interviews on seasons. Later on, Carpenter et al. (2020) used deep learning-based methods that could enhance feature representation of their data on students reflections in a biological hazard scenario. Related to LLMs, many researchers showed that these can successfully solve science problems to some extent, however, with severe limitations that we will detail later (Kieser et al., 2023; Krupp et al., 2023; West, 2023; Kortemeyer, 2023). Let's dive into some of these NLP methods more closely and see how they can be applied in science education research.

7.2 An Applied Example

Context

We start by merely focusing on the surface level of texts, i.e., what words appear in them (we put aside the question of what a word exactly is, which is more difficult than one might naively think. The different words that, say, students use in their written responses to a question or task are certainly linked to the ideas that students want to express, even though they are no one-to-one mapping. We will call a constructed response by a student a document, which is an established term in NLP. Focussing on the document level, we can begin by building a vocabulary (i.e., all different words that appear throughout documents), and then counting the frequency of each

Table 7.1 Term-document matrix for student responses

Student	Ball	Force	Gave	Gravitational	Hand	Energy	Gravity	Upward	Throw	Just
Student 1	1	0	2	0	1	3	0	0	1	0
Student 2	2	3	2	0	1	0	1	0	1	1
Student 3	3	3	1	0	1	0	3	0	1	2
Student 4	7	6	0	5	1	1	0	3	0	0

word in each document. In short: we build a term-document matrix. Imagine students answered on an item posing the following physics question: "Imagine you throw a ball vertically upward (on earth): Describe what happens to the force on the ball? Neglect air resistance" (Gregorcic and Pendrill, 2023). The (synthetic) responses could read[2] as follows:

Student 1 (Confusing force and energy [own generation]): When you throw the ball up, the energy you gave it to go up keeps it moving for a while. Even after it leaves your hand, I think that the energy is still pushing it upwards, and gets weaker. When it's coming down, all the energy you gave it is all gone.

Student 2 (Impetus Preconception): When you throw the ball up, the force you gave it to go up keeps it moving for a while. Even after it leaves your hand, I think that force is still pushing it upwards, but it gets weaker and weaker until the ball stops and starts to come back down. When it's coming down, I guess the force you gave it is all gone, and it's just gravity that's pulling it back to the ground.

Student 3 (Correct Concept, Less Sophisticated Language): Okay, so when you throw the ball into the air, the only force acting on it after it leaves your hand is gravity, which pulls it down toward Earth. There's no force pushing it up anymore; it just goes up because of the speed you gave it when you threw it. Gravity slows the ball down until it stops at the top of its path, then it makes it come back down. So, the force on the ball is always the same—it's just gravity.

Student 4 (Correct Concept, More Sophisticated Language): Upon releasing the ball with an initial upward velocity, the exertion of an upward force ceases the moment the ball loses contact with the hand. Subsequently, the ball is solely under the influence of a constant gravitational force directed towards the Earth's center. This gravitational force, which is equal to the product of the mass of the ball and the gravitational acceleration (mg), continuously opposes the ball's upward motion, thereby reducing its kinetic energy and decelerating it until it momentarily comes to rest at the peak of its trajectory. Thereafter, the same gravitational force accelerates the ball earthward. Throughout the entire process, the gravitational force remains constant in both magnitude and direction, acting as the sole force on the ball in the absence of air resistance.

Transforming the responses into a term-document matrix

A term-document matrix for three responses might read as follows (which can be easily generated with the `CountVectorizer` from the `scikit-learn` module in Python (see Python code below).

---

[2] Generated with GPT-4; conversation can be accessed here: https://chat.openai.com/share/ 5c6ad151-56a9-4b8e-9d2c-5c9abbc4575d.

Interestingly, even this simple term-document matrix of word co-occurrences can already provide insights. Note that, before calculating the term-document matrix, we removed common words such as "and" (called stopwords) from the documents. This step should be critically examined in your research, however, for illustration purposes this is reasonable to focus on the more important content words. Given that the frequencies are absolute counts and not normalized (e.g., relative frequencies with respect to overall document length), they give us an impression of the length of the responses. Student 4 apparently wrote more compared to the others. Response length is oftentimes an important proxy (though rather uninformative) for quality. You should make sure to examine length differences in responses, which is reported in many other (science) education studies that use NLP and analysis of text data more generally. Furthermore, the term-document matrix gives us an impression of the themes and writing quality in a more detailed sense. Note that students 2 and 3 used gravity, whereas student 4 used gravitational (force), which is more aligned with academic language rather than everyday language. Most importantly, student 2 used the word "gave" two times. This is consistent with the impetus preconception of forces, where an object is given a force that is then consumed or used up. As such, the term-document matrix can be of diagnostic value to examine students' ideas. Moreover, student 1 used energy instead of force, compared to the others. Note that instead of entire documents, the documents could also be split into sentences (e.g., with the spaCy module in Python) and these sentences could be used as documents which would allow a more nuanced analysis of ideas. Note also the limitations of merely analyzing documents based on frequency of terms. First, word order and context information is lost, which is an important feature for expressing ideas in natural languages. Therefore, such approaches are referred to as "Bag-of-Words" models. For example, student 3 also used the word "gave" once, however, in the context of velocity rather than force. Second, very long documents might make it difficult the see important differences between documents.

Weighting terms in the term-document matrix

The length dependence can be counteracted against with normalization. Typical normalization procedures for such count vectors are $L1$ or $L2$ normalization. The former refers to the procedure where all cells in a row are summed and each cell in a row (i.e., document) is then divided by this summed value. The latter refers to the common procedure where all values are squared and then summed. Finally, the square root is taken and each cell in a row divided by this value. Both these normalizations refer to the length of the vector (if the row is understood as a vector, where each cell maps to one dimension of this vector). We used $L1$ normalization and output the resulting normalized term-document matrix in a table (see Table 7.2). We now see that student 4 on average uses the term "force" as often as students 2 and 3, which seemed quite differently in the frequency-based term-document matrix. The term "force" might not be as informative after all. Student 2 with the impetus preconception used "force" equally often, and the term "force" was also given in the problem statement.

Table 7.2 Normalized term-document matrix for student responses

Student	Ball	Force	Gave	Energy	Hand	Gravity	Weaker	Pushing	Leaves	Throw
Student 1	0.056	0.000	0.111	0.167	0.056	0.000	0.056	0.056	0.056	0.056
Student 2	0.071	0.107	0.071	0.000	0.036	0.036	0.071	0.036	0.036	0.036
Student 3	0.100	0.100	0.033	0.000	0.033	0.100	0.000	0.033	0.033	0.033
Student 4	0.104	0.090	0.000	0.015	0.015	0.000	0.000	0.000	0.000	0.000

Table 7.3 TF-IDF Term-document matrix for student responses

Student	Force	Ball	Energy	Gave	Gravity	Weaker	Leaves	Throw	Pushing	Just
Student 1	0.000	0.040	0.183	0.099	0.000	0.061	0.049	0.049	0.049	0.000
Student 2	0.092	0.050	0.000	0.061	0.038	0.075	0.031	0.031	0.031	0.038
Student 3	0.082	0.067	0.000	0.027	0.101	0.000	0.027	0.027	0.027	0.067
Student 4	0.064	0.061	0.013	0.000	0.000	0.000	0.000	0.000	0.000	0.000

We might furthermore weight the terms by their occurrence within the documents. The idea is that documents are similar that use specific terms, i.e., terms that only appear in some, rather than all, documents. A commonly applied technique in information retrieval to resolve this issue is term frequency inverse document frequency (tf-idf) transformation. As the name suggests, term frequency is traded off with how often the term appears throughout documents (inverse document frequency). Tf-idf transformation can be utilized through the `TfidfTransformer`, again from the versatile `scikit-learn` module in Python (see Python code below). In the first line after to dots, the `TfidfTransformer()` is initialized and applied to the list of documents, named `X`, in the final line. The argument `norm='l1'` specifies the normalization for the count vectorization, which is performed beforehand. There are then some other mathematical details (adding 1 to prevent division by zero, etc.) that are of minor importance here. The resulting matrix is displayed in Table 7.3. Now, the term "force" gets less weight, whereas, for example, "gave" appears more prominent in the responses from students 1 and 2. Moreover, values for "force" differ for students 2 and 3 (even though they have similar term frequencies), because the matrix is first normalized.

Python code: Determine term-document matrix and tf-idf matrix

```
vectorizer = CountVectorizer(stop_words='english')
X = vectorizer.fit_transform(docs)

(...)

tfidf_transformer = TfidfTransformer(norm='l1', ...)
X_tfidf = tfidf_transformer.fit_transform(X)
```

Representing documents in embedding space

In the context of information retrieval research, term-document matrices and tf-idf matrices can be thought of as vector spaces. In fact, the matrices are transformed (columns and rows are switched) first, and then each column (i.e., document) represents a vector for the respective document in a space that is spanned by the vocabulary. In the term-document matrix, each dimension (i.e., a term) is rather unbounded, because it represents the frequency. In the tf-idf matrix, the dimension is bound to 0 and 1, which is advantageous for many ML applications. We can represent this tf-idf matrix in a 2D vector space, where the tf-idf values for the words "ball" and "gave" span the axes (see Fig. 7.1).

More advanced NLP techniques allow us also to separate verbs ("gave") from nouns ("ball") which enables us to perform more fine-grained analyses of the constructed responses. You could use part-of-speech tagging for this with the spaCy library in Python. Furthermore, looking at these points in space in Fig. 7.1 might enable us to apply some clustering techniques as outlined in Chaps. 5 and 11. For example, clusters of students could be identified that use certain terms more frequently compared to others with k-means clustering, PCA, or t-SNE.

Term-term co-occurrence matrix

Analyses can also be done with the term-term or word-word matrix. This is also a co-occurrence matrix of words in contexts rather than documents. Here, oftentimes a context window around 4 words to left and right of the target word is considered. Then, a cell in this matrix represents the times a target word co-occurs with another word in this particular context window. Words then can have similar vectors, i.e., similar cell frequencies. We included code in the supplement on how to generate

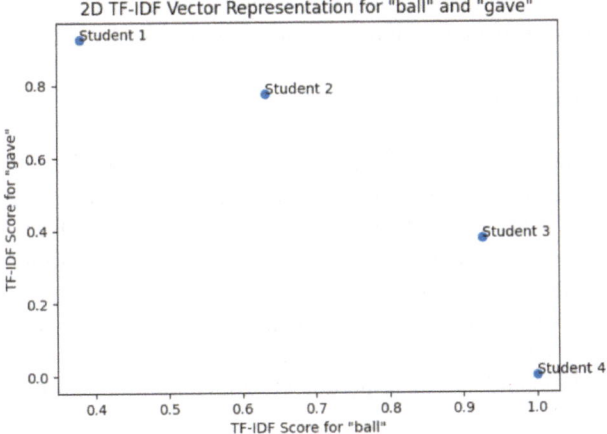

Fig. 7.1 2D representation of the tf-idf matrix for the terms "ball" and "gave"

this matrix. The matrix looks similar to the term-document matrix. Now, rows and columns refer to individual words, and documents have no meaning anymore in this matrix.

Python output: term-document matrix

```
array([[0, 1, 1, ..., 0, 1, 0],
       [1, 0, 1, ..., 0, 1, 0],
       [1, 1, 0, ..., 0, 1, 0],
       ...,
       [0, 0, 0, ..., 0, 0, 3],
       [1, 1, 1, ..., 0, 0, 0],
       [0, 0, 0, ..., 3, 0, 0]], dtype=int64)
```

A reasonable training corpus is then necessary to retrieve meaningful estimations, depending on your research goals. Again, vectors can encapsulate similarity of two target words, if they occupy a nearby location in the vector space, or have a similar direction. Cosine similarity is typically used here. Cosine similarity is the length-normalized dot product between two vectors. The problem with the dot product is that it favors long vectors. However, more frequent words also have longer vectors. Normalizing the dot product for length equals the cosine of the angle between the two vectors. The cosine similarity ranges from -1 (opposite direction) to 1 (same direction). Cosine of 0 refers to orthogonal vectors. Since frequencies are non-negative integers (in the term-term matrix), the cosine similarity then ranges from 0 to 1. In this term-term matrix, "force" is related to "energy" with .98, whereas "force" and "gets" are related with .81. Hence, the vector for "force" is more similar to the vector of "energy" as to the vector of "gets." We should not over-interpret these values, based on only four documents. Depending on your research goals, much larger corpora could be processed such as journal articles or students' writing.

Word meaning from distributional properties

Co-occurrence analyses of natural language hinge on the assumption that words that have a similar meaning appear in similar contexts. This is sometimes referred to as distributional semantics, or use theory of meaning (Manning, 2022): Meaning here merely is a function of context. Another important conceptualization is denotational semantics. Meaning, here, is a property of reference, i.e., the set of objects, processes, or phenomena in the world that a word, sentence, or similar refers to. It is crucial to remind ourselves that meaning of a word, phrase, or sentence is not static, but rather dynamic, context-dependent where no universal meaning for a word exists.

7.3   Advanced Language Modelling with LLMs

Early language models

Language modelling refers to the task of predicting the next word given previous words. More generally, language modelling seeks to "capture regularities of natural language for the purpose of improving the performance of various natural language applications" (Rosenfeld, 2000, p. 1270). You could try to generate massive look-up tables (rule-based systems) where for each possible sentence reasonable next words are suggested. However, there are infinitely many conceivable sentences making this undertaking futile. Language is characterized by "the infinite use of finite means" (Chomsky [1965] quoting von Humboldt). Finite means refer to the limited alphabet and set of conceivable words. Early language models simply sought to estimate the probability function of a sentence (i.e., a sequence of words). As with the term-term co-occurrence matrix, a window size can be defined and probabilities of co-occurrences can be estimated. For example, to build a language model researchers could seek to determine the conditional probability of a certain word given it's context of words. The context size partly determines the quality of the language model. A so-called n-gram model has a context size n. It was shown that even n-gram models of $n = 3$ generate reasonable sounding text. While these language models for a long time powered NLP applications such as speech recognition, they are challenged by the sparse estimation problem: observing frequencies of all possible 3-grams (i.e., trigrams) requires tremendously large (and in most contexts unattainable) text corpora.

Furthermore, ML algorithms such as decision trees have been employed to estimate probability distributions of words. Again, the space of possible combinations quickly becomes very large. For a 20 word sequence with a 100, 000 size vocabulary, there is a space of possibilities of the size 10^{100} (or: one Googol).

Static embeddings through ANNs

A means to tackle this explosion of complexity is the use of word/document embeddings. These embeddings help to generalize better and be more flexible compared to exact word matches, as used in n-gram models. Take the sentence: "I have to make sure when I get home to feed the ...". If during training only the sequence "to feed the cat" was seen, n-gram models would only predict 'cat' after "to feed the". However, embedding-based approaches make use of the fact that 'cat' and 'dog' appear overall in similar contexts, hence, 'dog' would also be presented as a reasonable candidate word.

In fact, such embeddings can be calculated with ANNs and LLMs. At the heart of many modern NLP techniques lie ANNs with sophisticated architectures. It has been found that task performance on many NLP tasks could be improved with ANNs and deep learning. In particular, language modeling proved to be of particular relevance to tackling language-related problems such as translating text between languages. ML-based language models then predict distributions of words by self-supervised training, and typically output scores or string predictions. For example, if a string is to be continued: "Throwing a ball into the ...", the language model proposes likely candidates for continuation, such as "air," "water," etc. Training LLMs hinges on the availability of Big Data (such as the Common Crawl of the Internet), and computing infrastructure (resources). Moreover, if this training data is biased or lacks examples from certain domains (e.g., chemistry-specific text), then the LLMs may struggle to generate reasonable words for chemistry-specific text completion tasks. At least, it is highly questionable to what extent they would be able to do so, and we will come back to these issues when considering the limitations of LLMs.

To train embeddings for words, sentences, paragraphs, or even documents, ANNs proved to be versatile tools. A feed-forward neural network can be utilized to train word embeddings. Training the feed-forward neural network is done by calculating a loss function (e.g., given a predicted word and the observed word) and using stochastic gradient descent to update the weights in the matrix. Stochastic gradient descent is a variation of gradient descent, which can be described as finding the low point in a valley by always following the steepest ascent (note that in complex landscapes, it is not assured that the global optimum can be found this way–you might get stuck in a high valley). Stochastic gradient descent utilizes subsamples of the data to approximate the gradients. After training, for each word we will receive a probability. This feed-forward neural network can then be used to auto-complete a sentence merely based on probabilities and the recent past context.

Such ANNs of various architectures with static embeddings power(ed) many NLP applications and enabled researchers to retrieve more performant vector space representations for words and documents. Important embeddings were, among others, Word2Vec, and GloVe (Global Vectors) that are sophisticated means of estimating embeddings. These are static embeddings. This means that a training corpus (as large and representative as possible) is required. Word2Vec has also been extended to include document vectors and topic vectors. Furthermore, it has been shown that reconstruction of text based on embedding vectors is also possible to some extent. Moreover, these embeddings can be calculated for specific data sets to adopt them to the purposes of different researchers. In science education, these embeddings could be used to enhance representation of students' constructed responses (Carpenter et al., 2020). However, before we mentioned that word sense is largely a function of context, hence: embeddings should also vary with context.

Contextualized embeddings with LLMs

Static embeddings were superseded with the introduction of the transformer-architecture for LLMs. These LLMs typically have a fixed vocabulary of 30K to 250K possible tokens, and each token is mapped to a fixed vector embedding. In the pre-training phase, these embeddings are learned. The embeddings are then passed through the transformer architecture, which "results in a 'contextualized' representation for each token" (Chang and Bergen, 2023, p. 4), which is then used to generate output tokens. LLMs in this form are typically trained with 500 to 2000 token input sequences, or even much longer.

Masked and auto-regressive language models

Researchers differentiate between masked and auto-regressive LLMs, otherwise also known as encoder and decoder LLMs. Encoder models such as BERT (we will utilize this in Chap. 12) represent an input sequence as an embedding, which can then be utilized in classification tasks or similar tasks. Decoder models such as GPT-3 or GPT-4 (we utilize ChatGPT which is based on GPT-4 in Chap. 3 to simulate data) can be utilized for text generation with specific goals (see below). Both types of LLMs are trained in a similar, self-supervised way, namely to predict masked and upcoming tokens in a sequence. In particular, auto-regressive LLMs seek to maximize the probability for a next word in a sequence so as to be utilized later for generative purposes. Auto-regressive LLMs such as GPT are typically much larger compared to masked LLMs. They are trained using gradient descent (see Chap. 4) with a text corpus of up to 1.5 trillion tokens. Training data set size should increase according to increases in model size, and some 100K to 4M tokens per batch (optimization step) are seen by the model to update token embeddings and weights. It was found that with sufficient pretraining and model size certain task-related capabilities in LLMs only begin to emerge. Moreover, scaling laws were identified that link performance of LLMs with data set size and model size. Also, the quality of the training data was found to be of crucial importance. Given the size and the versatility, such pretrained LLMs were also called "foundation models" (Bommasani et al., 2022). Typically, these foundation models are only trained once and then further fine-tuned and adjusted for specific purposes (transfer learning).

Auto-regressive LLMs such as ChatGPT are also further trained on non-task-specific tasks, such as curated human-written examples. These approaches are called supervised fine-tuning. Model parameters can then be updated through gradient descent. LLMs such as Bidirectional Encoder Representation for transformers (BERT) were versatile tools for being fine-tuned for specific tasks such as classification. This incited a research field called "BERTology" (Rogers et al., 2020), where many different findings regarding the linguistic "understanding" of such models and even world knowledge were performed. Other models such as GPT were found to be capable of systematically generalizing an abstract pattern that they were given as inputs, which is considered a capability that is also crucial in human reasoning and thinking (Lake & Baroni 2023).

Generative LLMs

Auto-regressive, generative LLMs output a probability distribution of next words. At inference time, auto-regressive LLMs can be utilized to continue a given input sequence by iteratively sampling next tokens, e.g., by choosing the most probable token, by providing a temperature parameter to introduce more randomness, to sample from top-k tokens (i.e., the k highest ranking tokens), or some other ways (Chang and Bergen, 2023). Given these generative capabilities, auto-regressive LLMs such as GPT are referred to as generative AI or generative LLMs.

Prompting and prompt engineering

Generative LLMs such as GPT were found to improve performance on tasks when provided specific context, e.g., examples on how a task is to be performed (few-shot prompting, or in-context learning): Few-shot prompting then refers to the case "where the model is given a few demonstrations of the task at inference time as conditioning [..], but no weight updates are allowed" (Brown et al., 2020, p. 6). Prompting is essential to effectively and efficiently interact with LLMs, and enforce the outcomes to have desired qualities. A simple prompt was providing examples on what solving a task would look like. This was called few-shot prompting. With few-shot prompting, these LLMs were found to perform better at tasks that they were not explicitly trained on. A few-shot prompt could look as follows:

Few-shot prompt (see: Brown et al., 2020)

```
Translate English to French # task
sea otter => loutre de mer  # 3 examples
peppermint => menthe poivrée
plush girafe => girafe peluche
cheese =>                   # prompt
```

Even prompting LLMs by simply adding "Let's think step by step" could help ChatGPT (based on GPT-4) to improve performance, distantly resembling what has been documented in research on "thinking slow" for humans. Based on this, chain-of-thought[3] prompting forces LLMs to generate several short sentences that lay out the involved reasoning steps in a problem, e.g., instead of providing the example "You have 3 apples; 2 are taken away; how many apples do you have? The answer is 1" we provide it "You have 3 apples; 2 are taken away; how many apples do you have? You have 3-2=1 apples. The answer is 1." This can be further improved by considering multiple reasoning paths and checking for self-consistency. In self-consistency prompting the LLMs is required to produce several solutions with chain-of-thought prompting, and cross-check them before producing a final answer. This has also been documented to improve performance in many well-defined tasks. Moreover, tree-of-thought prompting leverages chain-of-thought prompting and evaluates multiple reasoning paths. With all limitations of LLMs in mind, these techniques can also be used to automatically code data (Tan et al. 2024).

---

[3] We emphasize the critique uttered by Polverini and Gregorcic (2024) for the inadequate use of cognitive terms such as "thought" or "reasoning" in many NLP contexts. This critique dates back at least to the beginnings of the field of AI (McDermott, 1976, p.4), dubbed "wishful mnemotics": "A major source of simple-mindedness in AI programs is the use of mnemonics like 'UNDERSTAND' or "GOAL" to refer to programs and data structures."

Research with LLMs in science education

Generative AI offers many potentials for education in research and teaching. It was shown that knowledge of AI and LLMs is important in enabling students to meaningfully engage with generative LLMs. Krupp et al. (2023) observed that students in physics tended to accept the outputs of LLMs as is, and Zamfirescu-Pereira et al. (2023) found that students with no knowledge about LLMs used prompts as if they were talking to another human, which is less effective compared to other strategies that were outlined above.

LLMs have been used from the time of their inception, either masked and auto-regressive LLMs. It was shown that LLMs in the fine-tuning paradigm could improve classification and generalization performance for pre-service physics teachers' written reflections (Wulff et al., 2022a), and provide researchers opportunities to share their models across research sites and further train existing models (Sorge et al., XXX). Moreover, various researchers experimented with opportunities to utilize auto-regressive LLMs alongside with prompting to showcase useful features for science education research.

Also, prompt engineering has been extensively used in science education research (Gregorcic and Pendrill, 2023). Polverini and Gregorcic (2024) synthesize guidelines on effective prompt design in physics education research contexts and show that specifying the domain or specifying how to act (e.g., answer like a physics teacher) can improve LLM outputs for a physics problem. They also mention limitations of prompting with examples where the LLM attends to superfluous details and gets slightly off track when answering. For example, Kieser et al. showed that ChatGPT could simulate (answer as-if) students' preconceptions in mechanics (Kieser et al., 2023). Furthermore, West (2023) and Kortemeyer (2023) showed that ChatGPT (based on GPT-4) was capable of solving open and closed response conceptual questions in physics. However, problems with generating incorrect information or failing to evaluate the generated problem solutions were also highlighted (Gregorcic and Pendrill, 2023; Kieser and Wulff, 2024). Wan and Chen (2024) showed that specific prompting for feedback generation for physics problems improved feedback to levels that were considered equally correct and even more useful compared to human expert feedback. Prompt-engineering offers science education researchers many novel ways to extract information from students' responses and automatically generate adaptive guidance. However, with regards to prompting strategies it remains largely unclear to what extent these strategies also improve LLM performance in more complex reasoning tasks as they are typical for science education research (Polverini and Gregorcic, 2024). Moreover, for chain-of-thought prompting it was also critically mentioned that basically the humans-in-the-loop do much of the planning/problem solving (Valmeekam et al., 2023). Just like the horse Hans[4] that allegedly learned to count by recognizing signals from the human investigator, the LLMs leverage the humans' questioning to arrive at meaningful solutions.

---

[4] This story is beautifully told in Crawford (2021).

7.4   Applying LLMs in Science Education Research

Extracting static embeddings for science terms

Imagine you want to compute the similarity of terms in physics based on a large data set, such as Wikipedia. Pre-trained embeddings can then be a resource for you, which can be accessed through R or Python. A famous library to access static embeddings such as Word2Vec is `gensim`. This library allows researchers to efficiently either retrieve word vectors that have been trained on large language data sets such as the Internet, Wikipedia or book corpora.

Python code snippet: Utilizing `gensim` to retrieve word embeddings

```
import gensim.downloader as api wv =
api.load('word2vec-google-news-300')

# display word vector:
wv['current']
# calculate similarity for two word vectors: w1 = 'current'
w2 ='voltage' wv.similarity(w1, w2)

w1 = 'current' w2 = 'stream' wv.similarity(w1, w2)
```

The word vector here is a 300D vector with floating point numbers. We find that the average similarity (cosine similarity) between "current" and "voltage" is 0.13, and between "current" and "stream" is 0.02. At first sight, this seems reasonable: current should be positively linked to voltage. And it should be linked closer to voltage than to stream. Interestingly, word vectors as retrieved from Word2Vec or GloVe also encapsulate features such as representing analogies (king is to queen, as man is to women). These can be visually displayed in 2D space. However, closer inspection of these word vectors already yields problems. To have a comparison for the similarity values, we randomly selected 200 words from the models (`word2vec-google-news-300`) vocab and calculated the average similarity (SD) was 0.13 (0.11). As such, voltage is no more closely related to current as other randomly chosen words. This is not as expected. In physics current and voltage are conceptually closely related. When we search for the most similar words for "force", we get: forces (0.52); Faulcon_resigned (0.46); Mohammed_Majah (0.42); Nato_Isaf (0.41) and professional_nonpolitical_militarily (0.41). This highlights a problem with these word vectors when applied in science education: the training data sets are not sensitive to our contexts and specific training with discipline-based data sets would be necessary. There are probably millions of tweets and other texts in the training data singling out the joint forces in political situations, which has nothing to do with the scientific concept of force.

Moreover, calculating analogies with embeddings ("king is to queen, as man is to women") seems convincing at first. However, the symmetrical relation between them is quite different from how humans infer on analogies in reality. For example, human processing of analogies is intricately related to knowledge, e.g., average humans see North Korea as similar to China, but not vice versa. Moreover, these pretrained embedding vectors are static. I.e., the embedding vector for a word is fixed, irrespective of the context it appears in. Yet, the meaning for "bank" in "river bank" and "bank robbery" differs categorically. Hence, approaches are necessary, where the context is considered as well.

Utilizing LLMs to calculate embeddings (encoder)

LLMs can solve some of these problems and provide researchers with contextualized embeddings for the words. A famous LLM became known as BERT. BERT processes the entire input before forming a context-sensitive word vector. For the sentences: "The force may be with you." and "Gravitational force acts on a mass in the gravitational field." we would expect different word vectors for "force". Here (the code can be found in the Supplement), the cosine similarity between both embeddings for force is 0.38. Thus, they do not inhabit the same spot in embeddings space and indicate differences in meaning. Comparing vectors for individual words without context is not as meaningful as for static word embeddings from Word2Vec, hence we will omit it here. As with static embeddings, however, one can represent words or sentences now in the same embedding space. For example, if we add another sentence to the above responses, say: "When you throw a ball up, the force you gave it to move up keeps it moving for a while. Even after it leaves your hand, that force is still pushing it, but it gets weaker and weaker until the ball stops and starts to come back. When it's coming down, the force you gave it is all gone, and it's just gravity that's pulling it back.", which is slightly modified from student 1's response, then you will see that they are at a similar location in embedding space:

This, again, could be used to cluster similar responses with unsupervised ML techniques (see Chaps. 5 and 11) and thus identify patterns for answering a certain question such as the one above. For example, you could extract all similar sentences to the above sentence: "Gravitational force acts on a mass in the gravitational field." This will give you insights about the language used by your learners.

One intricate problem with embeddings in general and LLMs in particular is what these dimensions in the embedding space actually mean. For the simple encoding of the term-document matrix above it is clear: the dimensions refer to how often a certain word is used in a document. However, with the static and contextualized embeddings as generated through LLMs, this is more difficult. Researchers suggested that certain word embeddings could be correlated with the dimensions, which then would indicate that the dimension relates to the word. However, these approaches require substantial efforts and only approximate the true meaning of the dimension. This problem is comparable with interpretation of the dimensions in a PCA analysis. Substantial human expertise is required to make sense of the mathematical outputs.

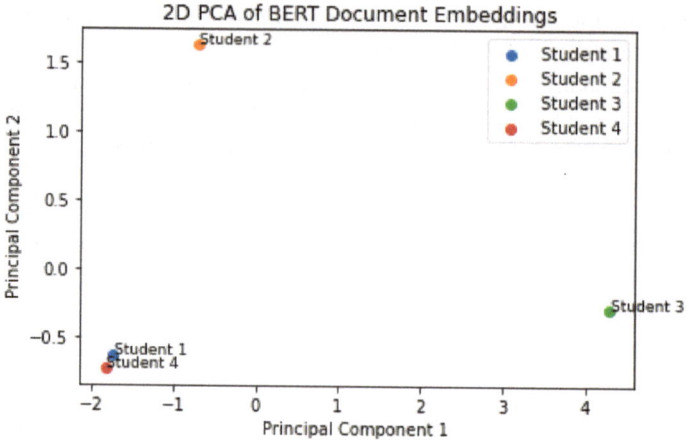

Fig. 7.2 Response embeddings through BERT for the four student responses

Utilizing generative LLMs

Besides embeddings we outlined the novel opportunities that arise from generative LLMs, such as synthetic data generation. For the task: "Imagine you throw a ball vertically upward (on earth): Describe what happens to the force on the ball? Neglect air resistance" synthetic responses such as the following were generated with the additional prompt "Student 1 has the impetus misconception, student 2 and student 3 have the correct for force concept, however, answer with differing degree of language sophistication":

Student 1 (Confusing force and energy [own generation]): When you throw the ball up, the energy you gave it to go up keeps it moving for a while. Even after it leaves your hand, I think that the energy is still pushing it upwards, and gets weaker. When it's coming down, all the energy you gave it is all gone.

Note in student 1's response that instead of force (which is typically associated with the impetus preconception), energy is used by the synthetic student. While energy is transformed when the ball moves up, it is inconsistent with physics instruction to attribute that the energy "pushes" the ball and "gets weaker". As such, these responses are quite fruitful instances for pre-service teachers in science education to reflect upon.

Generative LLMs such as ChatGPT could also be used to generate adaptive feedback to science education problems as illustrated by Wan and Chen (2024). In line with the prompt guidelines by Polverini and Gregorcic (2024), the authors show that well-designed prompts can enable GPT-3.5 to provide students adaptive feedback for a physics problem. In particular, the authors show that expectations of what the feedback should look like, the students response, examples of feedback, as well as an expert solution make a difference in feedback quality. The template the authors outline looks as follows:

Prompt design for adaptive feedback

```
#Context A physics instructor is rating students'
answers to the following physics problem:
## Physics Problem: A student pushes two boxes (...)
# The instructor rates students' answer and gives feedback
based on how similar it is to this expert answer:

## Expert Response: No, the 200 N force is not
'transmitted" to box B. (...)

## Physics principles involved:
1. The net force on an object is equal to the mass times
its acceleration.
2. ''Force'', ''energy'', and ''power'' (...)

# The feedback should start with whether the answer
is correct or incorrect, followed by a justification,
and then a follow up question for further thinking.
The feedback should not give away the expert answer.
Those physics principles should not all be repeated
in the feedback to students.

# Here are several examples of student answer and
instructor feedback: Answer: No because in order (...)

Answer: [new student response goes in here.]
Feedback:
```

This template shows that designing effective prompts in science education specific problems is a rather involved process and requires substantive domain expertise. However, once such a template has been devised, it can be shared across contexts and implemented at scale. Hence, carefully devising effective prompts for assessment instruments such as the force concept inventory could leverage resources to improve instruction.

7.5 Considering Limitations

We will provide a deeper look into LLMs in Chap. 12 and conclude this chapter with considering limitations of LLMs in particular. Chang and Bergen conlude in a review of LLM's capabilities that they "are still prone to unfactual responses, commonsense errors, memorized text, and social biases" (Chang and Bergen, 2023, p. 1). Let us explore some of these frailties in greater depth.

Explainability

Even though LLMs allowed new possibilities and showed promising results even in unseen (during training) tasks, it is oftentimes not possible to explain the model decisions (issue of explainability and black-boxes, see Chap. 2). Explainability is crucial for trust in AI systems and for researchers to advance their models (Lipton, 2018; Zhao et al., 2023). The goal for explainable AI is to understand model decisions in human-understandable terms (Zhao et al., 2023). By merely considering the size of most LLMs, and the size of the training data sets, it is understandable that no simple explanations will account for generated outputs. Rather, explanation models for the LLMs are required. Researchers devised approaches to understand model decisions and we will illustrate one such approach in Chapter 12 that can at least highlight which tokens in the input are important for the output generation. Particularly for LLMs, explainability for fine-tuned LLMs and prompting LLMs are differentiated (Zhao et al., 2023). Some procedures in the prompting paradigm relate to counter-factual reasoning and might be applicable for science education researchers. However, as of now, this is largely uncharted territory and progress is urgently needed in order to get a better understanding of LLMs. As of now, science education researchers probe generative LLMs and outline heuristics that eventually condition outputs so that informative outputs can be expected. It has to be constantly kept in mind that these outputs are statistical in nature (Polverini and Gregorcic, 2024).

Shortcut learning

One recognized problem in many fine-tuned LLMs such as BERT is called shortcut learning. For example, in the Argument Reasoning Comprehension Task, BERT was found to excel in tasks such as the following:

> Argument: Felons should be allowed to vote. A person who stole a car at 17 should not be barred from being a full citizen for life. Statement A: Grand theft auto is a felony. Statement B: Grand theft auto is not a felony.

The task here is to determine which statement is consistent with the presented argument, which in this case should be statement B.

However, instead of meaningful considerations, BERT simply picked up on some words such as "not" which are highly correlated with the correct answer. This is called shortcut learning, and is deceptive, given that humans cannot possibly understand why these unrelated words should have any bearing on the prediction. Hence, decisions are intransparent. BERT excels "by exploitation of spurious statistical cues in the dataset" (Niven and Kao, 2019, p. 4658). On an appropriately designed adversarial data set, these LLMs reach only accuracies that amount to random guesses. The underlying issue here relates to the fact that warrants in arguments are often left implicit, relating to world knowledge that is assumed to be shared. As such, BERT or other LLMs that are trained on some example argumentation comprehension tasks cannot possibly insert missing information, at least it would be rather unlikely for these models to have acquired such competencies.

Mimicry of examples in the training data

Similar to shortcut learning, also generative LLMs such as GPT-4 can stick to training examples too closely, which was called approximate (memory) retrieval. For example, while it was shown that GPT-3.5 could recognize the classic Monty Hall problem (choosing a goat behind three doors), slight modifications in the problem statement (that rendered the problem trivial) could not be recognized by GPT-3.5. The model sticked to the exemplary solution of the Monty Hall problem. This was also found for many other counterfactual tasks such as reasoning about chess positions where certain chess figures are swapped (and thus it was rather unlikely for the LLM to have seen these examples in the training data). Sometimes, such frailties seem to be fixed with increasing model size and training. However, it is also known that companies crowd source click-workers to fix such recognized flaws manually, rather than conceptually.

An interesting case in science education is presented by Polverini and Gregorcic (2024). They find a dialogue-based prompting enabled GPT-4 to provide reasonable solutions for an inclined plane problem. However, the presented solutions are not entirely satisfactory, given that the friction coefficient is not given in the problem but used in the solution (adding the prompt: "Isn't there another way to do it?" then solved the problem). The authors conjecture that this solution might be a relic from the fact that many textbook problems that might have been seen by the generative LLM during training likely included the friction coefficient and thus GPT-4's response mimicked these training examples. On the other hand, it was also found that being exposed to certain problems during training does not necessarily enable the models to recognize these problems (Macmillan-Scott and Musolesi 2024).

Unreliability and extrapolation

Moreover, generative LLMs commonly make up facts, i.e., they hallucinate knowledge, which refers to presenting false information as facts. Even in innocuous tasks such as textual summarization. Even though ChatGPT (based on GPT-4 and the multimodal version based on GPT-4V) seemed to have acquired many human-like capabilities, rigorous testing of abstraction and reasoning abilities show that such models are fundamentally constrained. The physicist Sean Carroll prompted Chat-GPT (based on GPT-4) with the problem of a toroidal (instead of square) chess board, where the edges and borders are connected and pieces can cross them seamlessly. This is an interesting problem since, given that White typically starts in chess, without rule-adaptation, it will instantly win over Black given that the king would be checkmated from the start of the game. However, he found that ChatGPT could not resolve this transfer problem (and other interesting problems), besides belaboring how different the game would be etc. etc. This all indicates that, after all, LLMs are capable of mimicking data that was seen during training and recompose it in well-crafted texts, however, emergent reasoning and thinking abilities require abstraction and synthetization of information as well which LLMs currently do not seem to have. It was concluded that "[m]any of these weaknesses [by LLMs] can be framed as over-generalizations or under-generalizations of learned patterns in text" (Chang and Bergen, 2023, p. 1). LLMs might seem creative to individuals who have not seen the entire training data (which is impossible for individual humans), however, if you knew the entire training data it might seem less creative.

There is evidence that test data set and task contamination is largely present in many LLMs, which went unnoticed by some of the developers of foundational LLMs. LLMs such as GPT-3.5 were tested in a few-shot or zero-shot paradigm and it was found that these models performed markedly better on data sets that were released prior to training the model, compared to data sets that became public only after the training of the model. This indicates that much of the acclaimed zero-shot learning might be interpolation of seen examples as well, and capabilities of LLMs to perform reasoning tasks and similar tasks are somewhat overestimated. All these studies point out that researchers need to be very cautious when investigating the performance of the utilized LLMs in their research. It oftentimes might merely mimic examples that were seen during training.

Understanding

It is often asked (e.g., in the context of AGI discussions) to what extent LLMs truely understand the world as humans do. This is certainly not the case. LLMs do not acquire world models as humans do (even very young children can abstract and generalize in ways LLMs cannot do). However, it is also probably not true that LLMs merely mimick what was seen in the training data. Some researchers suggest that LLMs learn models of semantic spaces and can do inference in these spaces. As of now, the limited access to other sensory experiences and interaction with the world prevent LLMs from developing human-like understanding of the world. What will become possible with regards to understanding the world when LLMs also "experience" ultrasonic and infrared, or UV radiation (even ionizing radiation) are exciting questions for future research.

Ethical and environmental issues

Finally, privacy concerns (as with other ML models as well) are also problematic, given that researchers can't train their own models (most lack the resources), nor can they inspect the models thoroughly. Given that these models are almost exclusively trained by private companies and conversation data in implementations such as ChatGPT are forwarded to the company, it raises the issue of making certain companies even stronger, potentially fostering monopolies. Companies such as Meta or Aleph Alpha started to open source their models (e.g., Llama or Luminous) which could raise transparency of why these models perform well or worse in certain tasks. Copyright issues with training data have already appeared in image models as well as language models.

Moreover, the substantial ecological footprint that training and accessing LLMs requires is another concern that should not be neglected when applying these models, as outlined in Chap. 2. We suggest that the presented conventional ML algorithms (see Chaps. 4, 5, 10 and 11) in many cases might be valuable alternatives, because they can be better inspected (transparency and explainability), they do not involve forwarding of data to private companies, and they do not bring along as negative ecological footprints.

7.6 Summary

Researchers called some generative LLMs "zero-shot reasoners" (Kojima et al., 2022) with "sparks of artificial general intelligence" (Bubeck et al., 2023). However, LLMs can be unreliable sources of information and should be treated with due caution, because "Slight changes in input word choice and phrasing can lead to unfactual, offensive, or plagiarized text" (Chang and Bergen, 2023, p. 2). As any other tool, LLMs should be applied in the contexts in which they were shown to work well and for which they are developed. In terms of practical applicability, LLMs have to be much improved in terms of explainability, ecological issues, as well as ethical issues in order to become useful at scale: "Although model performance [of LLMs] on broad benchmark datasets is relatively consistent for a given model size and architecture, responses to specific inputs and examples are not. This feature makes large language models tempting but unreliable to use in many practical applications" (Chang and Bergen, 2023, p. 2). Ganguli et al. (2022) outline the unpredictability of LLMs in practice, which needs to be solved before reliable implementation in high-stakes or sensitive decision making processes (besides the many issues related to privacy, personally identifiable information, and bias).

Science education researchers might engage in research related to the capabilities of (generative) LLMs for relevant science education problems. Good-practice examples and contexts in which these LLMs excel in science education should be outlined and eventually an understanding of LLMs for many science education problems can be developed. Even though models change and progress is rapid, larger and newer generative LLMs consistently outcompete older models, oftentimes without losing their specific capabilities. What is learned for one model will likely stand the test of time and be established as a body of knowledge in this entirely novel field of research. Moreover, the particular use-cases of generative LLMs, e.g., generating ideas without providing provably correct solutions to problems need to be critically examined. Utilizing these models under the wrong pretenses could otherwise introduce confusion and frustration for learners and instructors.

References

Bommasani, R., Hudson, D. A., Adeli, E., Altman, R., Arora, S., Arx, S. v., Bernstein, M. S., Bohg, J., Bosselut, A., Brunskill, E., Brynjolfsson, E., Buch, S., Card, D., Castellon, R., Chatterji, N., Chen, A., Creel, K., Davis, J. Q., Demszky, D., Donahue, C., Doumbouya, M., Durmus, E., Ermon, S., Etchemendy, J., Ethayarajh, K., Fei-Fei, L., Finn, C., Gale, T., Gillespie, L., Goel, K., Goodman, N., Grossman, S., Guha, N., Hashimoto, T., Henderson, P., Hewitt, J., Ho, D. E., Hong, J., Hsu, K., Huang, J., Icard, T., Jain, S., Jurafsky, D., Kalluri, P., Karamcheti, S., Keeling, G., Khani, F., Khattab, O., Koh, P. W., Krass, M., Krishna, R., Kuditipudi, R., Kumar, A., Ladhak, F., Lee, M., Lee, T., Leskovec, J., Levent, I., Li, X. L., Li, X., Ma, T., Malik, A., Manning, C. D., Mirchandani, S., Mitchell, E., Munyikwa, Z., Nair, S., Narayan, A., Narayanan, D., Newman, B., Nie, A., Niebles, J. C., Nilforoshan, H., Nyarko, J., Ogut, G., Orr, L., Papadimitriou, I., Park, J. S., Piech, C., Portelance, E., Potts, C., Raghunathan, A., Reich, R., Ren, H., Rong, F., Roohani, Y., Ruiz, C., Ryan, J., Ré, C., Sadigh, D., Sagawa, S., Santhanam, K., Shih, A., Srinivasan, K., Tamkin, A., Taori, R., Thomas, A. W., Tramèr, F., Wang, R. E., Wang, W., Wu, B., Wu, J., Wu, Y., Xie, S. M., Yasunaga, M., You, J., Zaharia, M., Zhang, M., Zhang, T., Zhang, X., Zhang, Y., Zheng, L., Zhou, K., & Liang, P. (2022). On the opportunities and risks of foundation models. *arXiv*.

Brown, T. B., Mann, B., Ryder, N., Subbiah, M., Kaplan, J., Dhariwal, P., Neelakantan, A., Shyam, P., Sastry, G., Askell, A., Agarwal, S., Herbert-Voss, A., Krueger, G., Henighan, T., Child, R., Ramesh, A., Ziegler, D. M., Wu, J., Winter, C., Hesse, C., Chen, M., Sigler, E., Litwin, M., Gray, S., Chess, B., Clark, J., Berner, C., McCandlish, S., Radford, A., Sutskever, I., & Amodei, D. (2020). Language models are few-shot learners. *arXiv*.

Bubeck, S., Chandrasekaran, V., Eldan, R., Gehrke, J., Horvitz, E., Kamar, E., Lee, P., Lee, Y. T., Li, Y., Lundberg, S., Nori, H., Palangi, H., Ribeiro, M. T., & Zhang, Y. (2023). Sparks of artificial general intelligence: Early experiments with gpt-4. *arXiv*.

Carpenter, D., Geden, M., Rowe, J., Azevedo, R., & Lester, J. (2020). Automated analysis of middle school students' written reflections during game-based learning. In I. I. Bittencourt, M. Cukurova, K. Muldner, R. Luckin, & E. Millán (Eds.), *Artificial Intelligence in Education* (pp. 67–78). Cham: Springer International Publishing.

Chang, T. A., & Bergen, B. K. (2023). Language model behavior: A comprehensive survey. *arXiv*.

Chomsky, N. (1965). *Aspects of the theory of syntax*, volume 11 of *Special technical report*. Cambridge: M.I.T. Press, Mass., 3. print edition.

Crawford, K. (2021). *Atlas of AI. Power, politics, and the planetary costs of artificial intelligence*. New Haven: Yale University Press.

Evans, V. (2006). Lexical concepts, cognitive models and meaning-construction. *Cognitive Linguistics, 17*(4).

Ganguli, D., Hernandez, D., Lovitt, L., Askell, A., Bai, Y., Chen, A., Conerly, T., Dassarma, N., Drain, D., Elhage, N., El Showk, S., Fort, S., Hatfield-Dodds, Z., Henighan, T., Johnston, S., Jones, A., Joseph, N., Kernian, J., Kravec, S., Mann, B., Nanda, N., Ndousse, K., Olsson, C., Amodei, D., Brown, T., Kaplan, J., McCandlish, S., Olah, C., Amodei, D., & Clark, J. (2022). Predictability and surprise in large generative models: Facct '22 (pp. 1747–1764).

Gregorcic, B., & Pendrill, A.-M. (2023). Chatgpt and the frustrated socrates. *Physics Education, 58*(3), 035021.

Jones, C., & Bergen, B. (2023). Does gpt-4 pass the turing test? *arXiv*.

Kieser, F., & Wulff, P. (2024). Using large language models to probe cognitive constructs, augment data, and design instructional materials. In M. S. Khine (Ed.), *Machine Learning in Educational Sciences: Approaches*. Applications and Advances: Springer Nature.

Kieser, F., Wulff, P., Kuhn, J., & Küchemann, S. (2023). Educational data augmentation in physics education research using chatgpt. *Physical Review Physics Education Research, 19*(2), 1–13.

Kojima, T., Gu, S. S., Reid, M., Matsuo, Y., & Iwasawa, Y. (2022). Large language models are zero-shot reasoners: 36th conference on neural information processing systems (neurips 2022).

Kortemeyer, G. (2023). Could an artificial-intelligence agent pass an introductory physics course? *Physical Review Physics Education Research, 19*(1), 15.

Krupp, L., Steinert, S., Kiefer-Emmanouilidis, M., Avila, K. E., Lukowicz, P., Kuhn, J., Küchemann, S., & Karolus, J. (2023). Unreflected acceptance—investigating the negative consequences of chatgpt-assisted problem solving in physics education. *arXiv.*

Lake, B. M., & Baroni, M. (2023). Human-like systematic generalization through a meta-learning neural network. *Nature, 623.*

Lipton, Z. C. (2018). The mythos of model interpretability. *Machine Learning.*

Macmillan-Scott, O., & Musolesi, M. (2024). (ir)rationality and cognitive biases in large language models. *arXiv.*

Manning, C. D. (2022). Human language understanding & reasoning. *Daedalus, 151*(2), 127–138.

Marcus, G., Rossi, F., & Veloso, M. (2016). Beyond the turing test. *AI Magazine, 37*(1), 3–4.

McDermott, D. (1976). Artificial intelligence meets natural stupidity. *SIGART Newsletter, 57*, 4–9.

Nehm, R. H., & Härtig, H. (2012). Human vs. computer diagnosis of students' natural selection knowledge: Testing the efficacy of text analytic software. *Journal of Science Education and Technology, 21*(1), 56–73.

Niven, T., & Kao, H.-Y. (2019). Probing neural network comprehension of natural language arguments: Proceedings of the 57th annual meeting of the association for computational linguistics.

Norris, S. P., & Phillips, L. M. (2003). How literacy in its fundamental sense is central to scientific literacy. *Science Education, 87*(2), 224–240.

Polverini, G., & Gregorcic, B. (2024). Performance of chatgpt on the test of understanding graphs in kinematics. *arXiv.*

Rogers, A., Kovaleva, O., & Rumshisky, A. (2020). *A primer in bertology: What we know about how bert works: Transactions of the association for computational linguistics., 8*, 842–866.

Rosenfeld, R. (2000). Two decades of statistical language modeling: Where do we go from here? proceedings of the IEEE.

Sherin, B. (2013). A computational study of commonsense science: An exploration in the automated analysis of clinical interview data. *Journal of the Learning Sciences, 22*(4), 600–638.

Sorge, S., Wulff, P., and Kubsch, M. (submitted). Using a pretrained language model to provide individualized feedback for pre-service physics teachers' reflective thinking.

Tan, Z., Beigi, A., Wang, S., Guo, R., Bhattacharjee, A., Jiang, B. et al. (2024). Large language models for data annotation: A survey. *arXiv.*

Turing, A. (1950). Computing machinery and intelligence. *MIND*, LIX(236), 433–460.

Valmeekam, K., Marquez, M., Sreedharan, S., & Kambhampati, S. (2023). On the planning abilities of large language models : A critical investigation. *arXiv.*

Wan, T., & Chen, Z. (2024). Exploring generative ai assisted feedback writing for students' written responses to a physics conceptual question with prompt engineering and few-shot learning. *Physical Review Physics Education Research.*

West, C. G. (2023). Ai and the fci: Can chatgpt project an understanding of introductory physics? *arXiv.*

Wulff, P., Buschhüter, D., Westphal, A., Mientus, L., Nowak, A., & Borowski, A. (2022). Bridging the gap between qualitative and quantitative assessment in science education research with machine learning—a case for pretrained language models-based clustering. *Journal of Science Education and Technology, 31*, 490–513.

Zamfirescu-Pereira, J. D., Wong, R. Y., Hartmann, B., & Yang, Q. (2023). *Why johnny can't prompt: How non-ai experts try (and fail) to design llm prompts: Chi '23, april 23–28, 2023* (pp. 1–21). Germany: Hamburg.

Zhao, H., Chen, H., Yang, F., Liu, N., Deng, H., Cai, H., Wang, S., Yin, D., & D. Mengnan (2023). Explainability for large language models. A survey. *arXiv*

Chapter 8
Human-Machine Interactions in Machine Learning Modeling: The Role of Theory

Christina Krist, Marcus Kubsch, and Peter Wulff

Abstract This chapter provides guidance for how to think about the role that humans play in conducting analyses that include ML as part of the workflow. It introduces a stance towards ML as "intelligence augmentation" rather than "artificial intelligence" and articulates four key questions for an analyst to consider when setting up analysis. It also introduces computational grounded theory as an alternative to automation-focused applications of ML.

8.1 Introduction

It can be easy to focus on the "thinking" work that a computer can do. After all, this book is focused on teaching you to understand the underpinnings of various computational models and how you might apply them in your own research. But there are two important assumptions that undergird our stance toward machine learning that we want to make explicit here:

1. ML is not always the appropriate or most useful tool; it may actually make your analysis more difficult and of lower quality.

We often see researchers who are beginning to wade into computational approaches drawn to ML out of a kind of fear of missing out, thinking that there exists some magical tool that will automate the "hard part" of their analysis and make their life and their research much better. This is false. No such tool exists. If anything, integrating ML will add more steps to your analytic workflow (as described in the previous chapters). In addition, because the "decisions" made in an ML model are black-boxed to varying degrees, it may make your results less interpretable. This

C. Krist (✉)
Graduate School of Education, Stanford University, Stanford, CA, USA
e-mail: stinakrist@stanford.edu

M. Kubsch
Freie Universität Berlin, Berlin, Germany

P. Wulff
Heidelberg University of Education, Heidelberg, Baden-Württemberg, Germany

© The Author(s) 2025
P. Wulff et al. (eds.), *Applying Machine Learning in Science Education Research*,
Springer Texts in Education, https://doi.org/10.1007/978-3-031-74227-9_8

is what we mean by saying it may make your analysis of "lower quality"—it may make it more difficult for you to construct a compelling claim from ML output.

2. Computers are not intelligent, critical, or innovative.

Despite the name "artificial intelligence" (a marketing ploy more than anything), it is important to remember that an algorithm is essentially a set of instructions. Computers are really good at following those exactly. And while more sophisticated machine learning models (e.g., ANNs; LLMs) appear to be "generative" in that they output something beyond specifically what they were given (e.g., solving a novel mathematics problem after being trained on a specific set of problems), these models are just identifying patterns that reflect the data they have to work with. These patterns may or may not make sense; the articulation of the abstraction is not the goal. Instead, functionally what occurs is a reproduction of patterns that exist within current data sets (Mitchell, 2023). This is the basis of many of the social critiques of AI: because even the most sophisticated ML algorithms are pattern identifiers and replicators, they reproduce existing social biases and inequalities (e.g., D'Ignazio and Klein, 2020; Noble, 2018).

In stating these assumptions, and after having read the descriptions of ML in the previous chapters, we hope to bring ML down from the pedestal, shrouded in mystery, that newcomers tend to place it on. ML is just a tool. And, like any methodological tool, it can be useful or it can be useless. What matters is when, how, and why it's used.

In this chapter, we provide some guidelines for how to think about whether and how using ML is right for your research project. As part of these guidelines, we focus equally on the role that humans can and should play as we do on the role that computational algorithms can play. After all, *you*[1] are also an important methodological instrument, with different strengths and weaknesses that you bring to any research task.

Thankfully, people who have worked on AI for decades, across multiple contexts, have also thought deeply about how humans and machines should interact and be integrated. For example, a recent report from the US Department of Education's Office of Educational Technology advocates for "Intelligence Augmentation" (IA) rather than "Artificial Intelligence" (AI) in order to "keep[] humans in the loop and position[] AI systems and tools to support human reasoning" (2023, p. 14). They describe that IA centers "intelligence" and "decision making" in humans but recognizes that people sometimes are overburdened and benefit from assistive tools. Intelligence Augmentation (IA) uses the same basic capabilities of AI, employing associations in data to notice patterns, and, through automation, takes actions based on those patterns. However, IA squarely focuses on helping people in the human activities of teaching and learning, whereas AI tends to focus attention on what computers can do. For instance, IA may help teachers make better decisions because computers notice patterns that teachers can miss. When a teacher and student agree

---

[1] We mean "you" to be understood in a broad sense here or elsewhere, i.e., including team members, collaborators, etc.

that the student needs reminders, an AI system may provide reminders in whatever form a student likes without adding to the teacher's workload.

8.2 A Guiding Metaphor for Centering "Intelligence Augmentation" Versus "Artificial Intelligence": Self-driving Cars

Because this is a relatively new approach in education, and because teaching and learning are massively complex tasks, we find the analogy to the ways AI has been used in cars to be helpful. For at least the past 100 years, engineers have sought to develop self-driving cars: machines that fully automate nearly all aspects of driving, so that humans only need to step in and intervene if anomalous situations arise. Despite the continuous enthusiasm (not to mention loads of money poured into their development), self-driving cars have yet to become a ubiquitous reality. Safety issues continue to halt any promising progress, even to this day.

So does this mean that the use of AI in cars is a failure? No, not necessarily. There are many very successful features in cars that utilize AI. These features, such as anti-lock brakes, air bags, parking assist, stay-in-your-lane warnings, and driver focus alerts are perhaps less flashy than a fully automated driving system, yet they significantly increase the safety of driving in vehicles equipped with these systems. These are examples of "intelligence augmentation"—tools that maintain the human activity of driving as the central goal, supported by specific tools targeted to notice and respond to specific patterns that humans can sometimes miss given the complex cognitive demands of driving. Most importantly, these tools are effective because they leverage the strengths of both humans and computers, identifying specific tasks where those strengths complement each other. In contrast, self-driving cars are a near-impossible engineering task because they are asking computers to mimic strengths of humans that they are simply not designed to be able to do (Fig. 8.1).

Consistent across the US DOE's (2023) recommendations for AI in education and the history of successes and failures in integrating AI into cars is the idea that it is crucial to think about humans and computers as a distributed system. With this perspective, we never consider *just* the strengths and limitations of an ML model in isolation, such as the type or structure of data it is best suited for or the particular biases of a specific algorithm. Instead, these considerations should always be contextualized within a broader ecosystem that includes the human researchers as part of the system. This extends these questions from, for example, "What type or structure of data is this ML model best suited for?" to, "What does utilizing this ML model for this type or structure of data serve to automate or augment, and is that automation or augmentation appropriate given the other tasks the human researchers will carry out as part of this analysis?"

The rest of this chapter elaborates the complexities of thinking through these contextualized considerations. In doing so, we highlight four guiding questions that

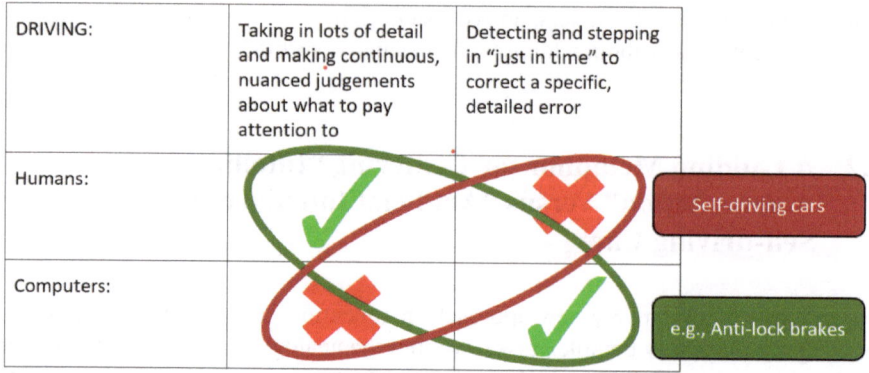

Fig. 8.1 How IA features such as anti-lock brakes take advantage of humans' and computers' strengths to complement weaknesses

are helpful for researchers to return to in deciding whether or how to use ML as part of a research project:

1. What is the degree of inference that I aim to draw from my data? Who is best suited to make those inferences, and how?
2. How will theory be used to guide interpretation of ML outputs, including identification of absences of features or patterns?
3. How is this ML analysis integrated and sequenced within a larger analytic workflow?
4. Is ML really augmenting my interpretive and analytic power as a researcher, or could this be done just as well (or better!) without it?

8.3   What are the Relative Strengths that Humans Versus Computers Bring to Conducting Research?

Before delving in to these four questions, we first want to make explicit some of the relative strengths of humans and computers when it comes to the task of doing research. Like driving, doing research is a complex and non-linear task. And, as with driving, humans are generally good at taking in lots of detail and making continuous, theoretically-informed judgements about what to pay attention to (Table 8.1). This does take practice: it requires developing an education-research-data-specific version of what Goodwin (1994) called "professional vision" coupled with deep knowledge of the existing literature to decide what is salient and relevant to contemporary questions and debates. It also takes time to learn to reduce the many visual and auditory inputs that a driver is bombarded with, cognitively automate the processing of that input, and direct attention to specific, salient aspects—a classical account of procedural learning (e.g., Corbett and Anderson, 1995). In other words, it's not that humans

Table 8.1 Relative strengths and weaknesses of humans versus computers for data analysis

Human strength/computer weakness for data analysis:	Computer strength/human weakness for data analysis:
Taking in lots of detail and making continuous, theoretically-informed judgements about what to pay attention to	Detecting patterns; performing the same specific, detailed operation many times

are naturally or innately able to take in lots of detail and make continuous judgements about it; but rather, with training and practice, humans are good at learning to do this in a specialized way.

In contrast, computers are distinctly not good at taking in lots of detailed information and making judgements about what to pay attention to, without specific, detailed instructions for what to pay attention to and how. These specific, detailed instructions may come "hard coded" in an algorithm, or they may come from emergent patterns based on a very large data set. In either case, what a computer *is* good at doing is detecting patterns, and at performing the same, specific, detailed operation many, many times (Table 8.1). Humans are distinctly bad at both of these things: we tend to pay attention to anomalies and outliers rather than patterns (one reason why statistical methods are so important as assistive analytic tools!), and we tend to become fatigued and make mistakes when performing repetitive tasks.

There are likely many more tasks or subtasks that we could articulate in terms of strengths and weaknesses, but we find these few broad comparative strengths to be sufficiently generative for getting started in thinking about how to distribute tasks.

What we hope to emphasize is that all of the intellectual work and decision-making that determines the quality of the ML analysis is done by the human. The computer simply carries out what the human tells it to do. (Remember Grounding Assumption 2: Computers are not intelligent, critical, or innovative). In other words, the appropriateness and quality of your analysis, and the kinds of claims that can be made from it, is entirely dependent upon *your* careful, critical thinking at each step of the process.

If this seems overwhelming and like a lot of responsibility, good! It is. This is part of what we mean when we say that using ML will definitely not make your research easier, and it may actually make it harder. We suggest going back and re-reading the chapters on supervised and unsupervised learning with this lens. Ask yourself, what are the key decisions I need to make? And what are the criteria I should use in order to make and justify those decisions? We have tried to make explicit these decision points and ways of thinking in context as we have walked through the previously presented examples, and future chapters will continue to do the same.

The next sections will attempt to tackle the four questions introduced above that make explicit a more general version of these decision points and criteria. In doing so, we will also provide some theoretical and methodological guidance on how to approach thinking about these questions.

8.4 What is the Degree of Inference that I Aim to Draw From My Data? Who is Best Suited to Make Those Inferences, and How?

As we split apart the work of the human from the work of the computer in the supervised/unsupervised machine learning workflows above, a feature of the task that was not made explicit but still influenced each stage of activity was the degree of inference required to transform the collected data into a variable that can then be visualized and used. To unpack this dimension, we will focus on one aspect of the workflow: assigning codes or numbers to data. The epistemic function of this aspect can be characterized by how well-defined the coding categories are a-priori. That is, do we know what we are looking for because there is an existing survey, coding manual, or rubric? Or, does part of the task involve developing emergent categories and codes (i.e., a grounded approach)? This distinction is similar to the difference between inductive and deductive analysis in qualitative content analysis (e.g., Kuckartz, 2014) and supervised versus unsupervised ML techniques.

Both deductive and inductive (or supervised and unsupervised) approaches can vary in the degree inference required. Some data transformations require very little inference. For example, whether a student continued on to Stage 2 of the Physics Olympiad can be very clearly mapped to a binary coding scheme (0 = did not continue, 1 = continued). Or, when using text data, when we are looking for the usage of technical language, the sequences of letters that signify key terms are clear. In contrast, mappings might be less well-defined and require large amounts of inference for constructs such as students' notions of uncertainty. Specific terms may not be clear signals, and the literature includes a wide range of definitions and operationalizations. In this sense, the human plays a much larger inferential role in deciding whether a given student response represents a particular epistemic belief and how to map that evaluation to a coding scheme.

Figure 8.2 presents these dimensions as four quadrants where each quadrant serves a different epistemic function.

8.5 How Will Theory be Used to Guide Interpretation of ML Outputs?

As mentioned previously, one of the key roles that a human plays is "seeing" data through the lens of theory. This includes identifying which theories might be relevant to interpreting data, both early on and throughout the analytic process. Of course, theory is useful for generating codes as described in the previous section. But there are also two additional roles for theory that we highlight here:

1. Theory guides interpretation of claims.
2. Theory helps analysts see what is not present in the data.

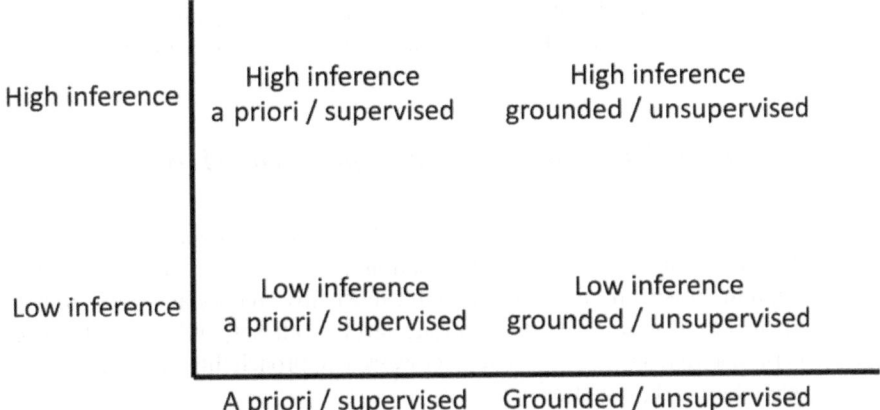

Fig. 8.2 Epistemic functions and level of inference of four different kinds of tasks in assigning codes or numbers to data based on characteristics of the data and question. Reprinted from Kubsch et al. (2023) which is licensed under CCA

In terms of guiding interpretation of claims, theory is what supports human analysts in being able to make decisions about whether a particular output—e.g., a way of parsing data into clusters—is meaningful or not. Here, "meaningfulness" means "contributing new insights." Does this way of dividing up the data tell me something new and interesting about the phenomenon I am studying? Does it help me answer my research question in some way? In other words, the theoretically informed human analyst is essential for interpreting computational outputs in ways that can lead to new insights.

What's important to keep in mind here is that the same data, clustered in the same way, may be meaningful or "junky" depending on the research question at hand and the theoretical lens applied. And, maintaining a theoretically informed lens helps prevent overstatement of claims or overattribution of certainty to computational results. Simply because a computer clustered data in a particular way, even with a great deal of statistical certainty, does not mean that those clusters are interpretable or meaningful for addressing a question of educational importance.

Theory is also essential for helping analysts to see what is not present in the data. This is true for both quantitative and qualitative analyses, and remains true for computational analyses as well. For instance, if a theoretical framework identifies four epistemic criteria (e.g., Berland et al., 2016) but computational outputs only identify or work for three, this is itself an interesting finding. Similarly, critical theoretical lenses are especially important here for identifying potential social biases in the data sets themselves, in the algorithms being applied to analyze the data, or even to evaluate the theoretical frameworks being used in the first place.

These high-level roles for theories are ones that guide how an analyst draws inferences from data. The details of how this is done varies by the type of analysis being performed, e.g., whether it tends to be more qualitative or quantitative in nature. We briefly provide some examples of what this work can look like in each case.

8.5.1 Theory in Quantitative and/or Supervised ML Analysis

In quantitative research, the work of drawing inferences from data outputs is traditionally shared by humans and computers functioning as a complementary system. In this system, humans specify relations based on substantive theory and use computational statistical tools to quantify the relations, test hypotheses, and draw inferences about generalizability. There are two distinct ways to approach this, serving different goals: explanation and prediction.

Traditional statistical models tend to be better for explanatory goals. A typical example of this would be to set up a regression model based on substantive theory and use statistical software to calculate the regression coefficients and conduct statistical tests of the model. The result of such an analysis is a descriptive and potentially explanatory model of the phenomenon under study.

In contrast, ML models tend to be better for predictive goals. Instead of letting humans specify the relations between the codes or numbers in the form of a statistical model such as a regression, humans may just specify which codes or numbers x is supposed to be related to codes or numbers y and let the computer find a function $f(x) = y$ that describes the relation. The result is a model, often uninterpretable to humans, that can predict y given x with a certain accuracy. This approach is typical of many supervised ML use cases. In essence, predictive accuracy is traded for the interpretability and causality of a traditional statistical model (Breiman, 2001; Pearl and Mackenzie, 2018; Schölkopf, 2019).

8.5.2 Theory in Qualitative and/or Unsupervised ML Analysis

In (grounded) qualitative research, the work of looking for relations and drawing inferences is traditionally done by humans. In evaluating "meaningfulness," a primary consideration that continually guides qualitative analysis is whether a pattern or theme is meant to be representative or anomalous, as both have value (Glaser, 2002).

Commercially available qualitative analysis software (e.g., NVivo, MAXQDA) increasingly provide a suite of tools to assist in consolidating, organizing, and visualizing qualitatively analyzed data. These tools range from quite simple, such as displaying the total count of each code across a set of documents, to more complex, such as network-type visualizations of code co-occurrences. The goal of such tools is not to produce statistical outputs, but rather to assist the researcher in conducting second-stage coding or building qualitative claims. In other words, these high-level patterns, such as the number of times a particular code occurred, are often an important starting point in supporting qualitative analysis, even though the counts of codes themselves are typically not the claim being made. Rather than the end (i.e., a claim), these patterns are the beginning of a qualitative analytic process.

Unsupervised ML approaches such as topic modeling or anomaly detection serve a similar role: they are not themselves generating a claim about the data, but they are surfacing a pattern from qualitative data that an analyst could then interrogate more deeply. Both theory and continued empirical work are needed to construct and evaluate the resulting qualitative claims.

8.6 How is ML Analysis Integrated and Sequenced Within a Larger Analytic Workflow?

The most common use of ML currently within the science education literature is for automation: to integrate a computer to do some of the work that humans would otherwise do. In response to the first two considerations, we have provided some guidance for how to think about which tasks to automate and how. With this consideration, we would like to provide an alternative purpose for ML: generating new insights. Rather than taking the human coding, or existing knowledge and processes, as a "gold standard" and training a computer to replicate human effort, ML can instead complement the human researcher towards a goal of developing the "gold standard" claim in the first place.

There are a few emerging examples of this in the science education literature. One example used unsupervised clustering techniques to triangulate human-conducted analyses of student interviews, using the clustering as an additional source of validity evidence for the qualitative claims made from the data (Sherin, 2013). Another example similarly used unsupervised clustering to refine a theoretically-informed coding scheme: in making sense of the clusters in light of the original coding scheme, Rosenberg and Krist (2020) identified a new type of code that led to a re-structuring of the original coding scheme. Their revised scheme led to better qualitative reliability between human coders as well as improved reliability of supervised automation of coding.

One common theme of these examples is the careful sequencing of human and computational approaches to analysis. Work in other fields has identified methodological frameworks for guiding such sequencing. We highlight one such framework here that has been foundational for our own work and for emerging work in science education: that of computational grounded theory.

Computational Grounded Theory (CGT) was originally developed by sociologist Laura Nelson (2020) to investigate the political logics in first- and second-wave feminist movements, primarily using documents produced by political organizations in New York and Chicago during each era. We highly recommend reading her original paper, as the findings themselves are very interesting! But for the remainder of this chapter, we focus on the methodological framework generally.

Broadly speaking, in CGT, one first leverages the power of computational techniques, especially unsupervised ML techniques, for pattern detection in large data sets—those of a size and scope that may prohibit human-driven analyses from the outset. Then, one leverages the interpretative power of humans to add quality and depth to the quantity and breadth of the first step. Finally, one uses computers to test the reliability and generalizability of the human refined pattern detection and interpretation from the second step. We use the following terms to describe the purpose of each of these steps:

- Step 1: Pattern Identification
- Step 2: Pattern Refinement
- Step 3: Pattern Confirmation

Importantly, in each of these three steps, assigning codes or numbers to data is followed by looking for relations and drawing inferences, which inform the next step leading up to the final sociological conclusions. This process is inherently cyclical, and the nature of the analytic work varies slightly at each stage. In other words, rather than assigning codes once and drawing inferences once, multiple rounds of analyses are chained together such that the inferences drawn from Step 1 are not final-form claims, but instead are used to guide the nature of the analytic work in Step 2, and so on.

Specifically, in Step 1, Nelson utilized unsupervised computational methods to identify both low-inference patterns (via lexical analysis, i.e., basic natural language processing techniques) and high-inference patterns (via topic modeling). In Step 2—the pattern refinement step—the analytic work remains grounded (and primarily done by the human analyst) but is predominantly high-inference. In this step, the human analyst uses a standard qualitative method—content analysis via guided deep reading—to interrogate the patterns that were identified using the computational techniques in Step 1, though only utilizing a small subset of the data set. In Step 3, pattern confirmation, the analytic work, can be characterized as high-inference, supervised/a priori. Nelson strategically selected supervised computational techniques that would examine the generality of the claims identified in Step 2 across the entire data corpus.

This framework demonstrates careful consideration of when and how to integrate computational methods to complement the work of the human analyst. By strategically sequencing methods, a CGT approach explores usages of ML in all but one of the four quadrants in Fig. 8.1, using computation in a way where it complements the role of the human analyst, aiming at "preserving the superior abilities to interpret text holistically provided by humans but incorporating the formal rigor, reliability, and reproducibility of computer-assisted methods." (Nelson, 2020, p. 8).

The use of CGT has recently gained some traction within science education research (e.g., Martin et al., 2023; Tschisgale et al., 2023). We will also provide examples of careful consideration of sequencing in Chaps. 6 and 15. In this way, we hope that the field of science education can continue to innovate in terms of how we are integrating ML into our analytic workflows in ways that move beyond automation and purely predictive applications.

8.7 Is ML Really Augmenting My Interpretive and Analytic Power as a Researcher, or Could This be Done Just as Well (or Better!) Without It?

Finally, it is always worth keeping this question in mind. What are the costs, both in terms of time spent learning a new technique, downloading and installing software, cleaning or re-formatting data, and in terms of the enviornmental impact and human labor costs of the computational infrastructure you are using (e.g., the cost of letting ChatGPT perform your tasks)? And—most importantly—can I make the case that integrating ML will produce a substantial payoff in terms of generating new knowledge that could not be generated without it?

As educators and science education scholars, we are always in favor of learning new techniques and tools. From that perspective, we encourage you to continue working through this book in order to be able to answer those questions in an informed way. We see value in gaining this knowledge, even if you end up never using it yourself. To us, that is decidedly not a waste of time! You will have expanded your intellectual and methodological toolkit in ways that can help you better assess your own and others' research. It may also give you new ideas and perspectives on your data or research questions to pursue, with or without ML tools. This is an important learning outcome.

We do hope that at some point, you are excited about a potential use for ML within your own research and see how a set of tools could augment your own interpretive and analytic power. This is the exciting part of new tools and computational developments! We see great potential for creativity and innovation as more and more science education scholars take up ML and other AI-based techniques in their research. At the same time, we do want to emphasize that just because emerging computational approaches are in the spotlight (now more so than ever) does not mean that you need to use them in order for your research to be relevant. Good questions, thoughtful insights grounded in the realities of practice, and careful interpretation and application of theory are evergreen—essential for all research, both with and without ML.

References

Berland, L. K., Schwarz, C. V., Krist, C., Kenyon, L., Lo, A. S., & Reiser, B. J. (2016). Epistemologies in practice: Making scientific practices meaningful for students. *Journal of Research in Science Teaching, 53*(7), 1082–1112.

Breiman, L. (2001). Statistical modeling: The two cultures. *Statistical Science, 16*(3), 199–231.

Corbett, A. T., & Anderson, J. R. (1995). Knowledge tracing: Modeling the acquisition of procedural knowledge. *User Modeling and User-Adapted Interaction, 4*, 253–278.

D'Ignazio, C., & Klein, L. F. (2020). *Data Feminism*. Strong Ideas Cambridge: The MIT Press.

Glaser, B. G. (2002). Conceptualization: On theory and theorizing using grounded theory. *International Journal of Qualitative Methods, 1*(2), 23–38.

Goodwin, C. (1994). Professional vision. *American Anthropologist, 96*(3), 606–633.

Kuckartz, U. (2014). *Qualitative text analysis: A guide to methods, practice and using software.* Los Angeles: Sage.

Martin, P. P., Kranz, D., Wulff, P., & Graulich, N. (2023). Exploring new depths: Applying machine learning for the analysis of student argumentation in chemistry. *Journal of Research in Science Teaching*, pp. 1–36.

Mitchell, M. (2023). Ai's challenge of understanding the world. *Science, 382*(6671), eadm8175.

Nelson, L. K. (2020). Computational grounded theory: A methodological framework. *Sociological Methods & Research, 49*(1), 3–42.

Noble, S. U. (2018). *Algorithms of oppression: How search engines reinforce racism.* New York: New York University Press.

Pearl, J., & Mackenzie, D. (2018). *The book of why: The new science of cause and effect* (1st ed.). New York: Basic Books.

Rosenberg, J. M., & Krist, C. (2020). Combining machine learning and qualitative methods to elaborate students' ideas about the generality of their model-based explanations. *Journal of Science Education and Technology.*

Schölkopf, B. (2019). Causality for machine learning. *H. Geffner et al. (eds): Probabilistic and Causal Inference: The Works of Judea Pearl, ACM, 27*, 765–804.

Sherin, B. (2013). A computational study of commonsense science: An exploration in the automated analysis of clinical interview data. *Journal of the Learning Sciences, 22*(4), 600–638.

Tschisgale, P., Wulff, P., & Kubsch, M. (2023). Integrating artificial intelligence-based methods into qualitative research in physics education research: A case for computational grounded theory. *Physical Review Physics Education Research, 19*(020123), 1–24.

Part II
Hands-On Case Studies

Chapter 9
Working with Data—Getting Started

Marcus Kubsch, Peter Wulff, and Christina Krist

Abstract In order to enable readers to implement the case studies on their computers, we will introduce essential information for implementing the necessary software. Comprehensive reviews and tutorials for the respective software can be found elsewhere and we will point to some of these resources.

9.1 Introduction

You have made it through the theory section of the book and arrived at the case study part where we start to work with data and do some actual machine learning in the context of science education. The following case studies contain code and we recommend executing the code as you progress through the case studies. At some points we only display important code, and full implementation details can be found in an online supplement, which can be freely accessed.

We decided to use two programming languages in this book: R and Python. While this may not seem helpful at first, we think it has value to know ones' way around in these two languages. Both, R and Python are open source and widely popular in the fields of data science and ML. Sometimes, however, specific challenges in ML are easier to tackle in one language due to the exclusive availability of packages, which consist of functions handling specific tasks within that language. Similarly, case studies or examples might be available in one language but not the other. In these cases, it is helpful if one is able to use both languages flexibly. Using different programming languages flexibly is in fact increasingly supported in popular tools like RStudio or Google Colab.

M. Kubsch (✉)
Freie Universität Berlin, Berlin, Germany
e-mail: m.kubsch@fu-berlin.de

P. Wulff
Heidelberg University of Education, Heidelberg, Baden-Württemberg, Germany

C. Krist
Graduate School of Education, Stanford University, Stanford, CA, USA

© The Author(s) 2025
P. Wulff et al. (eds.), *Applying Machine Learning in Science Education Research*,
Springer Texts in Education, https://doi.org/10.1007/978-3-031-74227-9_9

157

Before we get started with the actual case studies, we will provide a quick start guide to get both R and Python up and running and cover some of the most important commands and packages that we use throughout the following chapters. If you are already familiar with R and Python, you can skip the rest of this chapter.

> **Note of caution**
>
> If you run the analyses either in browser or especially on your own computer, you might get slightly different outputs. The outputs should not differ substantially, however, slight differences are always possible given that your hardware and software differs which can impact the calculations in the algorithms.

9.2   Getting Started with R

The objective of this section is to enable you to begin executing the code presented in the case studies. If you want to go more in-depth with R, Hadley Wickham's R for DataScience is an excellent and accessible resource, which can be found here: https://r4ds.hadley.nz (last access: Nov 2024).

9.2.1   Installation

To install R go to https://cran.r-project.org, download the version for your operating system and install it. Once you have R installed you could theoretically just write all code into the command line and execute it. However, that course of action is error prone and tiresome. A better way to use R is to work with scripts, i.e., you have a plain text editor in which you write down the code in the order you want to run it and then copy and paste it into the command line to actually run it. In this way, your analysis is automatically documented. We recommend using RStudio which is an open source development interface for R from Posit, that is, it is a program that gives you a text editor and command line in one application that comes with numerous helpful supporting functions like autocomplete for commands, an overview of the created objects, etc. You can download RStudio here: https://posit.co/downloads/.

9.2.2   Loading Packages and Data

Many of the functions we will use are not included in the basic R installation. To be able to use them you need to first install them and then load them every time you want to use them. Packages are installed with the install.packages() command, e.g., to install the required tidymodels package type install.packages("tidymodels") and run the command. Once the package is installed you can load it with the library() command, e.g., library(tidymodels) loads the tidymodels package so that the functions within it become available. In the following case studies, we will load packages as they are needed. If you try to load a package you have not installed a warning will appear. In that case install the package. Afterwards you should be able to load it.

The case studies utilize data and code accessible on either one of the book's accompanying websites (Google Drive: https://drive.google.com/drive/folders/ 1JNPLA7gu7-YeaTwYS0-dGDXmftYIb220?usp=sharing, GitLab: https://gitlab. com/ml_sci-ed/notebooks).[1] For the R based case studies, the website contains folders that include R projects files, scripts, and data. With RStudio, you can just use the project and then open the script. Paths to the data are configured automatically using the here package so that you should be able to execute the code without manual setup. If you do not want to use RStudio you can load the data using the load() function but you will need to specify a path to the respective datafile.

9.2.3 Important Operators: Assignment and Pipe

To assign data to variables, the assignment operator <- is used. The following example command stores the number 3 in the variable df:

```
df <- 3
```

To inspect the object df just type df and execute the command.

Another important operator is the pipe operator %>%. You can think of the pipe operator as pushing the output of one command or function to the input of the next command or function. The following code uses the pipe operator to get the maximum value of the variable df from the example above and then plot it:

```
df %>% max() %>% plot()
```

Lastly, the # is used in R to comment on code, i.e., code (and everything else) in a line that starts with # will not be interpreted by R (in Python this will be also #, and larger sections can be commented out in between """ . . . """).

9.3 Getting Started with Python

Python is a widely used programming language both in academia and industry, especially for AI and ML. Python provides researchers "efficient high-level data structures and a simple but effective approach to object-oriented programming." (see:

---

[1] Note that the Google Drive folder has also all pre-trained model in it, thus is relatively large in size.

https://docs.python.org/3/tutorial/, last access: Nov 2023). Readers who are interested in learning to program with Python are referred to resources such as https://learnpython.org/, or https://docs.python.org/3/tutorial/ (last access: June 2023). Also, AI resources such as Codex and ChatGPT (both by OpenAI) provide efficient resources that help you to implement code, as you can work with them using natural language. Say, you want to get help with calculating some statistic or plotting your data, simply prompt some general-purpose LLM with "Provide R/Python code to plot a histogram for variable X." These requests to LLMs such as GPT-4 yield helpful outputs for most common problems such as data visualization.

9.3.1 Installation

To install Python go to https://www.python.org/, download the version for your operating system and install it. A user friendly platform is Anaconda (https://docs.anaconda.com/). Anaconda is a platform that supports managing software such as Python and R. You will be able to run Python and R from Anaconda, and also Jupyter notebooks which are used in this textbook. As with R, once you have Python installed you could theoretically just write all code into the Python interpreter (e.g., through your terminal), by starting it through typing `python` into your console/terminal, and then engage in a so-called interactive mode to use Python. However (as with R), that is error prone and sometimes tiresome. A convenient way to use Python is to work with scripts, i.e., you have a plain text editor in which you write down the code in the order you want to run it and then copy and paste it into the Python terminal to actually run it. Or better, you run the scripts in script mode where you store your script and run it from the console/terminale via `>python name-of-script.py`. In this way, your analysis is automatically documented. If you want to write scripts and run them, we recommend using Spyder which is an open source development interface for Python which functions as an interactive development environment (IDE) (similar to RStudio for R) where you have a lot of command over you Python programming and data infrastructure. You can get it for free here: https://www.spyder-ide.org/.

Another way to execute Python code is through computational notebooks. In Python, Jupyter notebooks (or Jupyter lab) are a versatile way to use notebooks. They are more accessible compared to scripting Python code. However, there are also drawbacks that relate to performance issues, reproducibility, and debugging, among others. Given that we seek to provide readers easily accessible, simple applications of important ML algorithms and data processing procedures, Jupyter notebooks are nonetheless considered the optimal way to do so.

9.3.2   Loading Modules and Data

For ML in particular, many of the most powerful modules (this is Python lingo for "libraries", however, these terms are sometimes used interchangeably, and we use both terms in this book) were developed for Python such as scikit-learn (Pedregosa et al., 2011), tensorflow (and the closely related keras) (Martín Abadiet al., 2015), or pytorch (Paszke et al., 2019). These modules allow you to implement most of the available ML algorithms, and train and test them. Some of them are designed for efficient usability. For example, scikit-learn offers a variety of ML algorithms and data preprocessing functions, and it allows to access them very consistently with unified, thought-through functions (we will do so in Chap. 6). It is designed with the purpose to raise consistency, inspection, nonproliferation of classes, composition, and sensible default values (see further information on this library in Géron (2017)). Over time, many derived libraries emerged that will simplify your implementation of specific ML algorithms such as clustering and finding topics in language data, such as bertopic (which we will use in Chap. 13) or sentence-transformers.

For natural language processing, the spaCy library (Honnibal and Montani, 2017) has become a go-to reference. You can outsource many of your language preprocessing tasks to it such as removing redundant words, tokenization, part-of-speech tagging, etc. We will more deeply dive into these possibilities in chapters 7 and 12. Particularly for LLMs and data sets to download and use, the huggingface module present a rich resource (Wolf et al., 2020).

In this book, we will only use a selected range of the methods supplied by the abovementioned libraries. Readers who want to get a more comprehensive start with these libraries and learn about their capabilities are referred to the excellent tutorials provided by all major ML libraries nowadays, e.g., for pytorch see https://pytorch.org/tutorials/.

9.3.3   Implement Case Studies with Jupyter Notebooks

To implement the case studies and examples in Python, we provide the reader with Jupyter-Notebooks that can be downloaded and executed (Python version greater 3.8 is recommended). An important aspect in Python (or any other) programming is to control the versions of the libraries that you are using. This is done in Python typically through a so-called requirements.txt file (that you find in the GitLab supplement https://gitlab.com/ml_sci-ed/notebooks, or in the Google Drive repository https://drive.google.com/drive/folders/1JNPLA7gu7-YeaTwYS0-dGDXmftYIb220?usp=sharing), where modules and their versions are stored that will allow the user to re-run your project. A requirements file can be build with pip through pip freeze > requirements.txt, which stores a requirements file in your current working tree. Moreover, with pip check you can verify if any version conflicts of packages exist. In the requirements.txt file one line might read: scikit-learn==0.23.2, referencing the specific version of the scikit-learn library that is used for our notebooks. This then enables researchers in the future to replicate our outputs, given that later versions of `scikit-learn` might change some default parameters in the functions or the algorithms themselves. As such, it is crucially important to keep track of your used module versions for other to be able to reproduce your analyses. In fact, computational reproducibility was an issue as singled out in a review study (Kapoor and Narayanan, 2023).

R and Python develop rapidly, as does the field of AI research in general. This causes older projects with specific modules/libraries to sometimes fail to work in newer versions. It is therefore good practice to set up a virtual environment where all modules/libraries for this specific project are statically stored with the specific versions, or at least create a file where all versions of modules/libraries are stored (oftentimes named `requirements.txt`, see: https://learnpython.com/blog/python-requirements-file/).

9.3.4 Python Objects, Functions, and Data Types

Python relies on specific data types, functions, and other resources. Data types include integers, strings, lists, tuples, or dictionaries. The details do not matter here, and interested readers can find an accessible introduction here: https://docs.python.org/3/tutorial/datastructures.html (last access: 15 Oct 2023). As with R, you might assign a number a variable via `a = 1`. Instead of a number (integer), you could assign many different things such as strings or dictionaries to this variable and then access them through the variable name. You can even assign functions and classes to this variable (which then would not be called variable anymore). The Python rules (PEP) specify naming conventions for variables as: "should be lowercase, with words separated by underscores as necessary to improve readability".[2] Functions have similar naming conventions. A function can be very handy. It takes parameters, performs some computations, and (oftentimes) returns some output, or writes out output into an external file. A simple function that adds two numbers could look similar to what is found in Code 9.3.4. You additionally might want to assure that a and b are integers (or whatever), since Python would simply concatenate strings (and not return an error), if a and b are (unintentionally) strings.

Python code: Addition function

```
def addition(a,b):
    return a + b
```

---

[2] See https://peps.python.org/pep-0008/#introduction, last access: Nov 2023.

Python commonly uses objects to store functions and data, and make them flexibly accessible. A common object is the `class`. A `class` enables you to store data and functions together in one object. For example, you might create a class called `student` (see Code 9.3.4. This is a `class` object with two functions: `__init__`, and `give_name`. The `def __init__(self)` function is executed once the class is initialized, e.g., through `s = student()` in your Python script. This function then sets the age of the student to 16 (`age=16`), which is then attached to this object `s` via the `self.age`. The age can then be accessed later on. For example, if you run `s.age`, the number 16 will be returned. The second function enables one to give this student object a name, e.g., `s.give_name("Test")`, which can then also be accessed via `s.name` later on.

We will use class objects often when initializing ML algorithms such as `BERTopic` (see Chap. 13). We will then load the raw model from within a specific module via: `from bertopic import BERTopic` as a `class` object. We then initialize this class with specific parameters and assign it to a new object that inherits all data and functionality from the class: `topic_model = BERTopic( language = "english", ... )`. Once assigned and initialized, we can then train it with our data and visualize results (see Chap. 7).

Python code: implementing a class object

```python
class student:
    def __init__(self):
        self.age = 16
    def give_name(self,name):
        self.name = name
```

Further reading

Python introductions:

- Mark Lutz. Learning Python. 5th ed. O'Reilly Media, 2013.
- Fabrizio Romano. Learn Python Programming. 2nd ed. Packt Publishing, 2018.
- Chollet, F. (2018). Deep learning with Python (Safari Tech Books Online). Shelter Island, NY: Manning. Retrieved from http://proquest.safaribooksonline.com/9781617294433.

Using Jupyter notebooks:

Throughout the book, we will show Python program code excerpts that are part of larger Jupyter notebooks which can be accessed in the online supplement(s). If readers are unfamiliar with Jupyter notebooks, the following textbook presents an accessible introduction (see: Chap. 2 in Géron (2017)). If you want to use the notebooks in your browser, we suggest that you use Google colab (https://colab.research.google.com/, Google account required), and copy the notebooks into your Google drive, which can be mounted to colab (we included code for this in the notebooks).

9.4   How to Access the Notebooks for This Book

This textbook seeks to provide you with the first steps on how to get an ML research project started in the context of science education research. Hence, we open source all our computational analyses as notebooks and script files. You can access the notebooks via the companion website for this book via: https://drive.google.com/drive/folders/1JNPLA7gu7-YeaTwYS0-dGDXmftYIb220?usp=sharing (Google Drive). To run these notebooks, either R or Python are required, which are also freely available and open source. In the beginning of chapter 9 we introduced you to some basics on how to get started with R and Python in order for you to implement and run these notebooks.

9.5   (Some) Good-Practices in Scientific Programming

Let us finish with a list of recommendations (not conclusive) that can help you to efficiently and effectively set up science education research projects that involve running computer code:

- Set up a virtual environment for your project where you store all packages/modules (with their specific version) in order to be able to re-run analyses later on. This also aligns with open science standards for reproducible research.
- In your computations, whenever possible and meaningful, set a random seed so that the computations become more deterministic, which allows other researchers to check your computations.
- Try to comment your code (this is a thing LLMs such mostly do by default when they generate code). Later on in your project that might run several years, you will otherwise have a hard time to figure out what purposes and functions certain code had. Find guidelines here: https://realpython.com/python-comments-guide/ (last access May 2024).
- Use sanity checks in your code, as many algorithms are so complex that you cannot reasonably evaluate the validity of the outputs. A sanity check might be simply assuring, after scaling, that all means of your numerical features are zero.
- Especially for larger projects, modularization of code might be helpful, e.g., to write out own modules with often-used functions in it. Also, preprocessing of data

and analysis of data should be separated from each other. Oftentimes, preprocessing takes up a lot of time, and re-running preprocessing every time would basically waste your time.

- As in many analyses, you should consult with your fellow researchers to double-check your code and get ideas for improving it.

References

Abadi, M., Agarwal, A., Barham, P., Brevdo, E., Chen, Z., Citro, C., Corrado, G. S., Davis, A., Dean, J., Devin, M., Ghemawat, S., Goodfellow, I., Harp, A., Irving, G., Isard, M., Jia, Y., Józefowicz, R., Kaiser, L., Kudlur, M., Yuan, Y., ... & Zheng, X. (2015). Tensorflow: Large-scale machine learning on heterogeneous systems. arXiv.

Géron, A. (2017). *Hands-on machine learning with Scikit-Learn and TensorFlow: Concepts, tools, and techniques to build intelligent systems.* Beijing and Boston and Farnham and Sebastopol and Tokyo: O'Reilly.

Honnibal, M., & Montani, I. (2017). spacy 2: Natural language understanding with bloom embeddings, convolutional neural networks and incremental parsing.

Kapoor, S., & Narayanan, A. (2023). Leakage and the reproducibility crisis in machine-learning-based science. *Patterns (New York, N.Y.), 4*(9), 100804.

Paszke, A., Gross, S., Massa, F., Lerer, A., Bradbury, J., Chanan, G., Killeen, T., Lin, Z., Gimelshein, N., Antiga, L., Desmaison, A., Kopf, A., Yang, E., DeVito, Z., Raison, M., Tejani, A., Chilamkurthy, S., Steiner, B., Fang, L., Bai, J., & Chintala, S. (2019). Pytorch: An imperative style, high-performance deep learning library. In H. Wallach, H. Larochelle, A. Beygelzimer, F. d'Alché-Buc, E. Fox, & R. Garnett (Eds.), *Advances in neural information processing systems* (Vol. 32, pp. 8024–8035). Curran Associates, Inc.

Pedregosa, F., Varoquaux, G., Gramfort, A., Michel, V., Thirion, B., Grisel, O., Blondel, M., Prettenhofer, P., Weiss, R., Dubourg, V., Vanderplas, J., Passos, A., Cournapeau, D., Brucher, M., Perrot, M., & Duchesnay, E. (2011). Scikit-learn: Machine learning in python. *Journal of Machine Learning Research, 12*, 2825–2830.

Wolf, T., Debut, L., Sanh, V., Chaumond, J., Delangue, C., Moi, A., Cistac, P., Rault, T., Louf, R., Funtowicz, M., Davison, J., Shleifer, S., Platen, P. v., Ma, C., Jernite, Y., Plu, J., Xu, C., Le Scao, T., Gugger, S., Drame, M., Lhoest, Q., & Rush, A. M. (2020). Huggingface's transformers: State-of-the-art natural language processing. arXiv.

Chapter 10
Automation—Supervised Machine Learning

Marcus Kubsch, Christina Krist, and Peter Wulff

Abstract In this chapter we will apply ML with the purpose of building a reliable classifier for either classifying students into groups, or predicting test scores of students.

10.1 Introduction

In a recent study, Kenneth Holstein and colleagues (2019) asked teachers the following question "If you could have any superpowers you wanted, to help you do your job, what would they be?" One of the most desired superpowers turned out to be omniscience. In particular, teachers were eager for the ability to see students' thought processes including misconceptions, whether students were truly stuck, or whether they had nearly reached mastery. With this kind of information available, teachers argued, they would be in a better position to support struggling students. While teachers can get glimpses of students' thought processes in well-designed learning environments, it is almost impossible to do so for every student in a typical classroom. However, when students are learning in digital learning environments, their actions in these environments can reveal a lot about their thought processes. This information can then be relayed to teachers (and also to the students themselves).

This raises the question of how students' actions in a digital learning environment can be translated into information about their thought processes and learning. Theoretically, the data collected in a digital learning environment could be analyzed by a trained rater and then forwarded to the teacher or the students directly, e.g., in the form of feedback. However, this would not allow real-time feedback, and realistically, there are neither the financial resources nor the number of trained raters

M. Kubsch (✉)
Freie Universität Berlin, Berlin, Germany
e-mail: m.kubsch@fu-berlin.de

C. Krist
Graduate School of Education, Stanford University, Stanford, CA, USA

P. Wulff
Heidelberg University of Education, Heidelberg, Baden-Württemberg, Germany

© The Author(s) 2025
P. Wulff et al. (eds.), *Applying Machine Learning in Science Education Research*,
Springer Texts in Education, https://doi.org/10.1007/978-3-031-74227-9_10

available to do this. Thus, this challenge calls for automation, i.e., automating the process of rating the data. This brings us to the realm of supervised ML techniques. To briefly reiterate, supervised ML is the set of ML techniques where a computer is trained to predict an assigned code, label, or score based on a set of features. In terms of the above example, this would mean that a computer learns to predict human assigned labels, like identifying misconceptions, using students' interactions such as typed text or clicked buttons within a digital learning setting. Supervised ML techniques vary based on whether the computer learns to predict categorical aspects, such as identifying misconceptions, or continuous elements, such as counting the steps during problem-solving, labeled as classification and regression, respectively. This means that whenever we want to use ML to automate something—for example automatically coding students' misconceptions so that a teacher can see prevalent misconceptions in their class on a dashboard—the task we face is either a classification or a regression problem.

In this chapter, we will use supervised ML to tackle a classification and a regression problem in two case studies, including a discussion of the challenges regarding bias and fairness.

10.2   Supervised Classification

Our first case study in this chapter is situated in the context of the Physics Olympiad, a world-wide physics competition with selection processes in many different countries (Petersen et al., 2017). The selection process for the Physics Olympiad is a task-based science competition where students work individually on challenging theoretical and experimental problems. This selection process has four rounds in German (where we collected the data). While the initial two rounds involve tasks conducted by students at home or at school, primarily focusing on theoretical aspects, the third and fourth rounds incorporate a significant emphasis on experimental tasks. In each round, students have to meet a certain threshold of correct answers to proceed to the next round. While this selection process is primarily a competition, it is also an enrichment opportunity as it provides learning opportunities for students talented and interested in physics. Currently, there are ongoing discussions within the Olympiad about enhancing the emphasis on this enriching aspect by offering additional support to participating students. However, resources for providing individual supports are scarce. To make the most of these limited resources, i.e., make decisions about who can get what kind and what level of support, it is helpful to know how students are doing. The challenge is that students only hand in their exams at the end and, unlike in a regular classroom, there is no information about how students are doing as they work on the tasks. A potential solution to this issue is to use background information about students and build an ML model that predicts whether they will pass on to the next round. Predictions from the model could then be used to make decisions about who gets what level of support. In the following sections, you will build this model; a model that makes predictions about students passing on to the next round. As passing a round is a categorical or more specifically a dichotomous outcome (meaning there are two possible outcomes), this can be described as a classification task.

10.2.1 Getting to Know the Data Set

An important first step in any data analysis is getting to know the data set. We need to know what variables we have, what type of variables (numerical, factor, date) there are, whether there is missing data, and what distributional properties (normally distributed, long tailed, bi-modal, etc.) the variables have, as this will restrict our choices of ML algorithms and inferences we can make (see Chap. 2).

Variable types

Variables can describe many different things, e.g., students' gender, letter grades, or test scores. When we are doing machine learning, we are doing calculations with these variables. Which calculations are adequate, depends on the type of variable. Generally, three broad types of variables can be distinguished based on their measurement properties:

- *Nominal* variables distinguish between levels without information about order or distance between levels. Gender is a typical example of a *nominal* variable.
- *Ordinal* variables distinguish between levels and provide information about the order of levels. Letter grades are an example of *ordinal* variables as there is a defined order of the levels ($A > B > C > D > E > F$) but the distance between levels is not defined, i.e., the difference between A and B is not necessarily identical to the difference between E and F. Note that Likert type variables are also ordinal variables although they are often treated as if they were interval scale variables.
- *Interval scale* variables provide both—an order and equal distance between values. Appropriately scaled test scores are an example of interval scale variables.

Note that nominal and ordinal variables may (and at some point of working with data need to) be represented by numbers, e.g., 1 = Female, interest = 4. Which mathematical operations are appropriate with these numbers however needs to be critically considered.

Looking at distributions

Many machine learning techniques rely on calculated summaries of variables, e.g., the mean or the variance of a variable. These summaries may not always appropriately capture a variable if its' distribution is not easily captured by the summary. Consider for example the bi-modal variable plotted below (Fig. 10.1). It is questionable, that the mean of the variable is a meaningful summary.

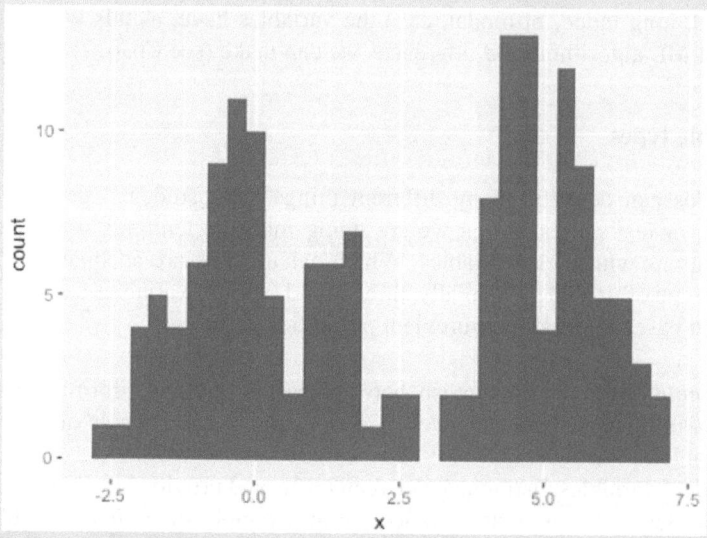

Fig. 10.1 Bi-modal variable

Therefore, it is important to consider the distributional properties of variables before working with them.

All this is important to know so that we do not make erroneous assumptions when modelling the data and interpreting our results later on.

In this chapter, we will use the statistical programming language R. The following code loads the packages we need for now and loads the dataset we want to work with.

R Code: Load packages and dataset

```
library(here)
library(tidyverse)
library(tidymodels)
library(GGally)
load(here("data","supervised_classification_data.RData"))
```

To take a first look at the data we run the command

R Code: Taking a first look at the data

```
head(df) (the glimpse() function is a good
alternative to head() as the output is returned
as a list - try it out) which outputs the first
couple rows of the data:
> head(df)
  grit mastery  sob apt prev_part success
1 3.25    4.00 3.40  NA         1       1
2 3.62    4.00 3.33  11         0       1
3 1.75    3.75 3.40  NA         0       0
4 2.62    3.75 3.47  NA         1       1
5   NA      NA   NA  NA      <NA>       0
6 3.12    3.75 3.07   4         0       0
```

The data set is organized so that one row of data represents one person and each column represents a variable. There are 6 variables in total. Success is a categorical variable with 1 indicating that a person continued to the next round and 0 indicating that a person did not. Prev_part is also a categorical variable and indicates whether a person has previously participated in the physics Olympiad (1 = has previously participated, 0 = has not previously participated). The four remaining variables represent constructs that are expressed on a numeric scale and treated as interval data. Apt represents participants' physics problem solving ability, sob represents participants' sense of belonging to physics, mastery represents participants' mastery goal orientation, i.e., their motivation to learn for the sake of understanding (Elliot & McGregor, 1999), and grit represents participants' "perseverance and passion for long-term goals" (Duckworth et al., 2007). For the sake of this case study, we assume these variables adequately represent the underlying constructs.

Now that you have an idea about the variables and the type of data that they contain, the summary() function provides an overview of the actual data:

R Code: The summary() function

```
> summary(df)

      grit           mastery           sob
Min.   :1.380   Min.   :1.750   Min.   :2.000
1st Qu.:2.500   1st Qu.:3.500   1st Qu.:3.070
```

```
Median :2.880    Median :3.750    Median :3.400
Mean   :2.822    Mean   :3.711    Mean   :3.304
3rd Qu.:3.120    3rd Qu.:4.000    3rd Qu.:3.530
Max.   :4.000    Max.   :4.000    Max.   :4.000
NA's   :13       NA's   :15       NA's   :21


     apt             prev_part  success
Min.   : 0.000    0   :111    0:150
1st Qu.: 5.000    1   :138    1:132
Median : 7.000    NA's: 33
Mean   : 7.369
3rd Qu.: 9.000
Max.   :14.000
NA's   :141
```

You should see the minimum, maximum, 1st and 3rd quartile, median, and mean for the numeric variables and the number of 0s and 1s for the two categorical variables. You also get information about the amount of missing data per variable (NA).

You can already see that all variables have variability. Variability is important because if a variable has only little variation, it is usually not helpful. It is not helpful, because variables with little variability do not allow to distinguish between cases. As an example, consider the extreme case of a variable having no variability, that is, all cases have the same value. Based on that variable alone, we will not be able to discern between different cases and that is not helpful if we want to build models that make predictions about those cases.

Another pattern to note is the degree of missing data. You can see that the variable apt has a lot of missing data with 141 NAs out of 282 rows of data (you can get the total number of rows in the dataset using nrow(df)).

While this is a good start, it is hard to get a sense of the distributional properties of the numerical variables from the output of the summary function alone. Plotting data is really helpful to better understand its properties. To take a look at single numeric variables, histograms are useful: df %>% ggplot(aes(x = grit)) + geom_histogram().

A quick look at the histogram (Fig. 10.2) for the variable grit shows that the data appears to be more or less normally distributed around its mean. By changing the variable name in the ggplot command, you can take a look at the other variables. Go ahead and try it.

Gaining an understanding of the distributional properties of a single variable is a good starting point. Additionally, comprehending the relationships among different variables in your data set is equally valuable. Here, a so-called pairs plot is helpful.

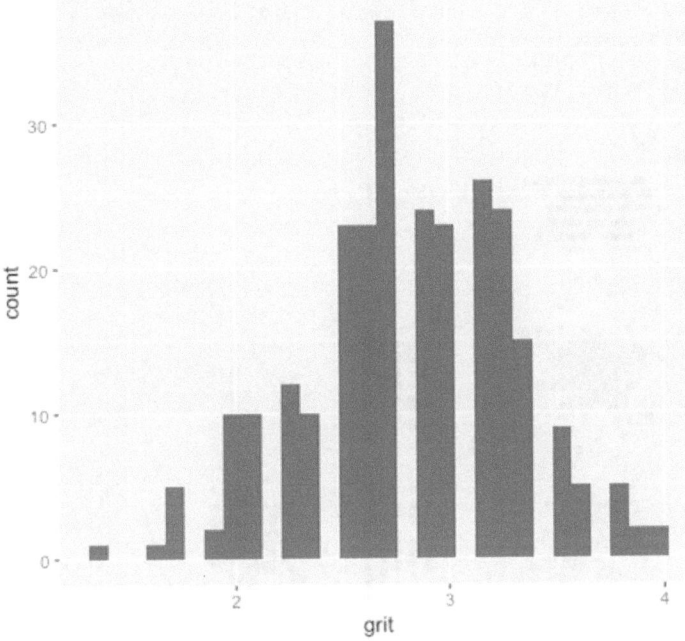

Fig. 10.2 Histogram for grit

A pairs plot (Fig. 10.3) shows the distributions of the variables in a data set and how they are related to each other. Take a look at the pairs plot of the numeric variables in the data set: `df %>% select(grit, mastery, sob, apt) %>% ggpairs()`.

Along the diagonal of the plot, you can see the distributions of the variables. Below the diagonal, you can see how the variables are related to each other and above the diagonal you get the respective correlations. When examining these plots, the objective is to obtain a comprehensive understanding of the interrelationships among different variables. Generally, as explained before, you do not want variables to be too highly correlated because then the information they hold is potentially redundant. In addition, the pairs plot also provides a quick overview of the overall variability of the data.

Across the panels below the diagonal, you can see how the variables covary and infer a small to medium amount of correlation between the variables. This is a positive observation. Additionally, you can see that the variable mastery has only low variability and thus might not be particularly helpful in making predictions.

Now that you are familiar with the properties of the data set, the actual machine learning process is nearly ready to commence. Prior to that, it's crucial to undertake data cleaning and transformation, which involves actions like removing, recoding, or rescaling data (see Chap. 2). Data cleaning and transformation at this stage describes

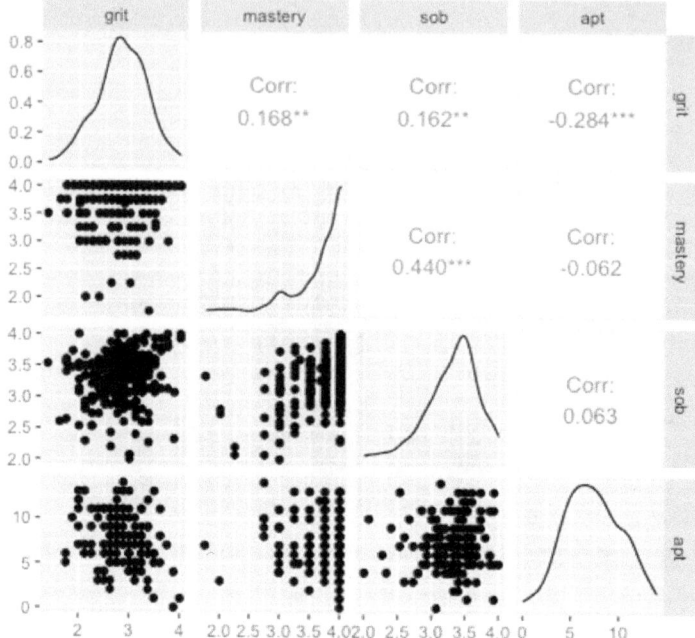

Fig. 10.3 A pairs plot of the variables in the data-set

all changes to the data that is either aesthetic (e.g., converting letter grades to numerical values) or inevitable because the data present irreparable flaws (e.g., eliminating rows containing exclusively missing data or erroneous entries, like negative ages resulting from data entry errors). Later during the ML workflow, further transformations on the data may be necessary but those reflect modeling choices and thus should be part of the actual ML workflow where the consequences of different choices in how data are transformed can be easily compared.

At the moment, data cleaning will be limited to excluding all rows where all the variables we use to make predictions—the predictor variables—are missing. While some missing data can be handled using techniques such as imputation (see Box in Sect. 10.3.2), imputing all data for a case can be misleading. Therefore, you should exclude cases where all data is missing: `df <- df[rowSums(is.na(df[, c("grit","mastery","sob","apt")])) != 4 ,]`.

Now, let's start building a model for predicting success in the selection process of the Physics Olympiad in Germany.

10.3   The Supervised ML Modeling Workflow

In supervised machine learning, the workflow entails three main steps: (1) data split-ting, (2) training the actual model or models and, (3) evaluating a model or set of models.

10.3.1   Data Splitting

First, the data is split into a training and a test set (this is called hold-out cross-validation). The training set is used to build and iteratively optimize a model; the test set is used for a final test of the most promising model developed using the training set.

So how do you split the data? You just split the data randomly with a proportion of about 70 to 80% for the training set and 20–30% for the test set, depending on the problem at hand and the available data. When splitting the data, you should consider whether the data set consists of any subgroups, e.g., students from different schools, students from different classes etc. When such subgroups are present, you need to make sure that when you split the data, the proportions of these subgroups are accurately reflected in the train and test set. Otherwise, you might accidentally train your ML model on one subset of the population which will (most likely) lead to low model performance when evaluating the model using the test set.

In the data set you are using now, `prev_part` is a variable that denotes a subgroup of students that have previously participated in the Olympiad. You can specify this as a variable to stratify by using the strata argument when splitting the data:

R Code: The strata argument

```
Set.seed(42) # makes the script reproducible
split <- initial_split(df, strata = prev_part)

df_train <- training(split)
df_test  <- testing(split)
```

You can use the following command to check that the proportions of `prev_part` in fact remain intact in the training and test set:

R Code: Check of the strata argument

```
> table(df_train$prev_part)/length(df_train$prev_part)
```

```
        0         1
0.4079602 0.5074627
> table(df_test$prev_part)/length(df_test$prev_part)

        0         1
0.4264706 0.5294118
```

With 51% of students having previously participated in the selection process of the German Physics Olympiad in the training set and 53% of students having previously participated in the selection process of the German Physics Olympiad in the test set, the split data adequately maintains the original proportions. Note that the percentages do not sum up to 100% due to some missing data.

As a last step before starting to train models you need to set up an object for cross-validation. Cross-validation basically allows us to get an estimate of how well a trained model will predict new data. Thus, we can use cross-validation to compare and evaluate different models with the training data and get a viable idea of how well they will perform when we eventually test our models against the test set. While there are different ways of doing cross-validation, they all follow the same principle which is based on resampling the data: partition the data into subsets, i.e., resample it, use one of those for training a model, use that model to predict the data in the other subset to evaluate the model, and iterate this using different partitions of the data. This provides us with a model evaluation for each partition and that information provides a better estimate of the model's predictive power than just using a single training and test set. Different methods of cross-validation differ in how exactly the data subsets are created. The code above used what is called k-fold cross-validation with $k = 10$ folds. K indicates the number of roughly equally sized subsets of data—the so-called folds (somewhat confusingly, the number of folds in vfold_cv function is set using the parameter v). Figure 10.4 illustrates the principle for 3-fold cross validation:

You can create the cross-validation data set using cv < − vfold_cv(df_train, v = 10, strata = prev_part).

Here, strata = prev_part makes sure, that across all folds, the proportion of students that have previously participated in the Physics Olympiad remains similar.

10.3.2   Training Models

In the tidymodels framework utilized in this book, the process of training a model involves various steps consolidated into what is termed a workflow. A workflow integrates model specification and data preprocessing into a "recipe" object, along with the model type intended for fitting into a separate "model" object.

We initiate the process by utilizing a basic recipe to predict student's performance in the initial round of the selection process of the German physics Olympiad based

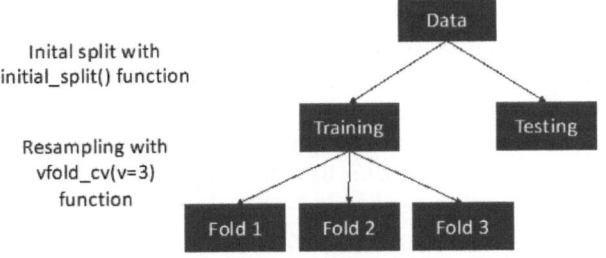

Inital split with
initial_split() function

Resampling with
vfold_cv(v=3)
function

Use fold 1 to evaluate model trained with folds 2 and 3.
Use fold 2 to evaluate model trained with folds 3 and 1.
Use fold 3 to evaluate model trained with folds 1 and 2.

Fig. 10.4 Cross validation

on the four predictor variables `grit`, `mastery`, `sob`, and `apt`. Due to notable missing data, especially in the `apt` variable, it becomes imperative to impute this data using bagged tree models by adding `step_impute_bag()` to the recipe:

R Code: Using bagged tree models

```
success_rec <- df_train %>%
recipe(success ~ grit + mastery + sob + apt) %>%
step_impute_bag(all_predictors()) # imputation based on all
predictor variables in recipe
```

Handling missing data

Encountering missing data is common. Hence understanding how to address it is essential. Although there are comprehensive textbooks dedicated to this subject (e.g., Flexible Imputation of Missing Data by Stef van Buuren), three fundamental steps should be taken into account.

1. Investigate: When data is missing, your first step should be to investigate why it is missing. Someone forgot to administer a survey in a class? Not ideal but probably not recurrent and causally related with any outcomes you are interested in. Standard methods to address this issue exist. Only students from a select demographic did not answer a question on a survey? There is possibly an underlying reason for this. Not accounting for this will probably distort your results and you should be careful in the interpretation of your results and clearly state the limitations.

2. Impute and Evaluate: A powerful technique to accommodate missing data is imputation. Imputation means that missing data is estimated based on available data and analysis are then carried out using these imputed values. While there are numerous considerations to make when imputing data (see recommended textbook above), the benefit of imputing data (and different imputation techniques) can be assessed. Thus we can empirically evaluate the effect of imputing data and report it.
3. Go back to design: When data is missing this often has an underlying reason—one that is often reflective of structural inequalities in our society. Understanding why data are missing, can lead to a better understanding of the educational context in which we are situated, helping us design more equitable approaches to education and improving outcomes.

Next, you define the type of model to fit. Let's start with something relatively simple, logistic regression, an extension of the traditional linear model for binary outcomes:

R Code: Set up model

```
model_log_reg $<$- logistic_reg() %>%
    set_engine("glm")

    # specifiy the specific package
    # for estimation, we use "glm" which is the default
```

Now, you can combine the recipe and model in a workflow and finally fit it on the training data:

R Code: Fitting on the training data

```
success_log_reg_wf <-
  workflow() %>%
  add_recipe(success_rec) %>%
  add_model(model_log_reg)
# fit model
success_log_reg_fit <- fit(success_log_reg_wf, df_train)
```

You can now view the fitted model by typing `success_log_reg_fit`. However, the model parameters are not really that interesting. After all, the model's job is to do predictions.

Models

You might have heard many things about what models are but, most importantly, models are tools. Tools, that help us explain past events and make predictions about the future based on abstractions of the world. Tools are always used with some intent. Thus, when we discuss models, the question is not whether we have the "right" model (remember: "All models are wrong[,] but some are useful" (Box, 1979, p. 2); rather we should reflect on whether the model is adequate for the intended purpose. Reflection on the equitability of the intended use should be a fundamental aspect to consider.

10.3.3 Evaluating Models

To evaluate the model, you use it to make predictions for the test set and then build an object that has the prediction from the model and the test set data so you can compare the model predictions with the actual data:

R Code: Predictions for the test set

```
success_log_reg_pred <- predict(success_log_reg_fit,
                          new_data = df_test)
success_log_reg_pred <- bind_cols(success_log_reg_pred,
                          df_test %>% select(success))
```

To evaluate classification models, a confusion matrix is a good start. A confusion matrix tabulates the true and predicted values. A perfect model would result in a matrix that has zeros everywhere but on the diagonal because the diagonal shows the correct classifications.

R Code and output: Confusion matrix

```
success_log_reg_pred %>% conf_mat(truth = success,
                          estimate = .pred_class)
          Truth
Prediction  0  1
         0 23 14
         1 16 15
```

The confusion matrix shows that 15 out of 29 students that are successful in the first round were correctly identified as successful and that 23 out of 39 students that do not continue were correctly identified as not successful. 14 students that were in fact successful in the selection process of the German physics olympiad were predicted not to be successful in the selection process and 16 students that are in fact not successful in the selection process were predicted to be. The confusion matrix can be helpful to diagnose why a model is not performing well.

The confusion matrix can also be summarized in multiple ways to allow for easier comparison between models. Accuracy, precision, recall, the F1, and AUC-ROC score are classification metrics commonly found in the literature. For now, we will just look at one of them—accuracy as it is intuitive and its definition does not require a reference case. Accuracy simply provides the proportion of correctly classified cases:

R Code and output: Example for the computation of accuracy

```
> accuracy(success_log_reg_pred, truth  = success, estimate
                                        = .pred_class)
# A tibble: 1 × 3
  .metric  .estimator .estimate
  <chr>    <chr>          <dbl>
1 accuracy binary         0.559
```

The resulting value is somewhat disheartening, as it indicates that the model is not really better than guessing. However, you also just walked through the ML modeling workflow successfully for the first time! Now, let's try again and do better in the second round.

Classification metrics

- Accuracy: Accuracy is the proportion of correctly classified instances out of the total instances in the dataset. It's a simple and intuitive metric, calculated as (True Positives + True Negatives) / Total Instances. However, it may not be the best metric for imbalanced datasets.
- Precision: Precision is the proportion of true positive instances among those predicted as positive. It measures how precise the model is in identifying positive instances. Precision is calculated as True Positives / (True Positives + False Positives). High precision means low false positive rate.
- Recall: Recall, also known as sensitivity or true positive rate, is the proportion of true positive instances among all actual positive instances. It measures the

model's ability to correctly identify positive instances. Recall is calculated as True Positives / (True Positives + False Negatives). High recall means low false negative rate.

- F1 Score: The F1 score is the mean of precision and recall, providing a balanced measure of both metrics. It ranges from 0 to 1, where 1 represents the best possible score and 0 the worst. The F1 score is particularly useful when dealing with imbalanced datasets. It is calculated as (2 * Precision * Recall) / (Precision + Recall).
- AUC-ROC is a classification metric used to evaluate binary classification models. It measures how well a model distinguishes between positive and negative instances by plotting the True Positive Rate (Recall) against the False Positive Rate for different thresholds. AUC is the area under this curve. AUC-ROC values range from 0 to 1. A value of 0.5 means the model is as good as random, while 1 indicates perfect performance. Higher AUC-ROC values show better classification

It is also important to keep in mind that the cost associated with wrong classification may differ across classes and applications. For example, automatically being given a hint in an online course if you do not need it is probably not as problematic as not getting the hint that is direly needed. This should always be considered and questioned when evaluating models with metrics.

10.4 Another Model

To get a better fitting model, you will now get in the weeds with a different class of models that is generally used with high success in educational contexts (e.g., Grinsztajn et al., 2022; Hilbert et al., 2021): tree-based models. The simplest tree-based model is a decision tree. A decision tree makes predictions by asking a series of yes or no questions. Imagine you are trying to identify a bird you see on a hiking trip. A tree-based model would ask a series of questions like "Can it fly?." If you answered "no" it would know that the bird is probably a penguin or ostrich. Now it would ask the next question such as "Can it swim?" and could infer that the bird is more likely to be a penguin than an ostrich and ask the next question to further narrow down the species. This process of asking questions continues until the model can make an educated guess about the bird you have in mind. In ML terminology, these yes or no questions which split the data into different branches are called nodes and the final nodes that do not allow any further splitting are called "leaf nodes." The model "learns" by analyzing data and finding the best questions to ask in order to make accurate predictions or classifications. Tree-based models are popular because they are powerful yet easy to understand and visualize, can handle both numerical and categorical data, and work well with large data sets. Note that they can be used to predict both categorical and numerical outcomes. Now let us dive into the thicket and train a model.

10.4.1 Training Models

As you will work with the same data, you can work with the same split of the data and recipe as before. You will need to specify a new model object however. The model you will specify is an extension of a simple decision tree model called a random forest (see also Box in Sect. 6.1 in Chap. 6). Instead of relying on a single decision tree, random forests consist of many decision trees. Each tree is built slightly differently due to randomness in the algorithm, making the "forest" diverse and better at capturing various patterns in the data. This addresses two key issues of simple decision tree models, that is, (1) their tendency to be sensitive to small changes in the training data and (2) their tendency to overfit. As random forests combine the predictions of multiple trees based on random subsets of the data, the overall model becomes less sensitive to small changes in the input data. The same mechanism also helps to reduce the tendency of overfitting the data so that random forests tend to generalize better to new, unseen data.

To set up the model, use the `rand_forest()` function. However, in addition to setting an engine to fit the model, you need to specify for more options. You need to set the mode which determines whether we want to predict a continuous or categorical outcome and you need to set three so called hyperparameters: `trees`, `mtry`, and `min_n`. `trees` describes the number of trees in the forest. Increasing the number of trees can improve the models' performance but it also increases the computational cost and can lead to diminishing returns. `mtry` controls the number of features considered for splitting at each node. A smaller value introduces randomness in the tree-building process, leading to a more diverse set of trees and better generalization. `min_n` is the minimum number of data points required to split a node further. Increasing this value can help prevent overfitting by ensuring that each split is based on a larger number of samples. These hyperparameters control fundamental aspects of the model and how well it will work. So, how do you determine their values? In fact, you do not directly assign them; instead, you allow the data to inform these values through a process known as hyperparameter tuning. To set up the model and prepare it for tuning, use the following code:

R Code: Specifying the model

```
model_rf <-
    rand_forest(mtry = tune(), min_n = tune(), trees = 1000) %>%
```

```
set_mode("classification") %>%
set_engine("ranger")
```

Note that this sets the trees hyperparameter directly and only prepares `mtry` and `min_n` for tuning using the `tune()` function. Setting trees to 1000 means that our forest will consist of 1000 trees which can be considered sufficient given the data and in the context of this case study, the computational cost is still acceptable.

10.4.1.1 Hyperparameter Tuning

To actually tune the hyperparameters means that you will fit many models with different combinations of hyperparameters, compare how well they perform at prediction using cross-validation, and then decide on a final set of hyperparameters for evaluating the model using the test-set. To do so, first create a new workflow with the random forest model:

R Code: Add Random Forest Model to Workflow

```
success_rf_wf <-
  workflow() %>%
  add_recipe(success_rec) %>%
  add_model(model_rf)
```

To save yourself some time and make use of the multiple cores modern processors have, load the `doParallel` library which allows to train multiple models at the same time during tuning.

R Code: doParallel library

```
library(doParallel)
doParallel::registerDoParallel(cores = detectCores())
# Then, the following codes does the actual tuning:
tune_res <- tune_grid(
  success_rf_wf,
  resamples = cv,
  grid = 40, metrics = metric_set(accuracy)
)
```

Here, grid tuning, which is the most straightforward way of tuning, is used: simply specify a number of hyperparameter combinations for which models are fit. You can supply these values by hand based on your knowledge of the data or—as done here—simply use the helper function that sets up a grid of 40 semi-random candidate parameter combinations. Further, `samples = cv` tells the `tune_grid()` function to use the cross-validation object to train and evaluate models on. Lastly, `metrics = metric_set(accurarcy)` specifies to use accuracy as a classification metric for evaluating the hyperparameter combinations. To get a sense of the results, plot them with the following command:

R Code: Grid tuning

```
tune_res %>%
  collect_metrics() %>%
  select(mean, min_n, mtry) %>%
  pivot_longer(min_n:mtry, values_to = "value",
               names_to = "parameter") %>%
  ggplot(aes(x = value, y = mean)) +
  geom_point() +
  facet_wrap(~parameter, scales = "free") +
  labs(y = "accuracy")
```

which gives this plot (Fig. 10.5):

From the plot, you can read off that there is indeed some variation with regard to model performance. With this kind of plot the goal is to look for patterns that indicate

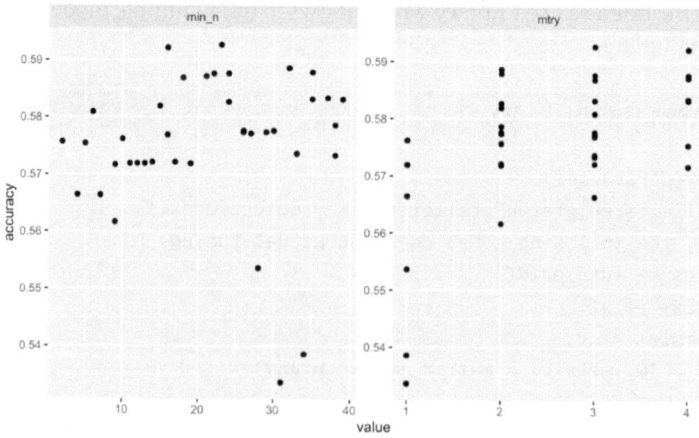

Fig. 10.5 Accuracy for different hyperparamter values

that the best values for the parameters are probably in the grid and whether the grid is sufficiently large. For `min_n` it seems that values of around 20 result in the highest accuracy, with accuracy decreasing both at higher and lower values. For `mtry` you can see model performance increasing from one to three but not further increasing at four. Thus, we can conclude that the best set of parameters is probably within the grid. Now, select the best set of hyperparameters using the `select_best()` function

```
best_acc <- select_best(tune_res, "accuracy")
```

and modify the workflow object to reflect these hyperparameter values using the `finalize_workflow()` function:

```
success_rf_wf_final <- finalize_workflow(model_rf, best_acc).
```

With this updated workflow you can now fit the random forest model with the best set of hyperparameters on the training data, use that model to predict the test data and are then ready to find out to what extent the random forests model performs better than the logistic regression.

R Code: Fit the model

```
        # fit model
success_rf_fit <- fit(success_rf_wf_final, df_train)

        #use model for prediction
success_rf_pred <- predict(success_rf_fit, new_data = df_test)
success_rf_pred <- bind_cols(success_rf_pred, df_test %>%
                            select(success))
```

10.4.2   Evaluating Models

With the fitted model at hand, let's find out how well the random forest did. First, take a look at the metrics again before digging deeper into the model using explainable machine learning techniques and asking the big question: should you use this model?

10.4.2.1   Comparing Metrics

Start by taking a look at the confusion matrix and compare it to the confusion matrix of the logistic regression model:

R Code and output: compare with logistic regression model

```
success_rf_pred %>% conf_mat(truth = success,
                            estimate = .pred_class)
          Truth
Prediction  0   1
         0 21   6
         1 18  23
```

If you compare this with the confusion matrix you got from the logistic regression model above you can see a stark improvement when it comes to correctly predicting success (23 correct predictions with random forests compared to just 15 with logistic regression) and there are far fewer false negatives (6 vs. 14), i.e., cases where the true outcome is "success" but the model predicted no "success." However, the model did not really get better at avoiding false positives. Both models, random forest and logistic regression, classify quite a few students as "successful" although they are not classified as "successful" in the data set.

Now, take a look at what this means in terms of the accuracy metric. Running `accuracy(success_rf_pred, truth = success,` `estimate = .pred_class)` returns a value of 0.647. This is a noticeable increase compared to the accuracy of 0.559 we got from the logistic regression!

10.4.2.2   Beyond Just Metrics

What else can you do to evaluate a model besides looking at how well it does at prediction? You can look at the model parameters and to what extent they align with our theoretical and substantive understanding. In the case of the logistic regression model we ran before, this is straightforward (well, as straightforward as the interpretation of parameters from logistic models gets). The following command extracts the model parameters from the model and transforms them to so that you can interpret them (the odds column gives the transformed model parameters):

R Code and output: extracting and transforming the parameters

```
> tidy(success_log_reg_fit) %>% mutate(odds = exp(estimate))

# A tibble: 5 × 6
  term         estimate std.error statistic p.value  odds
  <chr>           <dbl>     <dbl>     <dbl>   <dbl> <dbl>
1 (Intercept)     -2.30      1.89     -1.21   0.225 0.100
2 grit            0.384     0.324      1.19   0.235  1.47
3 mastery         0.201     0.455     0.442   0.659  1.22
```

| 4 sob | -0.0465 | 0.435 | -0.107 | 0.915 0.955 |
| 5 apt | 0.0781 | 0.0660 | 1.18 | 0.237 1.08 |

The transformed parameters are what is called odds ratios (OR). If the OR of a parameter is greater than one, the parameter is positively associated with the outcome, i.e., an increase in the variable increases the chances of the outcome to be of value 1. Respectively, if the OR of a parameter is less than one, the parameter is negatively associated with the outcome, i.e., an increase in the variable decreases the chances of the outcome to be of value 1. This means that `grit`, `mastery`, and `apt` are all positively associated with success in the physics Olympiad while `sob` is negatively associated. From a substantive perspective, only the negative association of `sob` is surprising. At the same time the large standard errors (`std.error` column) also suggest that the parameter estimates generally lack precision, making any interpretation questionable.

Logistic Regression

We can think of logistic regression as an extension of linear regression models to handle binary outcome variables. A simple linear regression model can be expressed as $y \sim a + bx$ where y is the continuous outcome, x the predictor variable and a and b are coefficients. a is the intercept and b is the slope. The intercept tells us the value of y when $x = 0$. The slope tells us how y changes when x changes. A positive slope means that y increases when x increases and a negative slope means that y decreases when x increases (for $x > 0$). In logistic regression, the outcome either takes the value 0 or 1. We can transform this into a continuous value by looking at the probability of the outcome y being equal to 1 which is $p = \frac{P(Y=1)}{P(Y=0)}$. This expression is also called the odds. Now, taking the natural logarithm ln of the odds which we write $logit(p) = \ln\left(\frac{P(Y=1)}{P(Y=0)}\right)$ which is also called the *log-odds*. Taking the natural logarithm is helpful because it allows us to establish a linear relationship between the input variables and the odds of the outcome. This makes it easier to estimate the model coefficients or weights using standard linear regression techniques. This means we can write a logistic regression as $logit(p) \sim a + bx$. In consequence, the coefficients are on the scale of the log-odds which is hard to interpret. To make the coefficients more interpretable, we need to transform them with the exponential function, e.g., exp (b). The resulting value will give us how the odds of the outcome being 1 will change. If for example exp $(b) = 1.27$ this means that there is a 27% increase in the odds of the outcome being one for each unit change in b.

Now, take a closer look at the random forest model. Since the random forest model is an aggregate of many tree-based models, there are no parameters that can directly be interpreted in the same way in which you just looked at the logistic regression model. Rather, you need to apply some explainable machine learning techniques, i.e., methods that try to bring some light into the more black-box like machine learning models.

Model interpretability vs. model complexity

A model with only a few parameters is easy to interpret. As an example, consider Ohm's law which (in its integral form) is a linear model that describes the relation between voltage U, current I and resistance R in an electric circuit. It can be written as $U = R \times I$. The model predicts that for a fixed voltage, the current is proportional to the resistance, i.e., the higher the resistance, the lower the current and vice versa. However, the model has serious limitations when we want to apply it in real world contexts, e.g., the resistance may be time dependent because with passing time a circuit may heat up which in turn may influence the resistance. Extending the model so that it fits such additional affordances of the real world will make the mathematical structure of the model harder to interpret. The same is true for the models we use when we apply machine learning techniques in science education research. A regression model is easier to interpret than a random forest model but the random forest model will in most cases fit better. To navigate this tension, it is helpful to think back to the reasons for which we are using the model and whether what we may gain in better predictions is worth the loss in interpretability.

First, we want to better understand how the random forest comes to its prediction for individual cases. To do so, you will use the DALEXtra package and prepare an explainer object with the following code:

R Code: DALEXtra

```
library(DALEXtra)
train_baked <- success_rec %>% prep() %>% bake(new_data = df_train)
# prepare explainer object
eval_success_rf <- explain_tidymodels(success_rf_fit,
  data = train_baked[,c("grit","mastery","sob","apt")],
  y = as.numeric(df_train$success),
  label = "random forest",
  verbose = F
)
```

Note, that first a new data object train_baked is created. This object contains the transformed version of the data stored in df_train which the recipe() provided in the previous steps. Now you can take a look at an individual case with

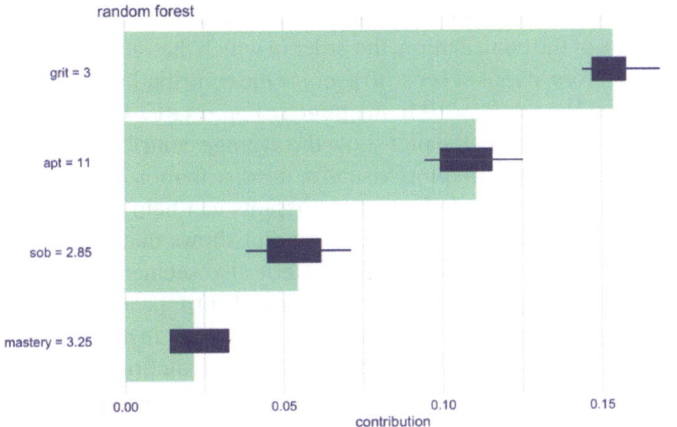

Fig. 10.6   Plot of SHAP results

`train_baked[7,]` (The "7" in the brackets selects the 7th row of the dataframe. Different numbers will return different rows.)

R output: Example of an individual case

```
    grit mastery    sob   apt success
   <dbl>    <dbl> <dbl> <dbl>   <fct>
 1    3     3.25  2.85    11 1
```

The output shows that the student was ultimately "successful." Now, use the `predict_parts()` function to calculate Shapley Additive Explanations (SHAP, Lundberg and Lee, 2017) and plot them (Fig. 10.6):

R Code: Calculate Shapley Additive Explanations

```
predict_parts(
  explainer = eval_success_rf,
  new_observation = train_baked[7, c("grit","mastery",
                                     "sob","apt")],
  type = "shap",  B = 10) %>% plot()
```

SHAP values show how much each variable contributes to a particular prediction. For the calculation of the contribution, the order in which the variables are considered is important. Therefore, it makes sense to average the contributions of variables under different orderings. We set $B = 10$ in the `predict_parts()` function to look at 10 random orderings. The bars in the plot show the average contribution of each feature for this case and the overlaid boxplots visualize the distribution of contributions from the 10 orderings that were used. Thus, the boxplots can help to get an idea of the variation of the contribution of a variable. The plot shows that grit and apt had the greatest contribution in the prediction. While grit also seemed to have the biggest influence in the logistic regression, the contribution of apt appeared relatively small in the logistic regression. Instead, mastery appeared important in the logistic regression model which only plays a relatively small role in this prediction. This is a good hint at why the random forest model is more successful as it can capture non-linearity that the logistic regression cannot.

Understanding how a model comes to its predictions for single cases is useful when you want to fine tune the model. Looking at cases where the prediction was wrong can be especially enlightening. However, you may mistake the forest for the trees when you focus too much on single cases. Therefore, we now want to find out which variables contribute most to driving the predictions overall, i.e., aggregated over the whole training set. To do so an idea that goes back to the seminal paper of Breiman (2001) is helpful: variable permutation. The idea behind variable permutation is that if a variable is important for getting the prediction right, the model should make worse predictions if we were to shuffle (permute) the values stored in that variable. The following command does just that and thus helps to get an idea of the importance of the variables:

R Code and output: Variable permutation

```
> model_parts(eval_success_rf, loss_function
            = loss_default("classification"))
        variable mean_dropout_loss           label
1  _full_model_          0.6302189 random forest
2          grit          0.5461380 random forest
3           sob          0.5801448 random forest
4           apt          0.5838872 random forest
5       mastery          0.6178502 random forest
6     _baseline_          0.4974882 random forest
```

The first line gives the accuracy of the full model, the last line gives the accuracy of a base model that just guesses. The other lines provide the accuracy if the respective variable is permuted. You can see that a permutation of mastery only leads to a rather small drop in accuracy (from 0.63 to 0.62). Permuting grit however leads to a quite large drop in accuracy (from 0.63 to 0.55). Thus, overall grit is more important for making predictions than mastery for the random forest model. In this way, variable importance can provide helpful information when deciding which variables to include in a model. Imagine that a variable about protected information such as gender had little to no importance in the prediction or that a variable was hard to measure— variable importance can help to make decisions in such cases.

To provide you with even more information when making the decision about which variables to include in a model there is a last tool to explore in this chapter: partial dependence profiles. Partial dependence profiles show how the predicted outcome of a model changes as variables change their values. This is especially helpful in discovering non-linearity and boundary cases. Using the `model_profiles()` function, you can calculate profiles for all variables across all students (N = NULL) in the data set and then plot them:

R Code: Partial dependence profiles

```
model_profile(eval_success_rf, N = NULL,
    variables = c("grit","mastery","sob","apt")) %>% plot()
```

The plot (Fig. 10.7) reflects what you could already see in the variable importance results: mastery seems of little importance in making predictions as the profile more or less resembles a horizontal line. Further, the profiles of apt, grit, and sob show highly non-linear behavior; supporting the hint you got from the single case analysis

Fig. 10.7 Partial dependence plot

with SHAP that the random forest model may do better than logistic regression because it can capture non-linear effects.

10.4.2.3 Should You Use It?

You have come a long way in this chapter. You trained two models and extensively investigated one of them. A question that still awaits is whether you should use the model in the wild. This question will come up again and again, especially if you consider using machine learning for automation. Automation brings scale and with scale both the potential benefits and the risks increase. In this case the best performing model with an accuracy of around 0.65 (the random forest model) does not seem good enough to be used with real students. The risk of misclassification appears just to be too high. Especially since there was a high number of false positives, i.e., cases where the model predicted "success" but there was no "success" in the data. This means that many students that could profit from additional support during the selection process of the physics Olympiad would not get that support if decisions were made based solely on the model. This shows that after all, ML is no magic bullet.

10.5 Supervised Regression

In the second case study you will take a look at students' learning about energy. The outcome you will look at is a continuous score so we move from supervised classification to supervised regression. After working through the previous parts, the general procedure should seem familiar. However, you will explore another type of model, and finally combine models to improve fit.

10.5.1 Getting to Know the Data

Start with loading the here and tidyverse libraries and loading the data:

R Code and output: loading data and libraries

```
# loading packages
library(here)
library(tidyverse)
# load data
load(here("data","learning.RData"))
Next, let's get an overview of the data using glimpse():
```

```
Rows: 490
Columns: 16 (truncated)
$ total        <dbl> 4, 73, 28, 38, 46, 2, 18, 44, 3, ...
$ P1_M.Radiant <dbl> 1, 1, 1, 0, 1, 1, 1, 1, 1, ...
$ P1_M.Electric <dbl> 0, 0, 0, 0, 0, 0, 0, 0, 0, ...
$ P1_T.Process <dbl> 0, 0, 0, 1, 1, 0, 0, 0, 0, ...
$ P1_LP.M      <dbl> 0, 0, 0, 0, 0, 0, 0, 0, 0, ...
$ P1_LP.T      <dbl> 0, 0, 0, 0, 0, 0, 0, 0, 0, ...
$ P1_P_expl    <dbl> 2, 1, 2, 2, 2, 1, 2, 2, 2, ...
$ P1_Practice  <dbl> 2, 1, 2, 2, 2, 1, 2, 2, 2, ...
$ P2_M.Radiant <dbl> 0, 5, 1, 3, 3, 0, 3, 3, 0, ...
$ P2_M.Electric <dbl> 1, 4, 1, 3, 1, 0, 1, 3, 0, ...
$ P2_T.Process <dbl> 0, 3, 1, 1, 1, 0, 1, 1, 0, ...
$ P2_LP.M      <dbl> 0, 1, 0, 0, 0, 0, 0, 0, 0, ...
$ P2_LP.T      <dbl> 0, 1, 0, 0, 0, 0, 0, 0, 0, ...
$ P2_P_expl    <dbl> 0, 1, 0, 0, 0, 0, 0, 0, 0, ...
$ P2_P_ana     <dbl> 0, 2, 0, 2, 1, 0, 2, 2, 0, ...
$ P2_Practice  <dbl> 0, 3, 0, 2, 1, 0, 2, 2, 0, ...
```

Some background on the data set is needed to understand what is visible here: the data comes from a digital learning environment where students engaged in a unit on learning about energy that followed project-based learning pedagogy. Specifically, the unit was built around the driving question "What is the best way to set up solar cells?" During the five parts of the unit, students learned about electric and radiation energy and how those two forms of energy may transform into each other in solar cells and related phenomena. As the course followed project-based learning pedagogy, students engaged in scientific practices like constructing explanations or analyzing data. All tasks that students engaged in were scored by a set of experienced scorers (the scores can actually be reproduced automatically using machine learning and natural language processing, for details see Gombert et al., 2022). The tasks were coded relative to three distinct categories: (1) evidence for specific knowledge elements such as electric energy or radiation energy presented in an answer, (2) evidence for engaging in a scientific practice such as analyzing data, and (3) evidence for successfully engaging in a learning performance, i.e., integrating elements of scientific practices and scientific knowledge in a way that demonstrates knowledge-in-use (e.g., Harris et al., 2019; Kubsch et al., 2019). With this background, you can further unpack the variables.

The variable total contains the sum score for each student across all tasks and codes in the unit, in this way it indicates students' learning over the course of the unit. All other variable names starting with PX refer to sum scores across the respective knowledge elements, practices, and learning performances in the part of the unit indicated by X, that is, P2_M.Radiant refers to the sum score across all scores for the knowledge element "radiant energy" in part two of the unit. In essence, the variable "total" provides an indicator for learning across all five parts of the unit while the other variables provide indicators for students holding knowledge elements, engaging in scientific practices, and demonstrating knowledge-in-use about energy during the first two parts of the unit.

The goal will be to predict student learning at the end of the unit, based on how they do during the first two parts of the unit. That prediction can be the basis for providing numerous interventions to personalize learning so that all students can unfold their potentials, i.e., struggling students may be provided with additional feedback and scaffolds while high performing students could profit from less scaffolded tasks that are more challenging or get the chance to dive deeper into some topics.

Now that you have an idea of what the variables mean it is time to explore them. Start by taking a look at the variable "total" using a histogram again:

```
df %>% ggplot(aes(x = total)) + geom_histogram()
```

The resulting plot (Fig. 10.8) shows that the scores vary over a broad range. Especially the data point at around 120 is peculiar. This is because this value is actually not a part of the real data set. Rather it is the theoretical maximum score possible. Thus, remove the row containing this value from the data set using the `filter()` function and saving the resulting object: `df <- df %>% filter(total < 120)`. Finding such outliers is exactly, why taking a look at the data before starting the actual machine learning process is so important.

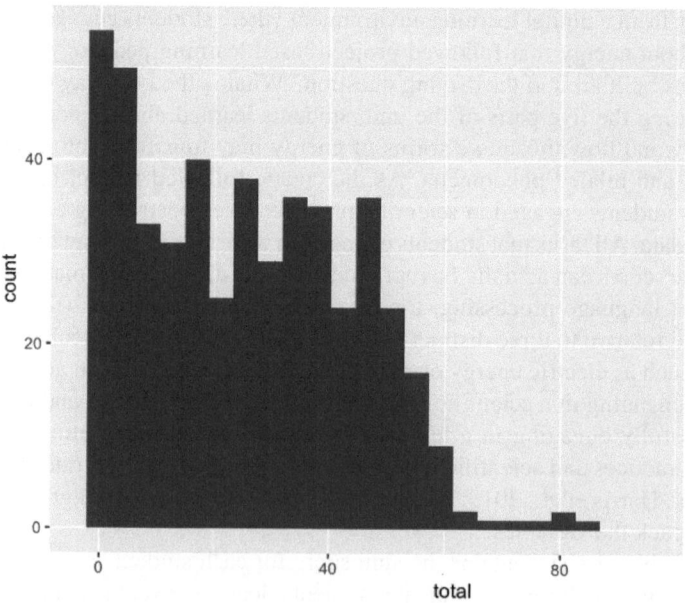

Fig. 10.8 Histogram of "total"

In the last case study, you only had a few variables so it was easy to look at the variables by looking at individual plots. This time, it is not that easy; there are more variables. To visually inspect large sets of variables in a convenient way, a stacked plot of those variables is helpful. Using the `ggridges` package allows us to create such a plot. The following code loads the library, transforms the data from wide into long format (which is the format required for the plot) using the `pivot_longer()` function and finally creates the plot:

R Code: Create the plot

```
library(ggridges)
df %>% pivot_longer(cols = starts_with("P"),
                    names_to = "variable") %>%
  ggplot(aes(y = variable, x = value))
  + geom_density_ridges(stat = "binline")
```

Long, Wide, and Tidy data

Long Data: Long data, also known as "tall" or "stacked" data, is a data format where each observation has its own row, and there are multiple rows for each subject or entity. In long data, each row contains a single data point, with columns representing the subject, variable, and value.

Wide Data: Wide data, sometimes called "unstacked" or "cross-sectional" data, is a data format where each row represents a single subject or entity, and each column represents a variable. In wide data, all information about a subject is contained within a single row, making it easier to see relationships between variables for a particular subject.

Tidy Data: Tidy data is a data format that adheres to a set of principles aimed at making the data easy to analyze, manipulate, and visualize. In tidy data, each variable forms a column, each observation forms a row, and each type of observational unit forms a table. Tidy data makes it easier to work with and analyze data using various statistical and visualization tools.

The plot (Fig. 10.9) shows that some variables like `P1_LP.M` have very little variance while others such as `P2_M.Radiant` have values spread across the whole range. Based on this information you may decide to remove variables from the data frame that essentially carry no information. The following command removes the bottom three variables of the plot from the data frame:

```
df <- df %>% select(!c("P1_LP.M", "P1_LP.T", "P1_M.Electric")
```

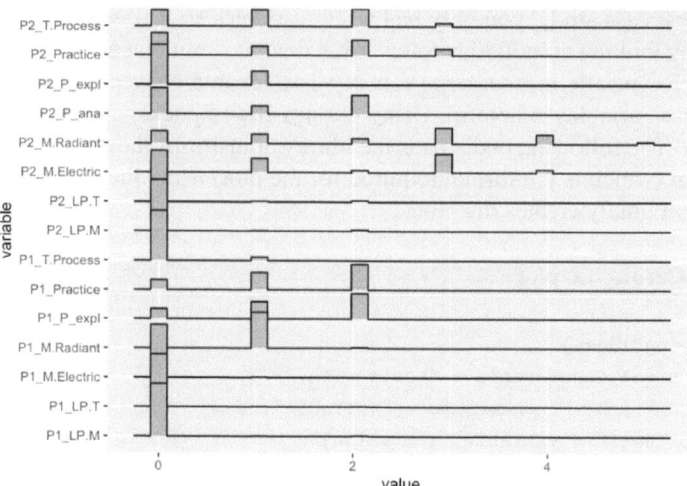

Fig. 10.9 Distribution of variable values

Before moving on to the data splitting, let's load the additional packages we will need for the remainder of the analysis.

R Code: Additional packages

```
library(tidymodels)
library(xgboost)
library(finetune)
library(vip)
library(kernlab)
library(GGally)
library(stacks)
```

10.5.2 Data Splitting

The data splitting progression in analog to the last case study. Note that only 5 folds for cross validation are used to decrease the time required for computation as you will use more computationally demanding models.

R Code: Data splitting

```
# data splitting ####
set.seed(42)
split <- initial_split(df)
df_train <- training(split)
df_test  <- testing(split)
# set up cross validation
cv <- vfold_cv(df_train, v = 5)
```

10.5.3 Training Models

Start with defining the recipe using the following code:

R Code: Defining the recipe

```
learn_rec <- df_train  %>%
  recipe(total ~ .) %>%
  step_normalize(all_numeric_predictors())
```

This time there are no missing values, however for some of the models you want to use, predictors should be normalized, i.e., they are transformed so that their mean is 0 and standard deviation is 1. If you wonder about suggested preprocessing required by different models there is a great overview in Tidy Modeling with R: https://www.tmwr.org/pre-proc-table.html.

Now you have come to the part where the models are defined. Last time you ran one model after the other and compared their performance. This is already somewhat tiresome with two models but what do you do if you want to screen many different model types? Luckily, the tidymodels framework has you covered. You can define a range of models and then combine them in a workflow set to fit, tune, and compare them simultaneously. You will define three increasingly complex models, starting with a decision tree:

R Code: Decision tree

```
model_tree <- rand_forest(mtry = tune(), min_n = tune(),
                          trees = 1) %>%
```

```
set_mode("regression") %>%
set_engine("ranger")
```

This is basically the random forest model you already know. However, you just fit one tree instead of a whole forest by setting the trees argument to 1.

Next, define a support vector machine.

R Code: Support vector machine

```
model_svm_r <- svm_rbf(cost = tune(), margin = tune(),
                       rbf_sigma = tune()) %>%
  set_mode("regression") %>%
  set_engine("kernlab")
```

The conceptual idea behind support vector machines (SVM) is to find the line that best separates different groups or classes in a given data set. Consider an example: suppose you have two different kinds of coins—nickels and pennies—in your wallet and empty them on a table. The task now is to find the line through the coins so that all nickels are on one side of the line and all pennies are on the other side. The coins are your data points and the type of coin—nickel or penny—refers to the class or group of the coins. To find the best line, an SVM tries to maximize the margin between the line and the closest nickels and pennies. The line is your decision boundary, and the margin is the distance between the line and the closest nickels and pennies on either side. The nickels and pennies that are closest to the line, those on the edge of the margin, are the most difficult to classify since they are the most similar. These are called support vectors. They support or define the decision boundary. This is how the name support vector machine comes to be. SVMs are often used for classification tasks but can be expanded to work with regression tasks as well. For regression problems, SVMs try to find the function that best fits all the data points and minimizes a margin of error around that line. Support vector machines are often effective when there are many predictors. However, they are not suitable for large data sets as they are computationally expensive. Thus, in the context of this example they are a versatile alternative to tree-based models.

Lastly, define an ANN as another widely used supervised (and unsupervised) ML algorithm.

R Code: Artificial neural network

```
model_mlp <- mlp(hidden_units = tune(),
                 penalty = tune(),
                 epochs = tune()) %>%
```

```
set_engine("nnet") %>%
set_mode("regression")
```

The ANN is the most complex model so far. ANNs are a class of models inspired by the human brain (see Chaps. 2 and 4). "Inspired" is important here because while there are some analogs to how brains work (at least to our current understanding) there are many differences between actual neurons in a brain and the neurons in a neural network. In ANNs, the neurons are organized into three kinds of layers: (1) the input layer, (2) the hidden layer or layers, and (3) the output layer. The input layer is the first layer where the model receives information—in this case the neurons in the first layer receive information from the variables in the dataset starting with P[X]. These neurons are then all connected to the neurons in the hidden layer. In the hidden layer, each neuron receives inputs from neurons in the previous layer, performs a calculation on these inputs, and then passes its output to neurons in the next layer. There may be more than one hidden layer. The name "deep learning" refers to neural networks that have many hidden layers, i.e., they are "deep" in terms of layers. The last layer is the output layer, where the model produces its final output. In this case this corresponds to the estimated total score. Each connection between neurons has a "weight," which is a numerical value that the model learns during training. When an input comes in, it gets multiplied by the weight. The neuron adds up all the weighted inputs and applies a mathematical function called an "activation function." The activation function decides whether and to what extent that input should progress further through the network. This allows neural networks to introduce non-linearity. During training, the model adjusts the weights based on the difference between its prediction and the actual value. The goal is to minimize this difference. The process of adjusting the weights is typically done using a method called "backpropagation." Research shows that neural networks typically perform well on complex data such as language data, image and video data, and time series forecasting. However, they typically require large data sets for training and are relatively hard to interpret (later in the book, there are examples of using pre-trained neural nets which circumvents some of these challenges). Further, on medium size tabular data—the kind of data we often encounter in science education contexts and in this case study—they usually perform worse or similar as tree-based models (Wilson et al., 2016; Grinsztajn et al., 2022; Küchemann et al., 2020). While there is some work on models that use an ANN architecture and work well on tabular data, these models either still require more data than we usually have in science education contexts (e.g.,TabNet (Arik & Pfister, 2021)) or are optimized for classification tasks (TabPFN, Hollmann et al., 2023). Still, you will use an ANN now; with the rapid progress in the field, ANNs should never be discarded entirely. Some word regarding the hyperparameters: hidden refers to the number of nodes in the hidden layers, epochs refers to the number of training iterations, and penalty refers to a parameter that aims at avoiding overfitting.

With the models at hand, it is time to combine them in a workflow set:

R Code: Keep the computation time within limits

```
wf_set <- workflow_set(preproc = list(learn_rec = learn_rec),
  models = list(rand_forest = model_tree,
                neuralnet = model_mlp,
                svm_r = model_svm_r)
  )
```

Now you can start the tuning process. To be able to use the tuning results in a combined model, set control parameters so that predictions are saved and the try to use parallel computation as much as possible:

R Code: Control parameters

```
tune_ctrl <-
  control_race(
    save_pred = TRUE,
    parallel_over = "everything",
    save_workflow = TRUE
  )
```

Given that you tune the hyperparameters of three models and that especially the ANN is computationally expensive, running the tuning command may take some time:

R Code: Tune the results

```
tune_results <-
  wf_set %>%
  workflow_map(
    seed = 42,
    resamples = cv,
    grid = 20
  )
```

Note that the grid of size 20 is probably too small to explore the whole parameter space given the many hyperparameters. Again, this small value is chosen to limit computation time.

When you have successfully tuned the model, you need to find out which model performed best.

R Code: Search for the best performing model

```
tune_results %>% rank_results(select_best = T) %>%
  filter(.metric == "rsq")
will provide you with a table and you can also
create a plot with
tune_results %>% autoplot(select_best = T, metric = "rsq")
```

The plot (Fig. 10.10) suggests that the SVM performed best, followed by the neural network and the random forest. However, the error-bars in the plot show that over the resampling process there was quite some variation so the differences in performance between the models are not necessarily robust. The performance metric `rsq` used here is short for r-squared (R^2). R^2 gives the variance explained, i.e., the part of the variance in the data that is explained by the model. The range you can see here is actually good in the context of education.

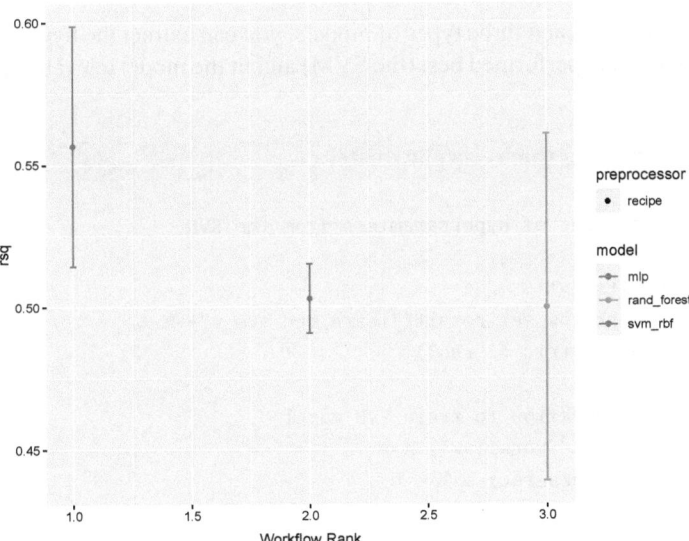

Fig. 10.10 Model performance

Regression metrics

In regression tasks, the aim is to predict a continuous value. The performance of regression models is typically evaluated using some form of metric which captures the difference between the model's predictions and the actual values. Commonly used metrics are:

Mean absolute error (MAE): This is the simplest and most intuitive metric. It calculates the average absolute difference between predicted and actual values. This metric is on the scale of the outcome.

Root Mean Squared Error (RMSE): This calculates the average squared difference between predicted and actual values and then takes the root. Because errors are squared in the process, this metric penalizes large errors more than small ones. This metric is on the scale of the outcome.

R^2: R^2 measures the measures the proportion of variance in the data that is explained by the model. It ranges from 0 (no portion of variance explained) to 1 (all variance explained). R^2 is often used in science education research in the context of regression models and their extensions. Thus, you might have an intuition for what is a "good" value for R^2. However, R^2 is not on the scale of the outcome so it can be hard to understand what the R^2 value means on the actual scale.

After having screened three types of models, you can extract the hyperparameters from the model that performed best (the SVM) and fit the model to evaluate it against the test data:

R Code: Extracting the hyperparameters

```
# select best set of hyperparamters from the SVM
best_rsq <-
  tune_results %>%
  extract_workflow_set_result("learn_rec_svm_r") %>%
  select_best(metric = "rsq")

  # set up a workflow to train SVM model
learn_wf_final <- workflow() %>%
  add_recipe(learn_rec) %>%
  add_model(model_svm_r)

learn_wf_final <- finalize_workflow(learn_wf_final, best_rsq)

# fit model ####
learn_fit <- fit(learn_wf_final, df_train)
```

```
#use model for prediction
learn_pred <- predict(learn_fit, new_data = df_test)

learn_pred <- bind_cols(learn_pred, df_test %>% select(total))
```

10.5.4   Evaluating Models

Now, you can evaluate the SVM model using the following command:

R Code and output: Evaluate the SVM model

```
>rsq(learn_pred, truth = total, estimate = .pred)
# A tibble: 1 × 3
  .metric .estimator .estimate
  <chr>   <chr>          <dbl>
1 rsq     standard       0.586
```

The output shows an R^2 close to .60 which is a good value in the context of science education research, especially when you consider that you predict students' outcomes after a unit consisting of five parts after the first two parts. However, you might want to do better and when you looked at the plot of the hyperparameter tuning results, you saw that all models performed similarly. Different types of models can capture different aspects of a phenomenon so when you have different types of models it can be worthwhile to explore if you can combine their predictions to make a better overall prediction. The conceptual idea of combining multiple models so that they can complement each other is known as multi-model thinking (Page, 2018) and in the machine learning and statistical literature known as model "stacking". Within the tidymodels framework, stacking is actually very easy so let's do it.

10.5.5   Training a Stack of Models

First, you create a stack object. For the sake of simplicity we will create a stack with just two models: the decision tree and the SVM. To do so, we will re-run some of the previous code but with a newly specified workflowset:

R Code: Stack object

```r
wf_set <- workflow_set(preproc = list(learn_rec = learn_rec),
                       models = list(rand_forest = model_tree,
                                     svm_r = model_svm_r)
)

# tune model hyperparamters
tune_ctrl <-
  control_race(
    save_pred = TRUE,
    parallel_over = "everything",
    save_workflow = TRUE
  )

tune_results <-
  wf_set %>%
  workflow_map(
    seed = 42,
    resamples = cv,
    grid = 20,
    control = tune_ctrl
  )

# display tune results
tune_results %>% rank_results(select_best = T) %>%
  filter(.metric == "rsq")

# plot tune results
tune_results %>% autoplot(select_best = T, metric = "rsq")

# extract paramters from best model
best_rsq <-
  tune_results %>%
  extract_workflow_set_result("learn_rec_svm_r") %>%
  select_best(metric = "rsq")

# set up a workflow to train model which had best results
learn_wf_final <- workflow() %>%
  add_recipe(learn_rec) %>%
  add_model(model_svm_r)

learn_wf_final <- finalize_workflow(learn_wf_final, best_rsq)
```

```
# fit model ####
learn_fit <- fit(learn_wf_final, df_train)

#use model for prediction
learn_pred <- predict(learn_fit, new_data = df_test)

learn_pred <- bind_cols(learn_pred, df_test %>% select(total))

# evaluate model ####
rsq(learn_pred, truth = total, estimate = .pred)

# using all the models with model stacking ####

# create a model stack object
learn_stack <- stacks() %>% add_candidates(tune_results)
```

Next, you create a "meta-model" by blending the predictions using a linear model with regularization. The regularization is used to prevent overfitting by making the meta-model prefer simpler models unless additional complexity leads to a substantial improvement in performance on the training set.

R Code: meta-model

```
# create a multimodel object
multimodel <- blend_predictions(learn_stack, times = 100)
```

The `times` argument controls the number of the resamples used to determine the model coefficients that determine the blending of the meta-model.

You can plot the resulting object using the `autoplot()` function:

`multimodel %>% autoplot()`

You can see (Fig. 10.11), that the `rsq` value stays relatively stable before increasing sharply at a penalty value of 0.1. This suggests that the default range was not helpful in our case. What you are looking for is a clear maximum value for `rsq` (or a clear minimum or RMSE). Rerun the `blend_predictions()` function with an extended range for the penalty value and plot (Fig. 10.12) the results again:

R Code: Blend predictions

```
multimodel $<$- blend\_predictions(learn\_stack,
    penalty = 10\^{}(seq(-3, 1, length.out = 40)), times = 100)
multimodel \%$>$\% autoplot()
```

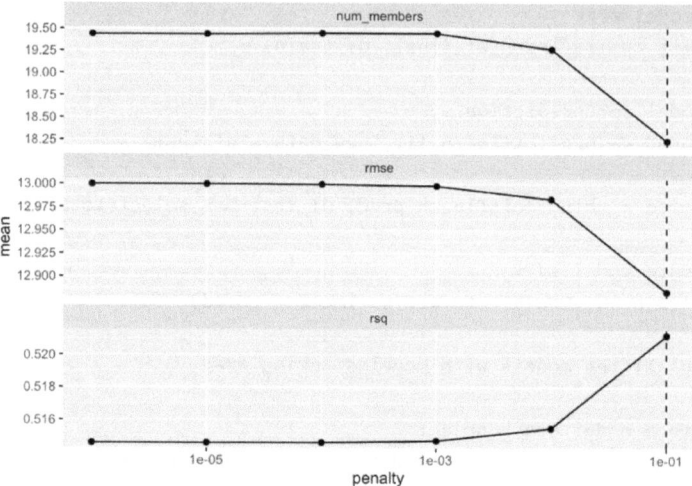

Fig. 10.11 Model metrics

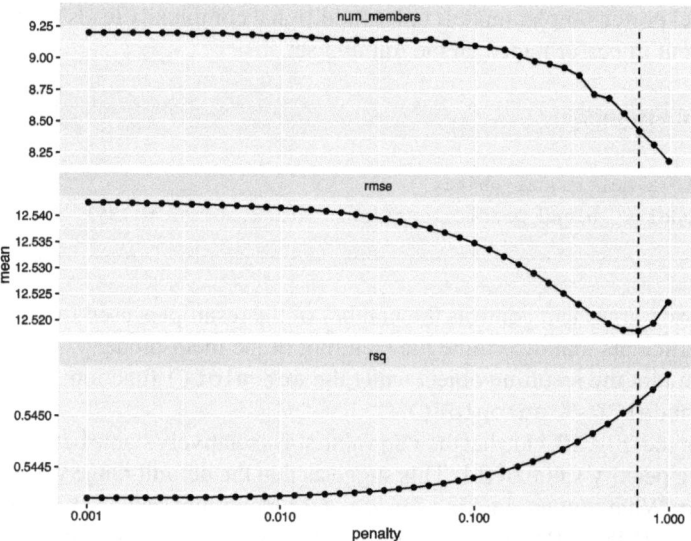

Fig. 10.12 Model metrics with adjusted range

Now, you can see that there is a minimum at the vertical line for the rmse value.
Further reducing the number of parameters also does not lead to a substantive increase

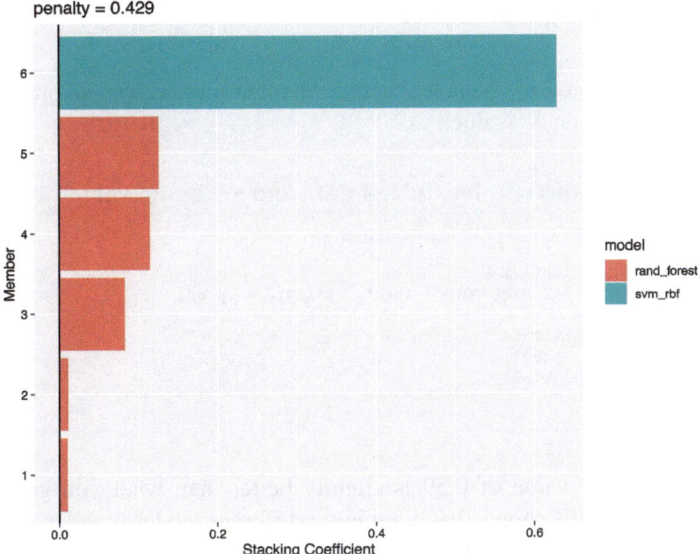

Fig. 10.13 Stacking coefficients

in rsq. You can also use the "weight" option in the `autoplot` command to take a look at the blending results:

```
multimodel %>% autoplot("weight")
```

The plot (Fig. 10.13) shows the coefficient for each model in the linear model that comprises the meta-model. Larger values suggest a larger contribution of the respective model. You can see that the SVM model which was already the best performing model alone also has the largest contribution here. However, the decision tree model is also added to the mix. Running `multimodel$equations` gives the actual equation of the meta-model:

R Output and code: The random forest model

```
2.21498499721433 +
(learn_rec_rand_forest_1_03 * 0.0103002372251664) +
(learn_rec_rand_forest_1_14 * 0.082561414300202) +
(learn_rec_rand_forest_1_05 * 0.125891863223886) +
(learn_rec_rand_forest_1_02 * 0.0111194179105971) +
(learn_rec_rand_forest_1_08 * 0.114711425173477) +
(learn_rec_svm_r_1_19 *  0.628839233188363)

# Now, we can use that formula to fit the meta-model:
multimodel <- fit_members(multimodel)
```

10.5.6 Evaluating the Model

Using the predict function you can now use the meta-model to make predictions for the test data and get an R^2 value:

R Code: Make predictions for the test data and get an R^2 value

```
predict(multimodel, df_test) %>%
+   bind_cols(df_test) %>% rsq(truth = total, estimate = .pred)
# A tibble: 1 × 3
  .metric .estimator .estimate
  <chr>   <chr>          <dbl>
1 rsq     standard        0.59
```

The resulting R^2 value of 0.59 is slightly better than what you got from just the SVM (0.586). This shows that combining different models can lead to better—although not by a large margin at all—predictions. In fact, the difference observed here is is within in the range of sampling variability and thus needs to be taken with a grain of salt. In general, a caveat of stacking models or ensemble methods is that with combining various models, interpretability is drastically reduced. Thus, one always needs to consider if a potential increase in predictive accuracy justifies a loss in interpretability.

10.6 Summary

In this chapter you have walked through two case studies where you used supervised ML with the goal of automation. The first case study aimed at predicting success in the selection process of the German physics Olympiad—a classification task. The second case study aimed at predicting students' final score in a unit on energy—a regression task. You have covered the basics of running supervised ML within the tidymodels framework. This includes ways of getting to know the data, the data splitting approach, different types of ML models and different ways to evaluate a model. You have seen how techniques of explainable ML can help you to understand how a trained model is coming to its predictions and how to screen many models at the same time and even combining many models into a meta-model for better predictive performance. In sum, this chapter should prepare you to start using supervised ML in your work and have provided you with the knowledge to follow current research and dive deeper into more specific aspects of supervised ML.

10.7 Tasks

Comprehension

1. Explain the concept of supervised machine learning and provide an example application in science education research.
2. What are the three broad types of variables discussed in the chapter? Give an example of each and discuss how this is relevant in the context of machine learning.
3. Why is it important to understand the distributional properties of variables before working with them in machine learning?
4. What is cross-validation, and why is it used in the supervised machine learning workflow?
5. What is hyperparameter tuning, and why is it essential in building machine learning models?

Application

1. Go back to the model that predicted success in the physics olympiad and choose another method of imputation. Compare the results with the original outcomes.
2. Evaluate the performance of a logistic regression model using accuracy, precision, recall, and F1 score. Write the R code to calculate these metrics.
3. Write R code to build a decision tree model and plot it. To do this, take a variable in the data set and split it into groups. Then define a decision tree model to predict the group membership. Plot the resulting decision tree. Use the rpart and rpart.plot libraries (you may need to install them) for this.
4. Create a new variable (also referred to as feature) that combines 'grit' and 'mastery' into a single variable called 'effort'. Write the R code to add this feature to the dataset and build a logistic regression model using this new variable.
5. Implement hyperparameter tuning for the random forest model predicting success using random search instead of grid search. Compare the performance with the grid search results.

References

Arik, S. Ö., & Pfister, T. (2021). TabNet: Attentive interpretable tabular learning. In *Proceedings of the AAAI conference on artificial intelligence* (Vol. 35, No. 8, pp. 6679–6687). https://doi.org/10.1609/aaai.v35i8.16826

Box, G. E. P. (1979). *Robustness in the strategy of scientific model building: Technical Report #1954*.

Duckworth, A. L., Peterson, C., Matthews, M. D., & Kelly, D. R. (2007). Grit: Perseverance and passion for long-term goals. *Journal of Personality and Social Psychology, 92*(6), 1087–1101. https://doi.org/10.1037/0022-3514.92.6.1087

Elliot, A. J., & McGregor, H. A. (1999). Test anxiety and the hierarchical model of approach and avoidance achievement motivation. *Journal of Personality and Social Psychology, 76*(4), 628.

Gombert, S., Di Mitri, D., Karademir, O., Kubsch, M., Kolbe, H., Tautz, S., Grimm, A., Bohm, I., Neumann, K., & Drachsler, H. (2022). Coding energy knowledge in constructed responses with explainable NLP models. *Journal of Computer Assisted Learning*, jcal.12767. https://doi.org/10.1111/jcal.12767

Grinsztajn, L., Oyallon, E., & Varoquaux, G. (2022). *Why do tree-based models still outperform deep learning on tabular data?* http://arxiv.org/abs/2207.08815

Harris, C., Krajcik, J. S., Pellegrino, J. W., & DeBarger, A. H. (2019). Designing knowledge-in-use assessments to promote deeper learning. *Educational Measurement: Issues and Practice, 38*(2), 53–67.

Hilbert, S., Coors, S., Kraus, E., Bischl, B., Lindl, A., Frei, M., Wild, J., Krauss, S., Goretzko, D., & Stachl, C. (2021). Machine learning for the educational sciences. *Review of Education, 9*(3).

Hollmann, N., Müller, S., Eggensperger, K., & Hutter, F. (2023). *TabPFN: A transformer that solves small tabular classification problems in a second.* http://arxiv.org/abs/2207.01848

Kubsch, M., Nordine, J., Neumann, K., Fortus, D., & Krajcik, J. (2019). Probing the relation between students' integrated knowledge and knowledge-in-use about energy using network analysis. *Eurasia Journal of Mathematics, Science and Technology Education, 15*(8). https://doi.org/10.29333/ejmste/104404

Küchemann, S., Klein, P., Becker, S., Kumari, N., & Kuhn, J. (2020). Classification of students' conceptual understanding in STEM education using their visual attention distributions: A comparison of three machine-learning approaches. In *Proceedings of the 12th international conference on computer supported education* (pp. 36–46). https://doi.org/10.5220/0009359400360046

Lundberg, S. M., & Lee, S.-I. (2017). A unified approach to interpreting model predictions: 31st conference on neural information processing systems (nips 2017), Long Beach, CA, USA.

Petersen, S., Blankenburg, J., & Höffler, T. N. (2017). *Challenging gifted students in science: The German science olympiads* (pp. 157–170). Abingdon: Routledge.

Chapter 11
Pattern Recognition—Unsupervised Machine Learning

Marcus Kubsch, Christina Krist, and Peter Wulff

Abstract In this chapter we will engage with a case study that utilizes unsupervised ML techniques to extract and interpret patterns in complex science education related data.

11.1 Introduction

Learning about science is a complex phenomenon. Many factors across multiple dimensions affect how students learn, e.g., students motivation (Hong et al., 2020), teacher's enthusiasm (Keller et al., 2017), or schools norms (Yeager et al., 2019). What makes issues even more complex is that many of these factors interact with each other as they change over time, e.g., students' motivation to learn at one time can influence to what extent they experience competence which in turn influences their future motivation to learn. To make sense of this complex phenomenon, science education researchers develop and test models and theories. A critical step in developing models and theories is to recognize patterns as our models and theories often start off as hypotheses that aim at explaining observed patterns. Alas, recognizing patterns is hard when phenomena are complex. Further, while humans excel at recognizing patterns in images, humans typically struggle with identifying patterns in large, tabular data. Luckily, computational approaches can help us with this. Specifically, unsupervised ML provides us with a set of powerful tools to recognize patterns.

Unsupervised ML techniques typically work on some variation of the following idea: to recognize a pattern means to find a set of cases in the data that are more similar—based on some metric of similarity—to each other than the remaining cases.

M. Kubsch (✉)
Freie Universität Berlin, Berlin, Germany
e-mail: m.kubsch@fu-berlin.de

C. Krist
Graduate School of Education, Stanford University, Stanford, CA, USA

P. Wulff
Heidelberg University of Education, Heidelberg, Baden-Württemberg, Germany

© The Author(s) 2025
P. Wulff et al. (eds.), *Applying Machine Learning in Science Education Research*,
Springer Texts in Education, https://doi.org/10.1007/978-3-031-74227-9_11

211

In this way, unsupervised ML techniques take data and provide us with subgroups of that data. Depending on where you look and the exact technique used, these subgroups are called clusters, groups, profiles or (latent) classes but the basic idea is the same. In this way, unsupervised ML is an exploratory approach and the resulting clusters require careful interpretation by the researchers.

In this chapter, you will use unsupervised ML to recognize patterns in how students' epistemic emotions change over the course of short units about energy in middle school.

11.2 Epistemic Emotions in Science Learning

The case study in this chapter is situated in the context of investigating students' epistemic emotions as they learn about energy in short, 4–6 lessons long units. Epistemic emotions are emotions that relate to the generation of knowledge (e.g., Pekrun et al., 2017). Examples are the joy people feel when they have solved a problem, the curiosity that is sparked by observing a phenomenon that escapes explanation, or the confusion that arises when contradictions between what is expected and what is observed cannot be resolved. Epistemic emotions can impact learning (D'Mello & Graesser, 2012) as modern and effective science learning explicitly engages students in the doing of science (e.g., National Research Council, 2012), i.e., students engage in scientific practices to learn about science ideas.

Over time, the emotions people feel and their intensity varies. Therefore, students were asked to report the epistemic emotions they felt during the units multiple times using Likert scales. With such data, you could now ask structural questions about the epistemic emotions such as "How is curiosity correlated with joy?". While this is a valid and interesting research question, let us take a more holistic approach here. The holistic approach centers on the individual student (see also person-centered approach (Magnusson, 2003); for an example in the context of science education see e.g., Hong et al. (2020)) and the set of epistemic emotions that they reported they felt at any of the measurements. This acknowledges the complex interactions between epistemic emotions within students.

11.2.1 Why Unsupervised ML

Analytically, this means identifying students that expressed similar sets of epistemic emotions, i.e., epistemic emotion profiles, that can then be describe qualitatively. In a next step, you can then try to find patterns in how students transition between profiles over time. Both of these tasks come down to recognizing patterns in complex data; a task well suited for unsupervised ML.

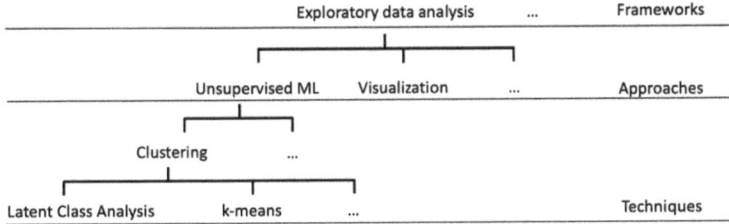

Fig. 11.1 Different frameworks for data analysis

Latent Profile Analysis vs. Clustering vs. Unsupervised machine learning

Depending on where you look and who you talk to, you might find the terms latent profile analysis, latent class analysis, cluster analysis, or unsupervised machine learning being used interchangeably or one being described as an example for the other, e.g., sometimes latent class analysis is labeled as an unsupervised machine learning technique as it is argued that a representation of the data is learned. Let's try and bring some order into all of this.

We consider Fig. 11.1 to be helpful to think about these terms. There are different Frameworks for data analysis such as exploratory data analysis or confirmatory data analysis. Within exploratory data analysis we have different approaches for doing exploratory data analysis—unsupervised machine learning and data visualization for example. Within different approaches we have different techniques; one of them is clustering. Finally, there are different ways to do clustering with latent class analysis being an example of model-based[1] clustering and k-means an example of a model free clustering approach.

11.2.2 Getting to Know the Data Set

Let us start with getting to know the data set again. You will continue to use R for this chapter. The following code loads the packages and data needed in this chapter:

R Code: Packages and data

```
# load required packages
library(here)
```

---

[1] Model based in the sense that there are relatively strong distributional assumptions involved, e.g., the assumption that the groups all have the same variance and only differ in the mean value is often made.

```
library(tidyverse)
library(tidyclust)
library(tidymodels)
library(dbscan)
library(umap)
library(GGally)
# load data
load(here("Data","unsupervised_learning_emotions.RData"))
```

Next, use the `glimpse()` function to get an overview of the variables in the dataframe `df`:

R Code and output: the `glimpse()` function

```
> glimpse(df)
Rows: 214
Columns: 11
$ userid      <chr> "1001", "1003", "1005", "1007", "1009...
$ joy         <dbl> 3.666667, 3.000000, 3.250000, 2.00000...
$ confusion   <dbl> 1.000000, 1.000000, 1.500000, 2.00000...
$ curiosity   <dbl> 4.000000, 3.333333, 3.000000, 3.00000...
$ boredom     <dbl> 2.000000, 1.500000, 2.333333, 3.00000...
$ anxiety     <dbl> 1.00, 1.00, 1.00, 3.00, 4.75, 1.00, 1...
$ frustration <dbl> 1.000000, 1.000000, 1.750000, 4.00000...
$ interested  <dbl> 4.000000, 3.666667, 3.500000, 3.00000...
$ control     <dbl> 4.333333, 3.000000, 4.250000, 4.00000...
$ value       <dbl> 3.666667, 2.666667, 3.250000, 4.00000...
$ total       <dbl> 22.0, 21.0, 34.0, 12.0, 35.0, 18.0, 3...
```

In the data, each row represents a student identified by the `userid` column and the other columns give the scores on different emotions (`joy, confusion, curiosity, boredom, anxiety, frustration, interest`), emotion related appraisals (`control, value`), and the total score students received during the unit based on scoring artefacts that were collected (`total`). Control and value appraisals are strongly related to understanding (epistemic) emotions (Pekrun, 2006; Pekrun & Linnenbrink-Garcia, 2014). Control appraisals describe to what extent someone feels able to influence or manage a situation. When students feel in control as they learn and conduct investigations in the science classroom, it influences their emotional responses and promotes positive emotions like curiosity. Value appraisals describe to what extent someone considers a task or topic relevant or important. If learners see a task as valuable or relevant to their goals and learning, it can trigger positive epistemic emotions, such as interest or enjoyment, that support deep learning and persistence. Conversely, if learners consider a task as unimportant, it can lead to negative emotions, like boredom or frustration. Thus, understanding control and value appraisals can provide insight into the emotional experiences of learners and potentially guide strategies for supporting learning and motivation. All scores are averages calculated from multiple measurements over the course of the unit.

The output also shows that the `total` variable and the emotion related variables are on different scales. The total variable that represents a measure of students overall learning during the unit ranges from 0 to 82 as `range(df$total)` shows. In contrast, the emotion related variables only ranges from 1 to 5 as the data was collected using Likert scales ranging from 1 (not at all) to 5 (very much). To check for missing data use `sum(is.na(df))` which returns 0. How does this work? When you apply the function `is.na()` to the data; the function checks for every cell in `df` whether it is NA or not and returns TRUE or FALSE respectively. Then you sum over the resulting object with R evaluating TRUE as 1 and FALSE as 0. Thus, the result of 0 tells you that there are no NAs. After checking this, take a closer look at the data using a pairs plot (Fig. 11.2):

```
df %>% select(-userid) %>% ggpairs()
```

Overall, there is nothing really striking; all variables show variance. Notably, anxiety and frustration and to some extent confusion show power law like (see Box below), long tailed distributions.

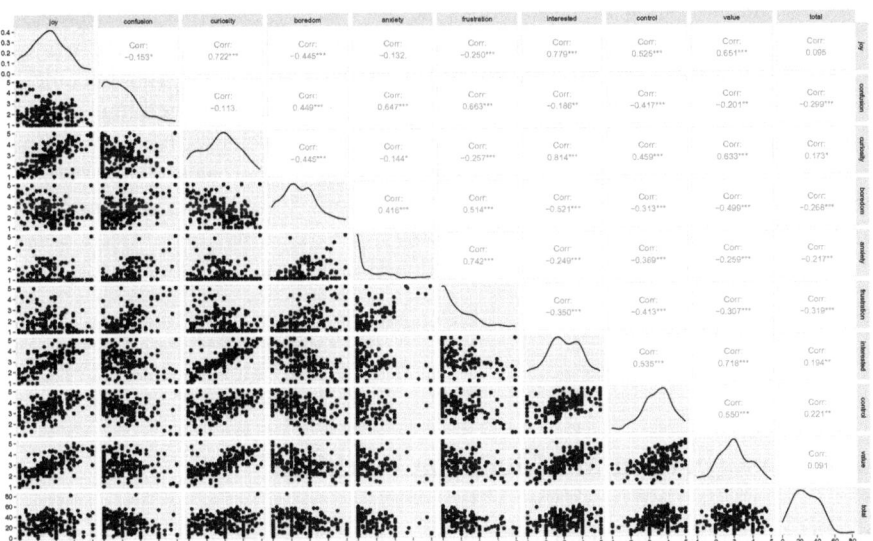

Fig. 11.2 A pairs plot

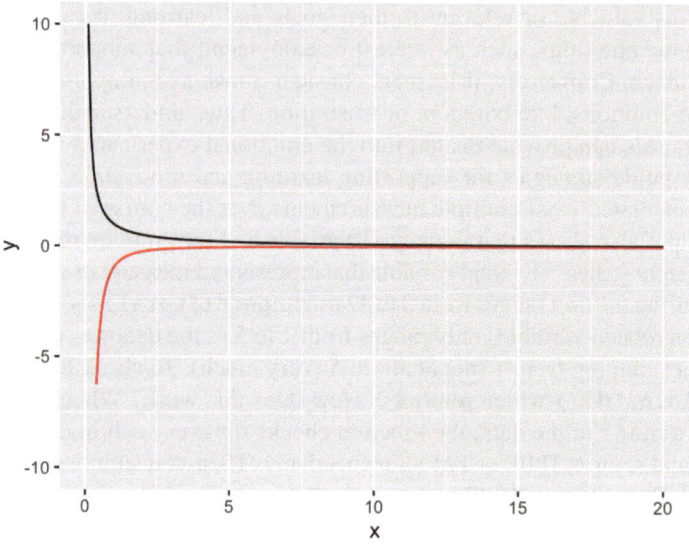

Fig. 11.3 Powerlaw function

Power law function

A power law is a function of the form $f(x) = x^k$. The black line in Fig. 11.3 shows how such a function looks like for $k = -1$. Data that is distributed in this way has the property that some values are very likely to occur while others are much rarer but have relatively similar frequencies. This becomes clear when you look at the red line in the plot (Fig. 11.3). The red line shows the derivative of the function $f(x) = x^{-1}$; the derivative describes how much the values of the function change. The red line quickly approaches 0, indicating that the values change very little. The distribution of words in languages is an example for data that follows a power law. Words like "the" or "a" occur very often while words like "source" or "choice" have much smaller but in fact similar frequencies.

11.3 Unsupervised ML Modeling Workflow

Now that you have a general understanding of the data, it is time to look for patterns using unsupervised machine learning. The general workflow for unsupervised machine learning has three steps: (1) pattern recognition, (2) qualitative pattern interpretation, (3) pattern validation.

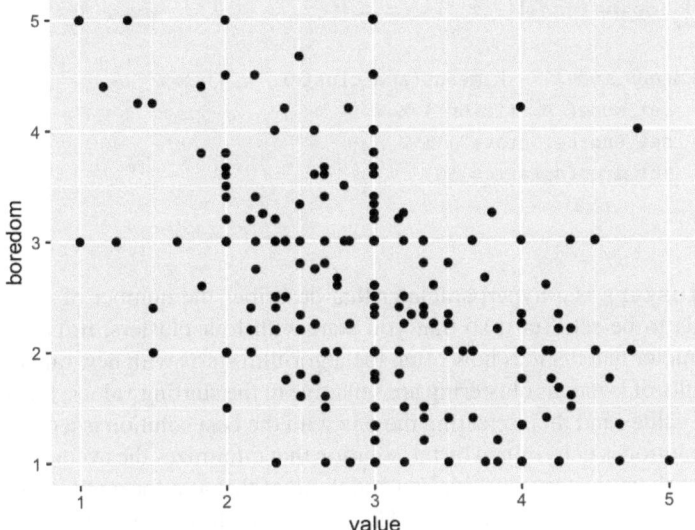

Fig. 11.4 Plot of boredom and value

11.3.1   Pattern Recognition

To start with something simple, look for patterns in just two variables—boredom and value—using clustering. Take a closer look at these variables by plotting them:

    df %>% ggplot(aes(y =boredom, x =value)) + geom_point().

Some parts in the plot (Fig. 11.4) certainly look more populated than others, suggesting that there might be something to uncover here. When there is two-dimensional data as in this case here, you can use simple clustering to look for patterns. One of the most used—and often quite effective—techniques is k-means clustering. The 'k' in k-means refers to the number of clusters the algorithm aims to partition the data into.

The underlying principle of k-Means is quite straightforward: it assumes that every data point belongs to one of k clusters, each represented by a centroid—the center of the cluster. Starting with randomly positioned centroids, the algorithm iteratively assigns each data point to the nearest centroid and recalculates the centroids based on these assignments until the clusters become stable, or a pre-specified number of iterations have been completed. The following code sets up a k-means model:

R Code: k-means model

```
kmeans_model <- k_means(num_clusters = 4) %>%
  set_mode("partition") %>%
  set_engine("stats") %>%
  set_args(nstart = 50)
```

num_clusters is a hyperparameter that describes the number of clusters. This value needs to be set. Set it so that you start with four clusters. nstart is also a hyperparameter that decides how often the algorithm starts with new random values. As the results of k-means clustering are sensitive to the starting values, trying a range of starting values and then selecting the one with the best solution is a good strategy. The best solution is determined by the solution that minimizes the average distance of all points to their cluster centroid. You can think of this as a form of hyperparameter tuning already built into the function. Having defined the model, you can now fit it:

R Code: Define the model

```
# fit model
set.seed(42)
kmeans_fit <- kmeans_model %>%
    fit(~., data = df[,c("boredom", "value")])
df_fit <- cbind(df, extract_cluster_assignment(kmeans_fit))
df_fit %>% ggplot(aes(y = boredom, x = value, colour = .cluster))
+ geom_point()
```

As k-means involves a random component with the starting values, a seed is set to ensure reproducibility. In the next line, the model is fit and then the extract_cluster_assignment() function is used to add the cluster assignment to the data and create a new object df_fit that stores this information. Finally, the clustering results are plotted (Fig. 11.5):

11.3.2 Qualitative Pattern Interpretation

You have now used k-means clustering to detect patterns—the clusters—in the data. However, what is the meaning of these patterns? How can they be characterized? Answering these questions is what the qualitative interpretation step is all about. While the computer can be used to find patterns in data, it cannot interpret what these patterns mean. This is where the theoretical sensitivity and substantive knowledge of

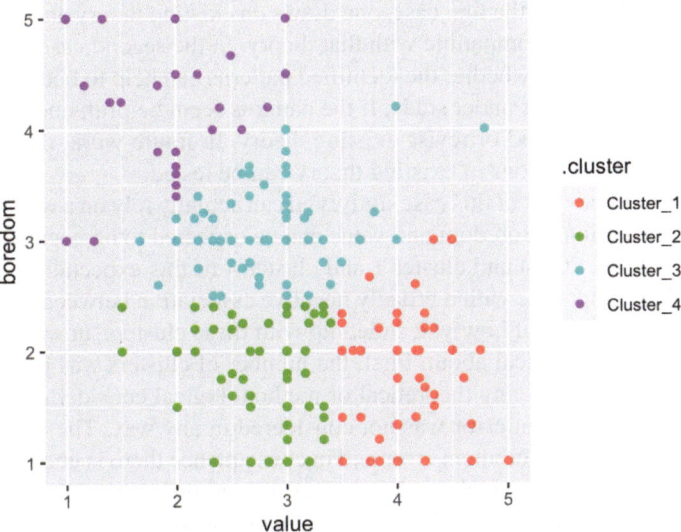

Fig. 11.5 Clusters in boredom and value data

the human analyst come into play. Cluster 1 contains students that have rather high value scores and low to average boredom scores. These students might be characterized as valuing the activities and being mostly not bored. Cluster 2 contains students that have average value scores and similarly low boredom scores as the students in Cluster 1. These students might be characterized as somewhat valuing the tasks and mostly not being bored. Cluster 3 contains students with value scores comparable to Cluster 2 but higher boredom and can be characterized as somewhat valuing the tasks while being also somewhat bored. Finally, Cluster 4 contains students that have high boredom scores and low value scores. Being bored and perceiving tasks as low value seems like a valid description.

11.3.3 Pattern Validation

Using k-means clustering, you just grouped students into four different groups that seem ... reasonable? What do you make of these four groups? Should you use them in further analyses? Answering these questions comes down to considering the validity of the patterns you found. Before going into further detail here, take a step back and reconsider what you are doing and why you are doing it.

What you are doing is exploratory data analysis. But why are you doing it? Actually, no reason was provided for doing it and there are different reasons for doing exploratory data analysis. It is helpful to differentiate two cases here: (1) finding patterns that can be expected within existing theory and (2) finding patterns for the sake of developing theory. In the first case, one draws on existing theory and asks to what extent the patterns are compatible with that theory. In the second case—developing theory—the question is whether the identified patterns can help to better understand whatever phenomenon is under study. If the patterns seem helpful, one may develop new or propose to expand or revise existing theory. In future work, the new theory or expansions and revisions of existing theory can be tested.

Within the framework of this case study, you can actually rely on a well-developed literature on emotions. Boredom and value can be expected to be negatively associated (e.g., Pekrun, 2006) and cluster 1 and cluster 4 fit this expectation. Clusters 2 and 3 are also compatible with a broadly negative association between boredom and value. When you think of how you ended up with these clusters, however, there are some things to be skeptical about. First, the number of clusters was just set to four in the algorithm without any theoretical or methodological consideration for doing so. Second, measurement error was not considered in any way. The variables used for clustering have measurement error and in consequence there is uncertainty in the cluster assignment. K-Mean clustering, however, ignores this uncertainty in cluster assignment. While there is little you can do about k-means clustering ignoring measurement error, you can go back and see how setting a different number of clusters affects the results. To do so, tune the respective hyperparameter.

11.3.4   Once More with Tuning

For tuning, follow the same steps as in the last chapter on supervised learning: set up a model and specify which hyperparamters to tune, build a workflow, create folds for cross-validation, set up a tuning grid for number of clusters ranging from 1 to 10, and finally tune the hyperparameters. The following code does all that:

R Code: Tuning hyperparamters

```
# tune hyperparamter
# define model
kmeans_model_tune <- kmeans_model %>%
  set_args(num_clusters = tune())
# build workflow
kmeans_wf <- workflow() %>%
  add_model(kmeans_model_tune) %>%
  add_formula(~.)

# set folds for cross-validation
cv <- df %>% select(boredom, value) %>% vfold_cv(v = 10)

# set tuning grid
```

```
grid <- tibble(num_clusters = 1:10)

# tune
system.time(tune_res <- tune_cluster(
  object = kmeans_wf,
  resamples = cv,
  grid = grid
))
```

When you look at the code closely, you will notice the function `system.time()` that was used on the tuning process and running the code, you will get a reading of the time it took to run the tuning process. It is included here to show how parallelization can save a lot of time. Whenever analyses that are independent of each other are run, you can use the multiple processors in your computer to run these analyses in parallel instead of one after another. Grid tuning is exactly such a case. The following commands load the `doParallel` library that provides an easy way to run analyses in parallel and specifies how many processor cores to use:

R Code: `doParallel` library for parallelization

```
# parallel computation
library(doParallel)
registerDoParallel(detectCores())
```

Now, let us run the tuning process again:

R Code: Parallel tuning process

```
system.time(tune_res <- tune_cluster(
  object = kmeans_wf,
  resamples = cv,
  grid = grid
))
```

You will notice that—depending on the specifics of your computer—processing time has decreased a lot. After this short detour, let us take a look at the results of the tuning process by running the following code:

R Code: Results of the tuning process

```
tune_res %>%
  collect_metrics()

tune_res %>%
  autoplot()
```

The resulting plot (Fig. 11.6) shows two metrics—sse_total and sse_within_total—and how they change as the number of clusters increases.

sse_total is the sum of squared errors about the centroid of a one cluster solution which just gives a sense of the overall variability of the data. sse_within_total is the sum of the sum of squared errors for each cluster. Therefore, sse_within_total is identical to sse_total for the model with just one cluster. Increasing the number of clusters generally reduces the sse_within_total. Try this for yourself using this code:

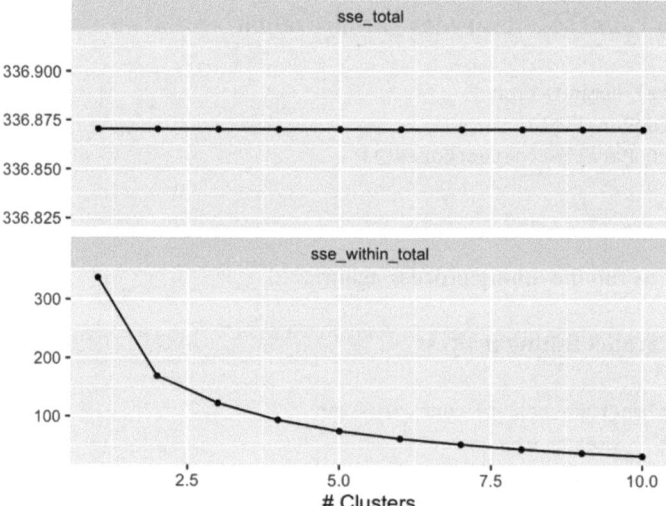

Fig. 11.6 Cluster tuning

R Code: Increase the number of clusters to reduce `sse_within_total`

```
#expand grid
# set tuning grid
grid_large <- tibble(num_clusters = 1:100)

tune_res_large <- tune_cluster(
  object = kmeans_wf,
  resamples = cv,
  grid = grid_large
)

tune_res_large %>%
  autoplot()
```

`sse_within_total` behaves like this because having more clusters allows more data points to be closer to the centroids of their clusters. This becomes intuitively clear when you take the number of clusters to its extreme and have as many clusters as you have data points. In that case, every cluster will reflect one data point and the squared error will be zero.

Squared errors

Why are errors squared? Because error can occur in different directions! Let us look at a simple example to see what this means. Imagine you have two data points a $x = 3$ and $y = 5$ and two respective predictions with $x_{pred} = 4$ and $y_{pred} = 4$. Now, if you just calculated the error by subtracting actual value from the predicted value you would get $x_{error} = 4 - 3 = 1$ and $y_{error} = 4 - 5 = -1$. Summing these two errors leads to a total error of $error_{total} = 1 + (-1) = 0$. This result is clearly not helpful because the predictions are erroneous. The solution is to square the errors before summing to get rid of negative numbers: $1^2 + (-1)^2 = 1 + 1 = 2$.

To select the number of clusters based on this plot, you can use the so-called elbow method. You might be familiar with this approach from exploratory factor analysis—a technique that is also sometimes referred to as unsupervised ML. Factor analysis is used to find a reasonable number of dimensions in a data set that captures most of the variation. The idea of the elbow method is to find the point where increasing the number of clusters starts leading to diminishing returns in terms of reduced error. This point is located at the "elbow;" the point where a linear decrease in error with

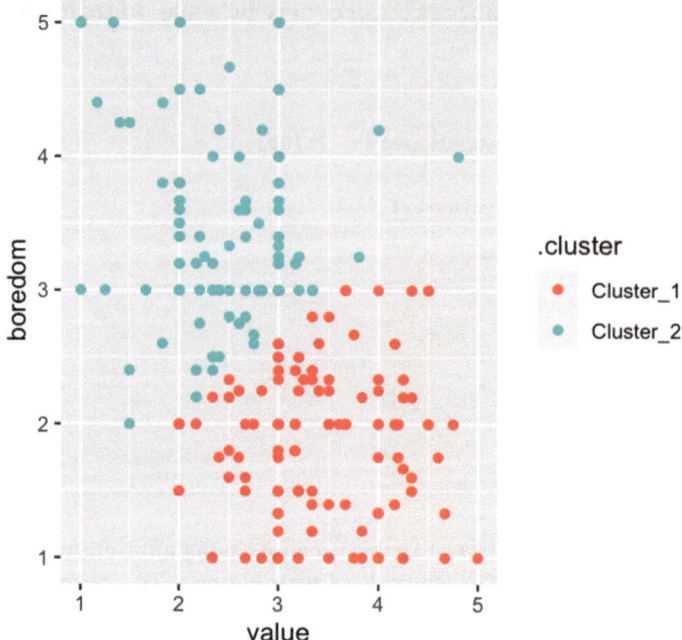

Fig. 11.7 K-means clusters of boredom and value

increasing number of clusters transitions to a curved relationship between decrease in error and increasing number of clusters. In this case, this happens when the number of clusters is two (Fig. 11.6).

Take that results and run a final model with two clusters and plot it (Fig. 11.7):

R Code: Selecting the two clusters and plotting it

```
# set-up final model
kmeans_model_final <- k_means(num_clusters = 2) %>%
  set_mode("partition") %>%
  set_engine("stats") %>%
  set_args(nstart = 50)
# fit final model
kmeans_fit_final <- kmeans_model_final %>%
  fit(~., data = df[,c("boredom", "value")])
# plot final model
df_fit_final <- cbind(df, extract_cluster_assignment(kmeans_fit_final))
df_fit_final %>% ggplot(aes(y = boredom, x = value, colour = .cluster))
+ geom_point()
```

A qualitative interpretation of the two clusters suggests that Cluster 1 represents high boredom, low value students while Cluster 2 represents high value, low boredom students. This aligns with substantive theory. In contrast to the last solution we looked at, we now have clusters that align with the substantive theory on emotions (e.g., Pekrun, 2006) and can be justified by a data driven procedure.

11.3.5 Adding Dimensions

After clustering two variables in the last section, let us now consider the whole range of emotions in the data set. Overall, there are a total of nine variables here: joy, confusion, curiosity, boredom, anxiety, frustration, interested, control and value. When looking at the pairs plot at the beginning of the chapter, you already saw relatively strong correlations between many of these variables—interest and curiosity for example are correlated with $r = .8$. When there are strong correlations between many of the variables this can lead to a range of issues such as redundancy and distortion of distance measures. Redundancy means that highly correlated variables often provide mostly redundant information. Essentially expressing the same underlying information, they may not add much value to the clustering algorithm but still increase the dimensionality and computational cost. Increasing dimensionality without a need to increase it leads to what is called the curse of dimensionality: in high-dimensional spaces, data tends to become increasingly sparse, and the distance between sample points starts losing its meaningfulness. Distortion of distance measures means that clustering algorithms often use some form of distance measure to assess the similarity between instances in the data. If variables are highly correlated, they may unduly influence the distance measure, leading to a distortion in the perceived similarity between instances. To avoid these issues, dimensionality reduction techniques can be a useful preprocessing step. In the following case study, you will explore the use of dimensionality reduction and use a different clustering algorithm that allows us to classify data as noise, i.e., instead of forcing clusters onto data like was the case with k-means clustering, the algorithm can decide not to classify data points.

11.3.6 Pattern Recognition

The packages that you will use in the following are not compatible with the tidymodels framework so the code will look somewhat different. As outlined above, there are two goals here—reducing the dimensionality of the data and then clustering it. For dimensionality reduction you will use Uniform Manifold Approximation and Projection (UMAP, McInnes et al. (2020)) and Hierarchical density based clustering (Campello et al., 2013) via the HDBSCAN package (McInnes et al., 2017) for clustering. UMAP is state-of-the-art dimension reduction technique that seeks to preserve the local structure of the data while also revealing the global structure, leading to improved performance of well-known clustering procedures (Allaoui et al., 2020). HDBSCAN is a hierarchical clustering algorithm that has three properties that make it attractive for us: (1) we do not have to specify the number of clusters, as the number of clusters is determined in a data-driven way, (2) it is conservative in the sense that it is willing to leave points unassigned rather than forcing them into clusters; grouping points together only when they truly form a cluster, (3) HDBSCAN is stable and predictive, i.e., when the algorithm is run multiple times with different random starting conditions the resulting clusters remain very similar and changes in the hyperparameters lead to relatively stable and predictable changes in the resulting clusters.

Hierarchical vs. non-hierarchical clustering

Hierarchical and non-hierarchical clustering are two prominent types of clustering algorithms used for unsupervised machine learning tasks. Hierarchical clustering builds a tree-like hierarchy of clusters, either by starting with individual points and merging them (agglomerative) or by starting with the complete data set and dividing it (divisive). Non-hierarchical clustering, on the other hand, typically works by optimizing an objective function (such as a distance metric) and requires the number of clusters to be set in advance, as seen in methods like k-means. The table below shows their primary strengths and weaknesses. Note that the choice between different techniques often depends on the specific use case and trying several approaches is often a good idea.

Hierarchical clustering		Non-hierarchical clustering	
Strengths	Weaknesses	Strengths	Weaknesses
No need to specify number of clusters	Computationally complex	Highly efficient	Sensitive to initial starting values
Distance measures can be chosen flexibly to fit the task	Sensitive to outliers	Robust to outliers	Issues with irregularly shaped clusters
Hierarchy in the data is preserved			Requires you to specify the number of clusters

UMAP and HDBSCAN both have a number of hyperparameters. To keep things simple, you will only tune one of the central hyperparameters for each algorithm and use defaults for the others (https://hdbscan.readthedocs.io/en/latest/parameter_selection.html and https://umap-learn.readthedocs.io/en/latest/parameters.html have excellent and extensive documentation of how major hyperparameters affect the results). For UMAP, this is n_neighbors. n_neighbors controls the balance between local and global structure by setting the maximum number of neighboring sample points to consider. For HDBSCAN, tune the primary parameter `minPts`. `minPts` is the minimum size (in terms of datapoints) a cluster can consist of. To tune the hyperparameters, some metric to judge the results is needed (for k-means clustering this was the `sse_within_total` measure). The HDBSCAN implementation used here (Hahsler et al., 2019) provides such a metric in the form of probabilities for the assigned cluster. The higher the probability, the more confident you can be in the cluster assignment. To find the combination of hyperparameters that provides the best average classification accuracy, use the mean of this probability. Depending on the use case, you could also define a different metric, e.g., trying to find the solution with the least cases where the probability is below 10%.

Now, it is time to do the actual tuning. The idea is to run UMAP for a range of n_neighbors values and for each of those values run HDBSCAN on the resulting projection (the result of the dimensionality reduction is called a projection) for a range of minPts values and store the results. The following code does just that:

R Code: Tuning

```
set.seed(42)

prob <- c()
for(n in seq(from = 5, to = 50, by = 5)){
  emo.proj <- df %>% select(all_of(emotions)) %>% umap(n_neighbors = n)
  for(i in 2:50){
    hclust <- emo.proj$layout %>% hdbscan(minPts = i)
    prob$prob <- append(prob$prob, mean(hclust$membership_prob))
    prob$minPts <- append(prob$minPts, i)
    prob$n_neighbors <- append(prob$n_neighbors,n)
  }
}
```

First, set a seed for reproducibility of the analysis (there is randomness involved in UMAP). Next, an object prob is created in which the classification probability as a metric is stored. Then a for loop is used. A for loop is a function that reiterates the commands in it for all values of a variable. In this case, the for loop will run all code within it once while the variable n sequentially takes each of the values in `seq(from = 5, to = 50, by = 5)`. `seq(from = 5, to = 50, by = 5)` creates a list of number from 5 to 50 in increments of 5, i.e., 5, 10, 15, 20, ...50. In the next line of code, variable n comes up:

```
emo.proj <- df %>% select(all_of(emotions)) %>% umap(n_neighbors = n)
```
the code takes the dataframe `df`, selects all of the emotion variables, applies the `umap()` function with `n_neighbors` set to the current value of n, and finally stores the result in the object `emo.proj`. This means that in the first iteration of the for loop,

the object `emo.proj` will store the UMAP projection of the emotions resulting from setting `n_neighbors` to 5. Note that `umap()` creates a two dimensional projection of the data by default. Next, we have another for loop within the for loop.

R Code: The inner for loop

```
for(i in 2:50){
    hclust <- emo.proj$layout %>% hdbscan(minPts = i)
    prob$prob <- append(prob$prob, mean(hclust$membership_prob))
    prob$minPts <- append(prob$minPts, i)
    prob$n_neighbors <- append(prob$n_neighbors, n)
}
```

In this for loop, the variable i sequentially takes all the values in 2:50—a sequence from 2 to 50 in increments of 1. In the first line of the loop, the two dimensional projection of the emotions stored in `emo.proj$layout` is taken, the `hdbscan()` function with the hyperparameter `minPts` set to i is applied, and finally the results are stored in the object `hclust`. In the next line, the mean classification probability is calculated and written to the end of the `prob` column within the `prob` object created before starting the outer for loop. The next two lines of code store the current values of i and n in the prob objects in columns named by the respective hyperparameter. This continues for all values that i can take. In essence, what this inner for loop does is clustering the UMAP projection for different values of `minPts` ranging from 2 to 50, calculating a metric, and storing the metric as well as the respective hyperparameter values in the prob object. After this inner for loop has ended, the outer for loop will start again for the next value that n can take. This continues for all values of n and you end up with an object `prob` that contains the average classification probability of HDBSCAN clusterings with `minPts` ranging from 2 to 50 of UMAP projections of the emotions data with `n_neighbors` values ranging from 5 to 50.

You can visually inspect these tuning results and get the best combination of hyperparameters using the following commands:

R Code: Visually inspect tuning results

```
# plot search results
data.frame(prob) %>% ggplot(aes(y = prob, x = minPts,
colour = as.factor(n_neighbors))) +
  geom_line()
# find best combination
data.frame(prob)[which.max(data.frame(prob)$prob),].
```

Now, use the returned hyperparameter values ($minPts = 4, n\_neighbors = 30$) and fit a final model and save it in the object hclust.final:

R Code: UMAP

```
emo.proj.final <- df %>% select(all_of(emotions)) %>%
                  umap(n_neighbors = 30)
hclust.final <- emo.proj.final$layout %>% hdbscan(minPts = 4)
```

Running hclust.final gives the following summary output:

R Output: Results of HDBSCAN

```
HDBSCAN clustering for 214 objects.

Parameters: minPts = 4

The clustering contains 3 cluster(s) and 0 noise points.

1   2   3
49  61  104

Available fields: cluster, minPts, coredist,
                  cluster_scores, membership_prob,
                  outlier_scores, hc
```

The summary shows that the clustering resulted in three clusters and 0 data points were classified as noise. Further, the sizes of the three clusters are displayed. You can also plot the hclust.final object using plot(hclust.final, show_flat = T).

The plot (Fig. 11.8) shows what is called a dendrogram; it illustrates the hierarchical relationships in the data. The x-axis does not represent any meaningful information; it shows the different partitions of the data. The y-axis represents how denesly clusters are packed (eps is a distance metric used in the algorithm), i.e., clusters that have smaller eps values are more dense. Lastly, the width of the vertical lines represents the number of data points in a cluster. The clusters with boxes labelled 1, 2, and 3 are clusters from the summary output. The dendrogram suggests that clusters 2 and 3 are more similar to each other than in comparison to cluster 1. The dendrogram can be a helpful tool in the qualitative interpretation step to triangulate decisions about merging or splitting clusters.

HDBSCAN*

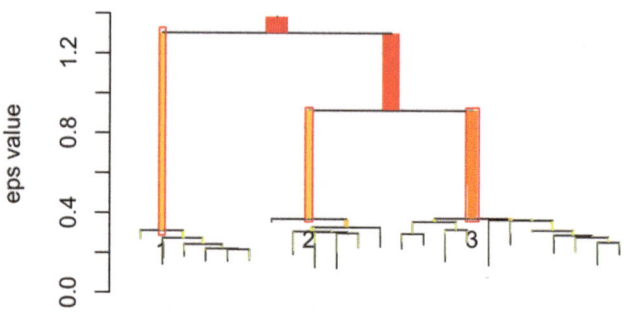

Fig. 11.8 Plot of the clustering results with HDBSCAN

You can also plot the clustering in the two dimensional projection from UMAP using the following code:

R Code: Clustering in the two dimensional projection

```
df.clust.emo <- cbind(df, hclust.final$cluster)
df.clust.emo <- cbind(df.clust.emo, unlist(emo.proj.final$layout))
df.clust.emo <- df.clust.emo %>%
    rename(cluster = 'hclust.final$cluster', dim1 = '1', dim2 = '2')

# plot solution
df.clust.emo %>% ggplot(aes(y= dim2, x = dim1,
                        color = as.factor(cluster))) +
  geom_point()
```

While this plot (Fig. 11.9) may look interesting, it actually provides very little information for us as there is no straight forward interpretation of the x and y axis. The two axes are the result of the projection from UMAP and cannot directly be linked to any construct. To learn more about what the clusters actually represent in terms of emotions, qualitative interpretation is needed.

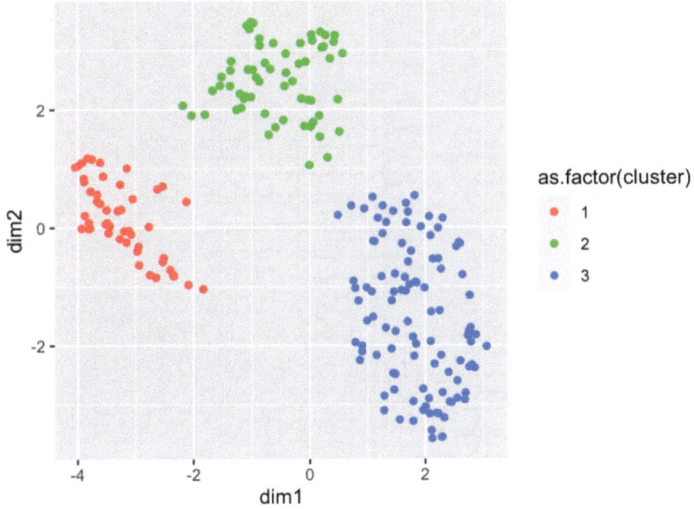

Fig. 11.9 Clusters in UMAP dimensions

11.3.7 Qualitative Pattern Interpretation

To interpret the three clusters qualitatively, they need to be related back to the constructs—the epistemic emotions. To do so, plot the epistemic emotions by cluster (Fig. 11.10):

R Code: Qualitative pattern interpretation

```
df.clust.emo %>% select(all_of(emotions) | cluster) %>%
pivot_longer(!cluster, values_to = "score",
             names_to = "construct") %>%
ggplot(aes(x = score, y = construct)) +
            geom_boxplot() + facet_wrap(.~cluster, ncol = 1) +
            theme(text = element_text(size = 15))
```

Cluster 1 can be described as average on all constructs. However, compared to the other clusters it has the highest values for negative emotions frustration, confusion, and anxiety. Thus one might describe cluster 1 as feeling *stuck* with reference to the model of affect dynamics by D'Mello & Graesser (2012). Cluster 2 shows low values on all constructs but control and boredom. This cluster may be interpreted as *disengaged* but following along with students that are not really touched by the material but probably doing at least ok so they feel in control but also bored as the material does not engage them. Cluster 3 is relatively high on most constructs with

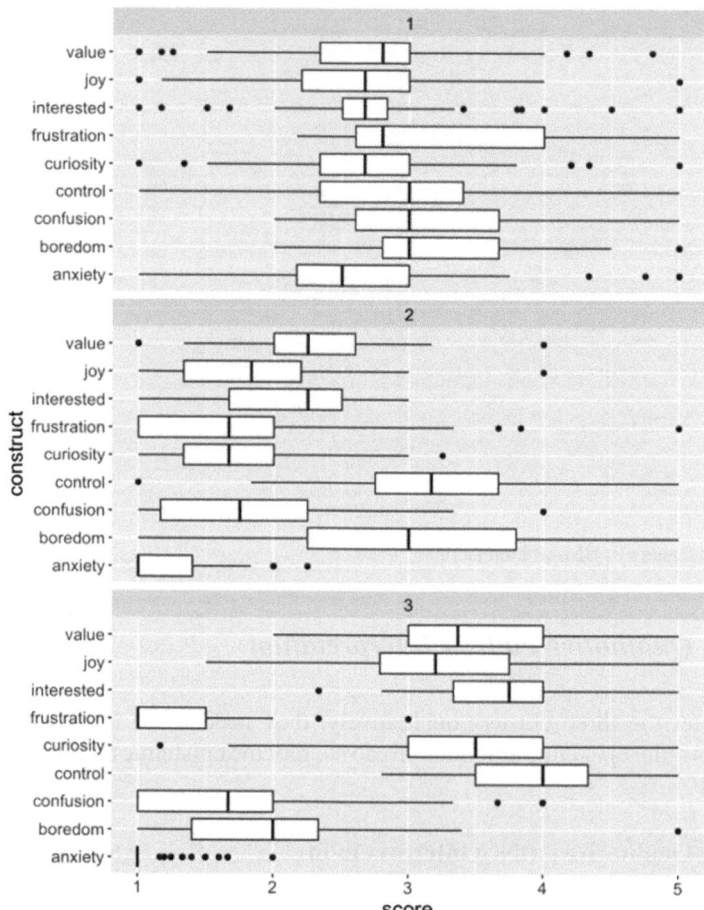

Fig. 11.10 Constructs by cluster

the exception of frustration, confusion, boredom, and anxiety. Thus, students in this cluster may be characterized as *engaged and motivated*.

11.3.8 Pattern Validation

To validate the three clusters, let us turn to substantive theory on emotions. From a theoretical perspective these clusters are sensible, e.g., high value and control in cluster 3 typically lead to positive emotions such as joy and are negatively associated with negative emotions such as confusion and anxiety. Further, other data can be used to look at criterion related validity (American Educational Research Association, 2014).

Criterion validity refers to the extent to which a measurement or a test is related to an outcome—the criterion—that it is expected to predict or correlate with. One would expect epistemic emotions to correlate with students learning. Based on the cluster interpretation, students in cluster 3—engaged and motivated—should show strong evidence of learning, students in cluster 2 are probably also relatively successful given that they feel in control but bored. For cluster 1 which we characterized as stuck relatively little learning can be expected.

Now, you can test these assumptions by looking at the `total` variable in the data set. Remember that `total` is a measure of students learning over the course of the unit. Plot students' scores by cluster and also run a respective ANOVA with post-hoc Tukey tests:

R Code: Clusters and learning

```
# relation between clusters and learning
df.clust.emo %>% ggplot(aes(y = total, x = as.factor(cluster))) + geom_boxplot()
df.clust.emo %>% aov(total ~ as.factor(cluster), data = .) %>% TukeyHSD()
# Tukey multiple comparisons of means
    95% family-wise confidence level
Fit: aov(formula = total ~ as.factor(cluster), data = .)
$'as.factor(cluster)'
        diff         lwr         upr      p adj
2-1 5.708933 -0.7203072 12.138173 0.0931013
3-1 8.655024  2.8479509 14.462096 0.0015388
3-2 2.946091 -2.4587964  8.350978 0.4043009
Residual standard error: 14.2 on 211 degrees of freedom
Multiple R-squared:  0.05543,   Adjusted R-squared:  0.04648
F-statistic: 6.191 on 2 and 211 DF,  p-value: 0.002438
```

The resulting plot (Fig. 11.11) generally supports the assumptions that the students in cluster 1 learned the least, and students in clusters 2 and 3 learned more. The results from the post-hoc Tukey test show that students in cluster 3 learned significantly more than students in cluster 1. The difference in learning between cluster 2 and 1 is marginally statistically significant while the difference in learning between clusters 2 and 3 is clearly not statistically significant with a p-value of .4.

Regarding the validity of the clusters, one can conclude that the clusters are theoretically sensible with further evidence from the relationship between the qualitative interpretation of the clusters and another variable (total learning) supporting this interpretation.

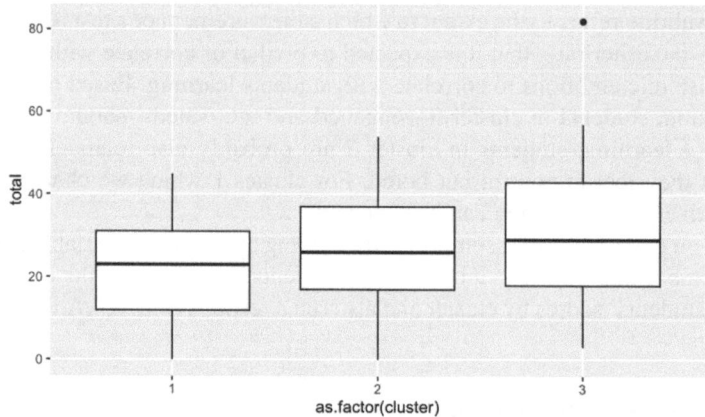

Fig. 11.11 Total score at the end of the unit by cluster

11.4 Exploring Additional Dimension with a New Technique

In the pattern validation step of the last section, you already looked at the relationship between students' epistemic emotions and learning. In this final case study of the chapter, you will to go one step further and look at patterns in students learning about numerous ideas about energy and a set of epistemic emotions together. To do so, you will explore another technique that is somewhat different from those two used before: Latent Profile Analysis[2] (LPA). What sets LPA apart is that it is model based. This means that LPA assumes a statistical model for the data and estimates parameters of the model to identify the clusters—these clusters are referred to as latent profiles within the context of LPA. The other techniques we explored, k-means and hierarchical clustering, are not model based as they rely on geometric or other forms of distance or similarity measures to assign data points to clusters. The benefit of a model-based approach is that it provides you with a whole range of statistical measures to help determine the optimal number of clusters and other aspects of the model specification (essentially, you can think of the number of clusters and details of the model specification as hyperparameters). Further LPA is very flexible regarding the kind of variables used together, i.e., it can handle a combination of categorical and continuous variables. LPA also allows soft-clustering, i.e., unlike in the case of k-means, cluster assignments are not absolute but LPA provides a probability of the assignment (similar to the one you used for tuning the hyperparameters of UMAP and HDBSCAN). These probabilities can then be used in follow-up analyses to provide more reliable and valid results. Further, you could also decide to limit

---

[2] Regarding terminology: Latent Profile Analysis and Latent Class Analysis (LCA) are both conceptually similar. When the variables that make up the clusters in the data are categorical this is often referred to as LCA, when the variables are continuous LPA is used.

follow-up analyses to cases where the probability of an instance belonging to a cluster is sufficiently high. Lastly, LPA models can be implemented as structural equation models (SEMs). SEM is a powerful modeling framework and when you use SEM software for estimating LPA models, you can incorporate LPA into larger SEMs. In this way, you can think of LPA as a technique that provides a bridge between unsupervised machine learning and more traditional statistical modelling techniques used in educational measurement and assessment.

Now, before diving into LPA, take a look at the data that you will use here (note that the output is abbreviated here due to space limitations):

R Code and output: Data

```
load(here("Data",
"unsupervised_learning_emotions_knowledge.RData"))
glimpse(df)
Rows: 160
Columns: 19
$ userid     <chr> "1001", "1003", "1005"...
$ joy        <dbl> 3.666667, 3.000000, 3.250000...
$ confusion  <dbl> 1.000000, 1.000000, 1.500000, 3.000000...
$ curiosity  <dbl> 4.000000, 3.333333, 3.000000, 2.500000...
$ boredom    <dbl> 2.000000, 1.500000, 2.333333, 2.000000...
$ anxiety    <dbl> 1.000000, 1.000000, 1.000000, 4.750000...
$ frustration <dbl> 1.000000, 1.000000, 1.750000, 4.250000...
$ interested <dbl> 4.000000, 3.666667, 3.500000, 2.500000...
$ control    <dbl> 4.333333, 3.000000, 4.250000, 3.500000...
$ value      <dbl> 3.666667, 2.666667, 3.250000, 1.500000...
$ total      <dbl> 22.0, 21.0, 34.0, 35.0, 34.0, 37.0...
$ LP.M       <dbl> 0.4, 0.2, 0.0, 0.0, 0.2, 0.0, 0.0...
$ LP.T       <dbl> 0.4, 0.2, 0.0, 0.0, 0.2, 0.0, 0.0...
$ Practice   <dbl> 1.0, 1.2, 1.6, 1.6, 1.6, 1.0, 1.2...
$ M.Radiant  <dbl> 2.0, 1.6, 4.0, 4.4, 3.2, 4.6, 1.6...
$ M.Electric <dbl> 0.8, 0.8, 0.8, 0.6, 0.8, 0.8, 0.8...
$ T.Process  <dbl> 0.2, 0.4, 0.4, 0.4, 0.6, 0.6, 0.4...
$ pre        <dbl> 1.43474564, -0.45652054, 0.44383939...
$ post       <dbl> 2.245116218, -0.002572408, 0.587175737...
```

The emotion variables and `total` are familiar variables by now. The new variables are `LP.M`, `LP.T`, `Practice`, `M.Radiant`, `M.Electric`, `T.Process`, `pre`, and `post`. Pre and post are test scores for the respective assessments before and after the unit. `M.Radiant` and `M.Electric` refer to evidence for knowledge about manifestations of radiation (`M.Radiant`) and electric (`M.Electric`) energy; `T.Process` refers to evidence for knowledge about energy transformation processes. Practice reflects students' engagement in numerous scientific practices throughout the unit such as constructing explanation or conducting investigations. Lastly, LP.M and LP.T reflect evidence of students applying their knowledge about manifestation of energy (LP.M) or energy transformation processes (LP.T) in the context of scientific practices. Just like the emotions, M.Radiant, M.Electic, T.Process, Practice, LP.M, and LP.T are averaged across the whole unit. The next step in getting to know the data would be to do a pairs plot again. However, a pairs plot with 18 variables does not come across well in a book so go ahead, type

df %>% select(-userid)%>% ggpairs(), and take a look at the plot on your machine. You will not encounter anything you have not seen before, so let us go ahead and continue with the pattern recognition step.

11.4.1 Pattern Recognition

Start by loading the tidyLPA package (Rosenberg et al., 2018) and define an object that stores the names of the emotions and knowledge related variables:

R Code: tidyLPA

```
library(tidyLPA)
emo_kn <- c("control", "value","joy","interested",
"curiosity", "confusion", "boredom",
"frustration", "anxiety","M.Radiant",
"M.Electric", "T.Process",
"Practice","LP.M", "LP.T")
# Now, you are already good to go for estimating
# latent profiles:
emo.kn.lpa  <- df %>% select(all_of(emo_kn))%>%
  estimate_profiles(1:10, models = c(1,3))
```

The code takes the data, selects all of the emotion and knowledge variables stored in emo_kn and estimates profiles. 1:10 specifies to estimate models with 1 to 10 profiles and models = c(1,3) tells the function to estimate these models with a range of different model specifications (1 and 3 refer to different model specification so in total 20 models will be estimated. There are also more specifications[3] but here we focus on two for the sake of brevity and accessibility. The different model specifications refer to the model implied assumptions. Model 1 makes strong assumptions that become more relaxed in model 3. The more assumptions a model makes, the less

---

[3] The mclust package (Scrucca et al., 2016) for R allows to specify an even wider range of models in a flexible way.

parameters need to be estimated which allows estimation with relatively little data. At the same time, model assumptions may not be met which may lead to uninterpretable profiles and / or not reflect the real underlying profiles. Relaxing assumptions can solve these issues but requires more data for precise estimation and may lead to overfitting.

With twenty estimated models, you now want to find the model solution that best fits the data. To do so, we use the `compare_solutions()` function.

Compare tidyLPA solutions:

tidyLPA solution

The best (least wrong) model according to BIC is Model 3 with 4 classes.

An analytic hierarchy process, based on the fit indices AIC, AWE, BIC, CLC, and KIC (Akogul & Erisoglu, 2017), suggests the best solution is Model 3 with 8 classes.

Model	Classes	BIC	Warnings
1	1	7153.788	
1	2	6479.586	
1	3	6137.853	
1	4	5962.805	
1	5	5923.172	
1	6	5900.532	
1	7	5857.432	
1	8	5729.532	
1	9	5756.237	
1	10	5790.178	
3	1	5395.248	
3	2	5319.694	
3	3	5353.034	
3	4	5253.824	
3	5	5325.760	
3	6	5257.404	
3	7	5303.932	
3	8	5285.852	
3	9	5340.181	
3	10	5359.552	

Fit indices

Fit indices are a method to compare the relative fit of different models. They balance how well a model fits the data against the complexity of the model, with the general rule being that simpler models are better unless a more complex model provides a significantly better fit. Lower numbers generally indicate better fit. Indices primarily differ in how exactly they penalize model complexity and how well they perform across sample sizes.

You get an overview of the fitted models with the BIC (Bayesian Information Criterion) values and a notification about potential warnings from the algorithm. Further, the function shows what model is best according to BIC and in addition also outputs what model is best according to the analytic hierarchy process (Akogul & Erisoglu, 2017) which is a procedure that uses multiple metrics to assess model fit. In this case, the analytic hierarchy process and the BIC suggest different solutions. So, what do you do now?

Deciding on the number of profiles is often challenging and in practice typically involves considering multiple solutions, comparing and contrasting them, and potentially running further follow-up analyses with both solutions in the spirit of a multiverse analysis[4] (Steegen et al., 2016). The key question that should guide this endeavour is: What do more profiles add to the understanding of the phenomenon at hand? Profile analysis is primarily done to look for differences in kind, not degree. Profiles that effectively reflect levels in the variables provides little insight (you knew that there were differences along the variables before) and if you are interested in differences in degree you can turn to related techniques (e.g., regression, latent growth curve, etc.). Thus, solutions with more profiles should add qualitatively new profiles. With this in mind, now take a look at the 4 profile solution suggested based on BIC:

R Code: Profile solution

```
get_estimates(emo.kn.lpa$model_3_class_4) %>%
filter(Category == "Means") %>%
  ggplot(aes(y = Estimate, x = Parameter,
  colour = as.factor(Class),
  group = as.factor(Class))) +
  geom_point() + geom_line() +
  theme(text = element_text(size = 15),
  axis.text.x = element_text(angle = 90,
  vjust = 0.5, hjust=1))
```

---

[4] The idea of a multiverse analysis is to follow along all forking paths where we make decisions in an analytical process and present all results together or average them. In this way, we can get an intuition for how analytical choices influence the results.

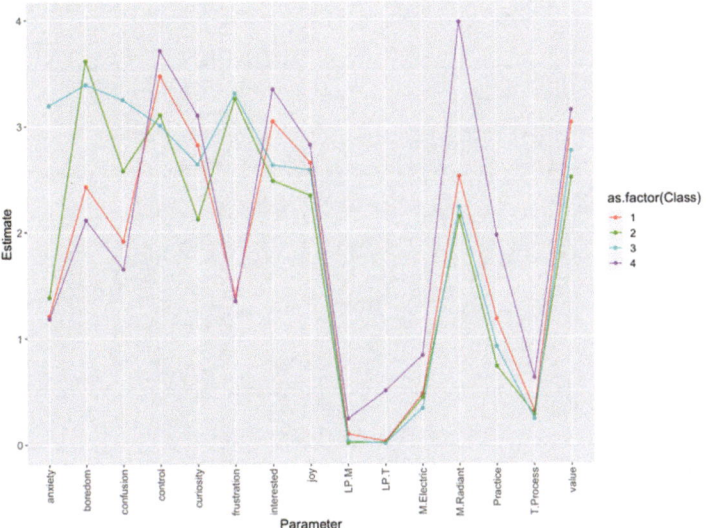

Fig. 11.12 Variable values for 4-profile solution

The plot (Fig. 11.12) shows that the four profiles show similar shapes across all variables from joy to value; profile four showing some difference in degree here. From anxiety to interested, profiles 1 and 4 show similar shapes and profiles 2 and 3 have shapes that are distinct from each other and also the shapes of profiles 1 and 2. This suggests you have four profiles with clear distinction here. Now, take a look at the suggested eight profile solution (Fig. 11.13):

R Code: Get estimates

```
get_estimates(emo.kn.lpa$model_3_class_8) %>%
filter(Category == "Means") %>%
ggplot(aes(y = Estimate, x = Parameter,
  colour = as.factor(Class),
  group = as.factor(Class))) +
geom_point() + geom_line() +
theme(text = element_text(size = 15),
  axis.text.x = element_text(angle = 90,
  vjust = 0.5, hjust=1))
```

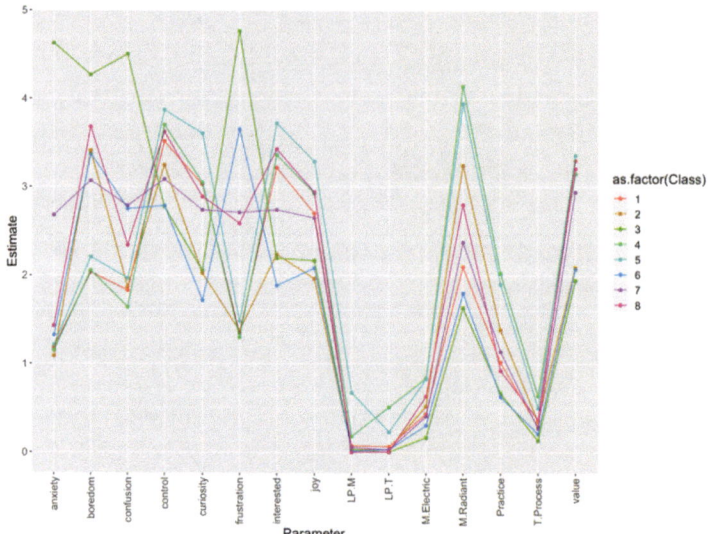

Fig. 11.13 Variable values for 8-profile solution

Again, there is little differentiation in kind to be seen on the variables on the right side of the plot. When you look at the left side, there is profile 3 which is high on anxiety, confusion and frustration which stands out. This profile is comparable in shape to profile 3 from the four-profile solution. The remaining profiles often show similar shapes such as profiles 1 and 4 and generally are similar in shape to the profiles 1, 2 and 4 from the four-profile solution with some added differences in degree.

There are not really any new shapes with the eight-profile solution compared to the four-profile solution. So far, the four-profile solution seems preferable. Let us also compare the comprehensive set of fit indicies and metrics for those solutions.

You can access the respective indices by storing the results of the `compare_solutions()` function in the object fit and then filtering for the two solutions that are of interest:

R Code and output: Storing the results

```
fit <- compare_solutions(emo.kn.lpa)
fit$fits %>%
filter(Model == 3 & Classes %in% c(4,8)) %>%
glimpse()
Rows: 2
Columns: 19
$ Model    <dbl> 3, 3
$ Classes  <dbl> 4, 8
```

```
$ LogLik    <dbl> -2128.142, -1969.723
$ AIC       <dbl> 4622.284, 4433.445
$ AWE       <dbl> 6798.430, 7371.367
$ BIC       <dbl> 5253.824, 5285.852
$ CAIC      <dbl> 5436.824, 5532.852
$ CLC       <dbl> 4258.218, 3941.336
$ KIC       <dbl> 4808.284, 4683.445
$ SABIC     <dbl> 4673.804, 4502.983
$ ICL       <dbl> -5262.106, -5309.282
$ Entropy   <dbl> 0.9671096, 0.9454112
$ prob_min  <dbl> 0.9761424, 0.8864891
$ prob_max  <dbl> 0.9972789, 0.9996584
$ n_min     <dbl> 0.11587983, 0.02575107
$ n_max     <dbl> 0.4806867, 0.3175966
$ BLRT_val  <dbl> 186.4271, 105.2969
$ BLRT_p    <dbl> 0.00990099, 0.00990099
$ Warnings  <chr> NA, NA
```

Using glimpse provides a vertical orientation of the output which is handy for viewing the data in the R console (if you have a large screen, running the command without `glimpse()` will provide a regular table). In the vertical orientation, the left number belongs to the four-profile solution. For AIC to ICL, lower values represent better fit.[5] Four indices favor the four-profile solution and four indices favor the eight-profile solution—a draw. Now, take a look at n_min next. n_min gives the size of the smallest profile in percent. With more profiles, the number of students assigned to each profile has to get smaller. For the eight-profile solution, the smallest profile only has 6 students and for the four-profile solution, the smallest profile has 27 students. The low number of students in the smallest profile of the eight-profile solution will be problematic for any follow-up analyses. Finally, prob_min is interesting to consider as it gives the lowest classification probability. The value should be as high as possible, otherwise you end up with classifications that are very uncertain themselves. Here, the four-profile solution is clearly superior to the eight-profile solution. In sum, the metrics favor the four-profile solution. In conclusion the four-profile solution will be in focus for the remainder of the chapter.

---

[5] For a more in depth discussion of the different fit parameters and also BLRT which we will not discuss here as there is no meaningful difference for this metric for the two solutions, we recommend Spurk et al. (2020).

11.4.2   Qualitative Pattern Interpretation

For the qualitative interpretation of the profiles, go back to the profile plot Fig. 11.12:

Profiles 1 and 4 are overall relatively similar. Values for control, value, curiosity, interested and joy are rather high, while the values for boredom, confusion, anxiety, and frustration are rather low. When it comes to the knowledge related variables, there is difference in degree favoring profile 4, especially for M.Radiant, M.Electric, and LP.T; profile 1 scores are mostly identical to those of profile 2 and 3. Thus one may characterize profile 1 as *overall engaged with average performance* and profile 4 as *overall engaged with high performance*. Profiles 2 and 3 show little differentiation regarding the knowledge related constructs between each other and in comparison to profile 1. What sets these profiles apart are the emotions. High values for boredom, and frustration set them apart from profiles 1 and 4; high anxiety further distinguishes profile 3 from 2. In consequence, profile 2 may be characterized as *frustrated and bored with average performance* and profile 3 as *anxiously frustrated and bored with average performance*.

11.4.3   Pattern Validation

For pattern validation, you will now again turn to an external criterion, specifically asking whether the profiles help to make sense of students' learning over the course of the unit as measured by the pre- and post-test. Based on our characterization of the profiles, it seems fair to assume that profile 1 learned the most, followed by profile 4. The rather negative affect of profiles 2 and 3 probably also influenced their learning; given the high anxiety for profile 3 probably more so for profile 3 than 2. The following code prepares a new object that has the profile assignment and the pre- and post-test data:

R Code and output: Preparing further analysis

```
# prepare further analysis
df.clust.emo_kn <-
cbind(df,emo.kn.lpa$model_3_class_4$dff$Class)
df.clust.emo_kn <-
df.clust.emo_kn %>%
rename( class = 'emo.kn.lpa$model_3_class_4$dff$Class')
```

Now, we run a regression model to see the influence of the profile on learning from pre- to post-test:

R Code and output: Regression model

```
df.clust.emo_kn %>%
lm(post ~ pre + as.factor(class), data = .) %>%
summary()
Call:
lm(formula = post ~ pre + as.factor(class), data = .)

Residuals:
     Min       1Q    Median       3Q      Max
-3.01950 -0.46777  0.08598  0.53326  1.65719

Coefficients:
                   Estimate Std. Error t value Pr(>|t|)
(Intercept)         0.11527    0.10146   1.136   0.2576
pre                 0.57791    0.06504   8.885 1.31e-15 ***
as.factor(class)2  -0.10708    0.22702  -0.472   0.6378
as.factor(class)3  -0.51328    0.20003  -2.566   0.0112 *
as.factor(class)4   0.18587    0.14901   1.247   0.2141
---
Signif. codes:  0 '***' 0.001 '**' 0.01 '*' 0.05 '.' 0.1 ' ' 1

Residual standard error: 0.822 on 159 degrees of freedom
  (69 observations deleted due to missingness)
Multiple R-squared:  0.3813,    Adjusted R-squared:  0.3657
F-statistic:  24.5 on 4 and 159 DF,  p-value: 8.3e-16
```

The output shows that (as can be expected) there is a strong relationship between prior knowledge (pre) and the post-test score. Further, there are effects for classes 2, 3, and 4. What about class 1? Class 1 is the reference class here (this is an automatic assignment happening with the lm() function). The estimated coefficients tell that classes 2 and 3 indeed learned less than class 1 (just as expected) and that class 4 learned more. You need to be careful with this interpretation however, because only the coefficient for class 3 is statistically significant. More data is needed to further substantiate these results.

However, keeping in mind the question of whether the profiles help to make sense of students' learning, one can also take a look at the R-squared. The pre-test and the four profiles together explain about 38% of the variance in the data. Compare that with the variance that the pre-test alone explains:

R Code and output: R-squared

```
df.clust.emo_kn %>% lm(post ~ pre, data = .) %>% summary()


Call:
lm(formula = post ~ pre, data = .)


Residuals:
     Min       1Q   Median       3Q      Max
-2.99270 -0.54185  0.06994  0.59617  1.86011


Coefficients:
            Estimate Std. Error t value Pr(>|t|)
(Intercept)  0.10621    0.07225   1.470    0.143
pre          0.60048    0.06615   9.077 3.66e-16 ***
---
Signif. codes:  0 '***' 0.001 '**' 0.01 '*' 0.05 '.' 0.1 ' ' 1


Residual standard error: 0.8429 on 162 degrees of freedom
  (69 observations deleted due to missingness)
Multiple R-squared:  0.3371,    Adjusted R-squared:  0.333
F-statistic: 82.39 on 1 and 162 DF,  p-value: 3.658e-16
```

The pre-test alone explains about 34% of the data. This means that the profiles help to explain 5 percentage points more of the variance—an increase of about 10%. You can even check whether this increase is statistically significant. To do so, store the results from the two regressions in respective objects and then run an ANOVA on them:

R Code and output: ANOVA

```
mod1 <- df.clust.emo_kn %>%
lm(post ~ as.factor(class) + pre, data = .)
mod0 <- df.clust.emo_kn %>% lm(post ~ pre, data = .)
anova(mod0, mod1)
Analysis of Variance Table


Model 1: post ~ pre
Model 2: post ~ as.factor(class) + pre
  Res.Df    RSS Df Sum of Sq      F  Pr(>F)
1    162 115.09
2    159 107.43  3    7.6671 3.7826 0.01175 *
---
```

```
Signif. codes:   0 '***' 0.001 '**' 0.01 '*' 0.05 '.' 0.1 ' ' 1
```

The output shows, that the increase in explained variance is statistically significant. In this way, the profiles do indeed help to better understand how students learn during the unit.

11.5   Summary

This chapter covered a lot of ground. You went from simple clustering using k-means to state of the art hierarchical clustering to latent variable models. You should now have a good idea of the landscape of different unsupervised ML techniques and know what questions to ask when someone presents you with such an analysis. How did they determine the number of classes? Are they accounting for uncertainty in measurement or the classification? Lastly, does the analysis add to the substantive understanding of the phenomenon?

11.6   Tasks

Comprehension

1. Explain the concept of unsupervised machine learning and how it differs from supervised machine learning.
2. Describe the difference between hierarchical and non-hierarchical clustering techniques, providing an example of each.
3. What research question in your field of study could be addressed using unsupervised machine learning? What are benefits and drawbacks of using unsupervised machine learning compared to the methods that are more commonly used?
4. Discuss the importance of dimensionality reduction in unsupervised machine learning.
5. Summarize the steps involved in the unsupervised machine learning workflow described in the chapter.

Application

1. Design a validation study to test the robustness of the clusters identified through unsupervised ML techniques. Describe the steps you would take and the criteria you would use.
2. Use k-means clustering to investigate the relationships between control, value, joy, and anxiety.

3. Use UMAP for dimensionality reduction on the provided dataset using all variables but total, control, and value. Explore what effect different settings of the hyperparameters have on the outcome.
4. Use HDBSCAN to look for clusters in the set of total, control, and value variables. Discuss the outcome with respect to the literature on emotions in education.
5. Develop a checklist of things you would like to see reported in a journal article that uses unsupervised machine learning so that you would feel comfortable to be able to reproduce the analysis.

References

Akogul, S., & Erisoglu, M. (2017). An approach for determining the number of clusters in a model-based cluster analysis. *Entropy, 19*(9), 452. https://doi.org/10.3390/e19090452

American Educational Research Association, American Psychological Association, & National Council on Measurement in Education. (2014). *Standards for educational and psychological testing*. American Educational Research Association.

Allaoui, M., Kherfi, M. L., & Cheriet, A. (2020). Considerably improving clustering algorithms using UMAP dimensionality reduction technique: A comparative study. In *International conference on image and signal processing* (pp. 317–325).

Campello, R. J., Moulavi, D., & Sander, J. (2013). Density-based clustering based on hierarchical density estimates. In J. Pei, V. S. Tseng, L. Cao, H. Motoda, & G. Xu (Eds.), *Advances in knowledge discovery and data mining*, (pp. 160–172). Berlin, Heidelberg: Springer.

D'Mello, S., & Graesser, A. (2012). Dynamics of affective states during complex learning. *Learning and Instruction, 22*(2), 145–157. https://doi.org/10.1016/j.learninstruc.2011.10.001

Hahsler, M., Piekenbrock, M., & Doran, D. (2019). dbscan: Fast density-based clustering with R. *Journal of Statistical Software, 91*(1). https://doi.org/10.18637/jss.v091.i01

Hong, W., Bernacki, M. L., & Perera, H. N. (2020). A latent profile analysis of undergraduates' achievement motivations and metacognitive behaviors, and their relations to achievement in science. *Journal of Educational Psychology*. https://doi.org/10.1037/edu0000445

Keller, M. M., Neumann, K., & Fischer, H. E. (2017). The impact of physics teachers' pedagogical content knowledge and motivation on students' achievement and interest: Physics teachers' knowledge and motivation. *Journal of Research in Science Teaching, 54*(5), 586–614. https://doi.org/10.1002/tea.21378

Magnusson, D. (2003). The person approach: Concepts, measurement models, and research strategy. *New Directions for Child and Adolescent Development, 2003*(101), 3–23. https://doi.org/10.1002/cd.79

McInnes, L., Healy, J., & Astels, S. (2017). hdbscan: Hierarchical density based clustering. *The Journal of Open Source Software, 2*(11), 205. https://doi.org/10.21105/joss.00205

McInnes, L., Healy, J., & Melville, J. (2020). *UMAP: Uniform manifold approximation and projection for dimension reduction*. http://arxiv.org/abs/1802.03426

National Research Council. (2012). *A framework for K-12 science education*. The National Academies Press. http://www.worldcat.org/oclc/794415367

Pekrun, R., Elliot, A. J., & Maier, M. A. (2006). Achievement goals and discrete achievement emotions: A theoretical model and prospective test. *Journal of Educational Psychology, 98*(3), 583–597. https://doi.org/10.1037/0022-0663.98.3.583

Pekrun, R., Lichtenfeld, S., Marsh, H. W., Murayama, K., & Goetz, T. (2017). Achievement emotions and academic performance: Longitudinal models of reciprocal effects. *Child Development, 88*(5), 1653–1670. https://doi.org/10.1111/cdev.12704

Pekrun, R., & Linnenbrink-Garcia, L. (Eds.). (2014). *International handbook of emotions in education*. Routledge, Taylor & Francis Group.

Rosenberg, J., Beymer, P., Anderson, D., van Lissa, C. j., & Schmidt, J. (2018). tidyLPA: An R package to easily carry out latent profile analysis (LPA) using open-source or commercial software. *Journal of Open Source Software, 3*(30), 978. https://doi.org/10.21105/joss.00978

Scrucca, L., Fop, M., Murphy, T. B., & Raftery, A. E. (2016). mclust 5: Clustering, classification and density estimation using Gaussian finite mixture models. *The R Journal, 8*(1), 289.

Spurk, D., Hirschi, A., Wang, M., Valero, D., & Kauffeld, S. (2020). Latent profile analysis: A review and "how to" guide of its application within vocational behavior research. *Journal of Vocational Behavior, 120*, 103445. https://doi.org/10.1016/j.jvb.2020.103445

Steegen, S., Tuerlinckx, F., Gelman, A., & Vanpaemel, W. (2016). Increasing transparency through a multiverse analysis. *Perspectives on Psychological Science, 11*(5), 702–712. https://doi.org/10.1177/17456916166586372

Yeager, D. S., Hanselman, P., Walton, G. M., Murray, J. S., Crosnoe, R., Muller, C., Tipton, E., Schneider, B., Hulleman, C. S., Hinojosa, C. P., Paunesku, D., Romero, C., Flint, K., Roberts, A., Trott, J., Iachan, R., Buontempo, J., Yang, S. M., Carvalho, C. M., & Dweck, C. S. (2019). A national experiment reveals where a growth mindset improves achievement. *Nature*. https://doi.org/10.1038/s41586-019-1466-y

Chapter 12
Automation and Explainability: Supervised Machine Learning with Text Data

Peter Wulff, Marcus Kubsch, and Christina Krist

Abstract In this chapter we revisit supervised ML and apply it to text data. We particularly utilize a LLM in the fine-tuning paradigm to showcase how these models can be used in science education research projects.

12.1 Supervised ML for Textual Data

12.1.1 Classifying Written Reflections of Science Teachers

In this case study we consider an example where NLP and supervised ML are utilized in science teacher education. It has been increasingly recognized that teaching practice in science teacher education programs is of great value to enable pre-service science teachers to apply their professional knowledge and reflect it. This helps them to develop action-oriented professional knowledge. To support the reflection of pre-service physics teachers teaching enactments, Anna Nowak and colleagues (Nowak et al., 2019) developed a reflection-supporting model (based on the ALACT reflection model by Korthagen et al. (1999)), which helps pre-service teachers to structure their written reflections. Among others, the authors differentiate several elements that constitute a full reflection. In this model, pre-service teachers are required to outline the circumstances of the lesson (e.g., goals), describe what happened during the lesson and evaluate specific situation. Finally, they devise alternative modes of action and consider consequences for their professional development. An instructors task could now be to apply this reflection-supporting model in practice, which requires her/him to read through the individual written reflections of the pre-service

P. Wulff (✉)
Heidelberg University of Education, Heidelberg, Baden-Württemberg, Germany
e-mail: peter.wulff@ph-heidelberg.de

M. Kubsch
Freie Universität Berlin, Berlin, Germany

C. Krist
Graduate School of Education, Stanford University, Stanford, CA, USA

© The Author(s) 2025
P. Wulff et al. (eds.), *Applying Machine Learning in Science Education Research*,
Springer Texts in Education, https://doi.org/10.1007/978-3-031-74227-9_12

teachers. This requires expertise, after all, the teachers make a lot of experiences that relate to various content-related and pedagogical issues such as students' preconceptions or classroom management. These substantive issues alongside the fact that even pre-service teachers teach a lot and instructors only have limited resources to assess the reflections and provide feedback. Also, the feedback situation in teacher training can be unbalanced as to power-differentials. Typically, instructors will also grade the teacher trainees, and the teacher trainees might be inhibited to disclose sensitive issues to their instructors, given that they know that they will grade and evaluate them. For example, teacher trainees might shy away from disclosing uncertainty about some content-related issues to not give the impression that they don't know about it.

As such, AI and supervised ML could be part of a solution to automate parts of the process to provide pre-service teacher guidance for their teaching and reflections. For example, if we had a supervised ML model that could classify teachers written reflections according to the reflection-supporting model we outlined, we could provide the students with some guidance on the structure of their reports (research found that written reflections tend to be overly descriptive and judgemental). As outlined in part I of this textbook and the preceding case studies, we will start with getting acquainted with the data.

12.1.2 Getting to Know Your Data

To train an ML model that can perform classifications according to a reflection-supporting model as outlined above (or any other model with distinct categories), we need data from pre-service teachers that actually reflected upon their or someones teaching. While some former research utilized rule-based approaches (e.g., defining specific words that are reflective), we seek to leverage the power of ML to detect patterns in complex data and utilize these patterns for classification problems. The data set[1] comprised 1091 segments. We segmented the data by sentences, which can be done with the spaCy module in Python, or by hand. As a first step, we split the data into train and test sets (cross-validation), and made sure to never touch the test data until the very last ML model was trained (which is sometimes difficult and most certainly an exciting moment in your project). The `train_test_split` function from the `scikit-learn` module in R and Python can be used. We did not stratify our data by students, which could be done and might be a wise decision in your project. After all, having sentences from the same student in train and test data might comprise unjustified data leakage and should be critically discussed (see Chap. 2). Stratification can also be done with the `train_test_split` function (for an application see Chap. 10). For illustrative purposes, we keep it simple and randomly split the entire

---

[1] The data set was collected in a research project, ARETE.KI, conducted by Marcus Kubsch, Stefan Sorge, and Peter Wulff.

data set into 872 sample sentences for training (and development) and 219 sample sentences for final testing.

Now we can start to explore our data. Just for you to get an impression of the kinds of sentences that are written and how they are classified by human raters, say a sample sentence reads: "The students take the materials and conduct the experiments." This would be classified as "description" according to the reflection-supporting model. A sentence that would be classified as evaluation could read: "The lesson overall went bad, because the students did not perform the experiment correctly."

Tokenization

Tokenization is an important procedure in NLP. It can be done on a word level (through `word_tokenizer` from the `nltk` library in R and Python), or through wordpiece tokenization. Wordpiece tokenization does not seek to split text into individual words, but rather into a finite number of wordpieces that can but must not be words. LLMs are typically restricted to a finite vocabularly (e.g., for BERT some 30,000 tokens) which enables the LLMs to partition any language input into recognizable units that do oftentimes not map directly to words, and has been found to improve LLM performance on some tasks. For example, the word "experiment" will be tokenized into: 'exper' + '##iment'.

To get an idea of the complexity of the data that we are dealing with we calculate the overall vocabulary size (i.e., a list of unique words) for the train data set, which is 2701. Vocabulary already provides you some idea on the complexity of your data (large number of unique words would correspond to greater variability in your data which might make classification more difficult). Another informative statistic is length of responses. Looking at the individual lengths of the sentences requires further wordpiece tokenization (see Box). We expect the wordpiece tokenized sentences to be longer compared to the word tokenized sentences, after all, wordpieces by definition mostly split words. This can be verified through the density plot (see Fig. 12.1). In this plot, the lengths and the respective proportions for both tokenizers are depicted.

All sentences were classified according to the reflection-supporting model by expert human raters. As well-established standard, one is here required to assure human interrater agreement, which can be calculated through Cohen's κ, if segmentation boundaries are clear (see Box). In our case, segmentation boundaries are sentences.

Segmentation

As we outlined in Chap. 7, language data typically comes unsegmented. However, in classification problems (or coding more generally) you typically aspire for a consistent coding unit, so that two raters get the same segments. How-

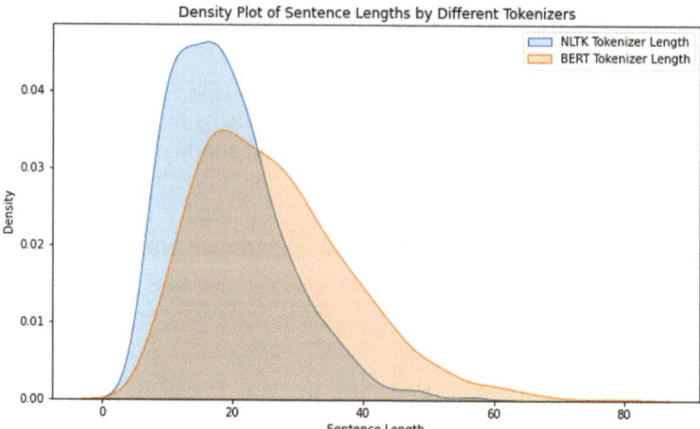

Fig. 12.1 Length of the train sentences as measured through word tokenization and wordpiece tokenizer

ever, this is not naturally the case and segmenting your data is quite an intricate problem. A more meaningful way to segment language data are elementary discourse units. Segmenting language data into elementary discourse units in itself is an intricate problem where human expertise is required and human interrater agreement should be assessed. Agreement metrics such as Cohen's κ (or Krippendorff's α) are of no help in this context. Other metrics such as γ have been proposed to estimate agreement on segmenting language data (or other unsegmented data) (Mathet et al., 2015).

Often, researchers in science education for convenience purposes use sentence segmentation. However, sentence boundaries, especially with dialogues, can be intricate to determine as well. In Chap. 7 we also showed how sentence segmentation can be performed with Python.

We should inspect the proportions of the sentences that are classified into each category (see Table 12.1). In fact, descriptions and evaluations are prevalent, which is not surprising, because one probably has to describe a situation at greater length in order to devise alternative modes of actions for it. Another important point to notice is the imbalance of samples in each category (unbalanced data set). This is quite common, rarely are categories in a data set balances and you might encounter it with your data as well (see some problems with unbalanced data, and strategies to address this problem in Box).

Table 12.1 Summary statistics for the reflection elements. The absolute count and relative proportion are displayed

Category	Count	Proportion
Circumstances	64	0.07
Description	407	0.47
Evaluation	263	0.30
Alternatives	99	0.11
Consequences	39	0.04

Dealing with unbalanced data

There are several techniques that can be applied to accommodate for this problem. After all, it is known in ML research that categories with more samples also typically are more accurately classified by the final ML model. ML is an inductive learning approach, and providing samples is crucial to enable the machine to learn. You could apply oversampling, where you balance the distribution of categories by replicating samples from the minority classes. You could draw samples randomly from the minority classes. You could also downsample the majority class. Generative LLMs might be of help to produce semantically similar sentences, rather than replicate existing sentences. Of course, due caution has to be given to bias and hallucinations that these generative LLMs might introduce. In any case, you should be aware that you artificially manipulate your training data and introduce novel assumptions that have to be critically examined alongside the model performance. You might cause overfitting, for example, when you draw multiple times the same sample and use it for training the ML model. You could also undersample (downsample) the majority class, however, then you constrain your valuable training data, which can be costly given that ML models typically perform better with larger sample sizes. Moreover, you might distort the underlying data distribution which can constrain the generalizability of the ML model. In this case study, we will leave the sample distribution in categories undistorted.

12.1.3 Applying Supervised ML to this Problem

After exploring your data, the next step is to decide on a supervised ML model in order to classify the data. We got to know many different classification algorithms (e.g., logistic regression or naive Bayes). In this case study, we will employ a more advanced NLP technique: Large Language Models (LMM). An accessible and widely used large language model is Bidirectional Encoder Representations for Transformers (BERT). Among the many ANN architectures, BERT is based on the so-called transformer architecture. Transformer models were found to particularly excel with NLP tasks, among others, because they rely on a self-attention mechanism that facilitates that for each word the relation to the other words is modelled. We won't bother with the details and accept that it works for certain tasks, given the performance on many NLP tasks, and the implementation of transformers in all sorts of models such as Generative Pretrained Transformers (GPT), in ChatGPT, etc. You can access the BERT model with the `transformers` library in Python as laid out in the Python code. This is a pretrained BERT model, which means that the model parameters were already trained through self-supervised training such that BERT provides meaningful contextualized embeddings (see Chap. 7).

OpenScience and LLMs

Given the intention of this book to encourage OpenScience (e.g., using OpenSource ML models on OpenSource software), we like to highlight that Python and R provide valuable resources to access transformer models on your own hardware and fine-tune (i.e., adjust model weights to be more suitable to your data) them to your data without any subscriptions and access fees, and without providing commercial, third-party companies your research data. Many of the models such as the BERT model in our present case study are by far not as capable as LLMs such as GPT-4, however, they are also mostly smaller and faster. On another note, you might consider that accessing ChatGPT each time you code written reflections billions of parameters are accessed and used for calculations. One request is estimated to average to 2.9 Wh of energy. This approximately equals the energy needed to lift a car (mass approx. 1 ton) one meter high! And this for only one average request. ChatGPT reached over 100 million users and of early 2023 had an estimated 195 million requests per day (de Vries, 2023). This all substantially taxes the environment, and should be critically reflected. Hence, training smaller models for specific purposes might be a reasonable way to solve this problem (of course, training costs a lot more energy - and these various goals have to be balanced).

Python Code: Loading pretrained BERT model

```
from transformers import BertTokenizer, BertModel
tokenizer =
    BertTokenizer.from_pretrained(
        'pretrained_models/bert-base-uncased')
model =
    BertModel.from_pretrained(
        'pretrained_models/bert-base-uncased')
```

For once, with only the pretrained BERT we can use it to represent our data. LLMs can output dense, contextualized embeddings (see Chap. 7), which can be visualized in 2D space. The embeddings can be accessed through the `forward()` function the `pytorch` models have, via `output.last_hidden_state`. These are the contextualized, dense, and high dimensional embeddings for our sample sentence. We can then use our knowledge on dimensionality reduction to visualize these embeddings in 2D. You could use functions such as PCA, UMAP, or t-SNE for the dimensionality reduction (see Chap. 5). We will actually perform similar analyses in the subsequent chapter and will jump to supervised classification here.

What the pretrained LLMs such as BERT are particularly capable for is fine-tuning and, with much larger LLMs such as GPT-4, few-shot and zero-shot learning, which means that they do not need specific training, but rather a prompt that exemplifies the task at hand to be performed once or zero times. Oftentimes you find different versions for certain LLMs. BERT "base" comprises 110 million parameters and 768 hidden layers, while BERT "large" uncased comprises 340 million parameters and 1024 hidden layers. Fine-tuning could be desirable for science education researchers, given that they for example developed a specific model in a research context and want the LLMs to be capable to code data according to this model. This is also conveniently done by using the `pytorch` library in R or Python. We will adhere to the implementation by Ostendorff et al. (2019) to wrap a simple ANN (e.g., FFNN) around BERT to essentially use BERT to represent the input data through dense, contextualized embeddings and use these as length-standardized inputs for the ANN that then can be trained to classify data according to the model of interest in the specific research context. The model can be initialized as depicted in the Python code.

Python code snippet: Implement LLM

```python
from torch import nn
import numpy as np

class BertMultiClassifier(nn.Module):
    def __init__(self, labels_count, hidden_dim=768, dropout=0.1):
        super().__init__()

        self.config = {
            'labels_count': labels_count,
            'hidden_dim': hidden_dim,
            'dropout': dropout,
        }

        # use BERT as pretrained LLM:
        self.bert =
            BertModel.from_pretrained("bert-base-german-cased")
        self.dropout = nn.Dropout(dropout)
        self.linear = nn.Linear(hidden_dim, labels_count)
        self.sigmoid = nn.Sigmoid()
```

```
def forward(self, encoded_input):
    pooled_output = self.bert(**encoded_input).pooler_output
    dropout_output = self.dropout(pooled_output)

    linear_output = self.linear(dropout_output)
    proba = self.sigmoid(linear_output)

    return proba
```

Again, we will not worry about the details here, but only say that in the middle the BERT model is implemented via `BertModel.from_pretrained("bert-base-german-cased")`, and embedded into a feed-forward neural net through `nn.Linear(hidden_dim, labels_count)`. Note that one can also access basically the same model via `BertForSequenceClassification` that does a similar job: a classification layer is added on top the encoder. In fact, you could add a decision tree or logistic regression classifier on top that you encountered in Chap. 4. As such, BERT can be trained with your specific data set. However, classification is only one among many NLP tasks for which LLMs could be utilized. A comprehensive overview and implementation details can be found here: https://huggingface.co/. As in other classification models, it is important to use cross-validation, monitoring of training and validation loss, and similar measures to keep track of the models' performance. A common flaw in ML research is overfitting and any measure to prevent it should be taken (see Chap. 2).

Our data set at hand is rather small. The strategy is now to utilize BERT to form contextualized embeddings for the sentences that can be used as inputs for a classifier, i.e., the feed-forward neural net with each category as an output node. Again, we had to specify hyperparameters such as the learning rate, the number of epochs, or the batch size input to the model. There are suggestions for these hyperparameters in prior studies. In your project, you might want to perform a grid search, i.e., systematically varying the hyperparameters and track performance on the validation data set. Remember that performance cannot be tested yet on the test data set, for otherwise our final model might overfit this test data set (we cannot stress this point enough, because otherwise your final model might be flawed and not perform as expected in practice).

The BERT model now runs multiple times through the training data set and we track the train loss, i.e., a measure of the discrepancy between actual prediction and gold standard, which mostly are human labels. A binary cross-entropy loss function was utilized as a convenient loss function for such a multi-class problem. Note that there are also multi-label classification problems. The difference is that in multi-class problems any sample x_i is assigned only one class, $y_i \in \{1...K\}$ (K is the number of classes). In a multi-label problem, each sample can be assigned multiple non-exclusive labels. Having finished the training (it might take some time, since even BERT is quite a large model) we now evaluate the model performance on the test data. A handy method is the `classification_report` function from the `scikit-learn` library. This yield the following output:

Python Output: Classification report

	precision	recall	f1-score	support
0	0.78	0.67	0.72	21
1	0.91	0.71	0.80	105
2	0.60	0.89	0.72	65
3	0.75	0.33	0.46	9
4	0.89	0.84	0.86	19
accuracy			0.76	219
macro avg	0.79	0.69	0.71	219
weighted avg	0.80	0.76	0.76	219

In this report you see the performance as measured through precision, recall, and f1-score (see Box in Chap. 10). Support refers to the number of samples in the test data set for each category (above) and overall (below). The numbers refer to the categories (alphabetically: alternatives, …). The lower three lines refer to aggregate performance. In unbalanced data problems, it is important to monitor the macro and weighted avg (average) for precision, recall, and f1-score, because the macro average treats all categories equally, and the weighted average weights the scores for respective support.

When interpreting this output, we first see that also the test data set is unbalanced (see last column, the support for each category). This also accounts for the discrepancy in macro and weighted average performances. Except category 4 (which would be circumstances), performance is linked to support. Most important are the last three rows, where we find the macro and weighted averages for precision, recall, and f1 score. BERT achieved a performance (macro F1) of .71, which translates into substantial agreement between computer and human rater. This value is dragged down by the only 9 samples of consequences and the low F1 value for it. We would now conclude to collect more data on consequences particularly to improve model performance for this category. Similar to our comparison with the addition example in Chap. 4 we also here employed a shallow learning ML algorithm to put performance of BERT into some perspective (you find the implementation details in the online supplement). A SVM classifier with the same data and no particular hyperparameter optimization reached a macro F1 of .52, which is substantially lower compared to the LLM. While this is no definitive proof that LLMs are always more capable in NLP task (in fact, sometimes simpler models are better), it is suggestive that LLMs are versatile and can provide valuable tools to try in your research problems.

BERT has been widely employed in science education research (Dood et al., 2022; Gombert et al., 2022; Winograd et al., 2021; Wulff et al., 2022). Our trained ML model can now be utilized in practice. Note that it is important to employ it in similar contexts in which it has been trained in. If learners of different linguistic background (e.g., younger age with different vocabulary) write in different style and with different words, this model might not perform well. This taps into the interesting question to what extent LLMs are good at extrapolation. To develop a better understanding of BERT's decision making, we will seek to utilize methods from explainable AI in the subsequent section. Before that, let us also say that another benefit of using this LLM approach over more shallow ML algorithms is that the BERT model can be conveniently further trained with novel samples and even in novel contexts (Wulff et al., 2023). You would use the trained model as the baseline model and further train it with data that you collected in a novel context.

12.1.4 Inspecting the LLM's Decisions

A key ingredient to trustworthy models and AI in general is explainability of model decisions (XAI, explainable AI).[2] LLMs are complex, deep learning-based ML models, and we already mentioned that explaining model decisions is quite tricky. It is not easily possible to reconstruct all the individual artificial neurons and establish an understanding what they are doing. Hence, we need an easier model that can be inspected. For example, one can examine the outputs of a complex model in depth and find similarities and differences in them. This could help to find features that are important for classification decisions.

When investigating explainability of ML models, researchers differentiate between local and global explanations. Local explanations would ask the question why a certain response of student A got the classification it got, and global explanations would ask why an entire set of students was attested with a certain preconception or similar. Both can indicate valuable information on why an ML model made a certain decision.

Integrated Gradients

A means to analyze local explanations is provided by integrated gradients. Integrated gradients work by calculating an attribution score for the importance of each input feature for a given output. This importance is measured against some (ideally uninformative) background (e.g., in image analysis an entirely black picture). Integrated gradients fulfill important properties of sensitivity and implementation invariance, i.e., they should attribute non-zero values to input features that are related to the output (compared to an uninformative baseline), and, if inputs and outputs of two differently implemented ANNs are different, attributions should be the same for the inputs. Say, you want to predict students' performance in an assessment based on their sense of belonging to science and their prior knowledge as input features. Assume that zero sense of belonging and zero prior knowledge would form the baseline. From this baseline, a stepwise procedure would then calculate an attribution score

---

[2] See also Chap. 2.

that indicates to what extent a certain value in sense of belonging or prior knowledge would predict performance.

We can illustrate this with our addition toy example from Chap. 4. Calculating integrated gradients in Python is enabled through the `captum` module. In this toy problem, a feed-forward neural network was trained to perform simple addition problems. Remember, a distributed representation for the inputs was used. For example, the number one was represented as the vector:

Distributed representation of number one

```
[0., 1., 1., 1., 0., 0., 0., 0., 0., 0., 0., 0.].
```

The input for a problem such as $1 + 1 = 2$ was then a 52 dimensional vector, i.e., the network has 52 input features and 36 output dimensions ($3 \times 12 = 36$). It was shown that simple addition could be learned perfectly, even for out-of-sample, unseen additions. We can now use integrated gradients to inspect the attribution scores for each of the 52 input features. We compare them to a baseline of zero as input. As such, all inputs that are zero will receive no attribution score per definition. We compare the trained model against an entirely untrained model to better understand what changes in the attributions (see Fig. 12.2).

While in the untrained model the attribution scores for the ones in the input are small and rather randomly scattered around zero, this changes for the trained model (see Fig. 12.2). The attribution scores become almost exclusively positive and large, indicating that these input features attribute to the actual output.

Integrated gradients can be used also for our text classification model. After all, text is oftentimes tokenized into a finite number of distinct tokens. Each token can then receive an attribution score for how much it contributed to the output, say a classification. In our above example, we categorized sentences into distinct categories (circumstances, description, evaluation, alternatives, and consequences). However, multi-class problems are difficult to implement.[3] To circumvent this issue you can simplify your model into a binary classifier. For example, you could simply retrain your model to predict if a sentence is an evaluation or not. To do this, you have to adjust your data into binary categories, adjust the `BertMultiClassifier` (i.e., set the labels indicator to 1), and re-train the model. The trained binary classification model even yields a better performance in terms of F1 score compared to the multi-class (5-way) classifier (see Python code). This is oftentimes the case: less categories to predict increase the model performance.

---

[3] See our conversation with the `captum` developers here: https://github.com/pytorch/captum/issues/355, last access 22 Nov 2023.

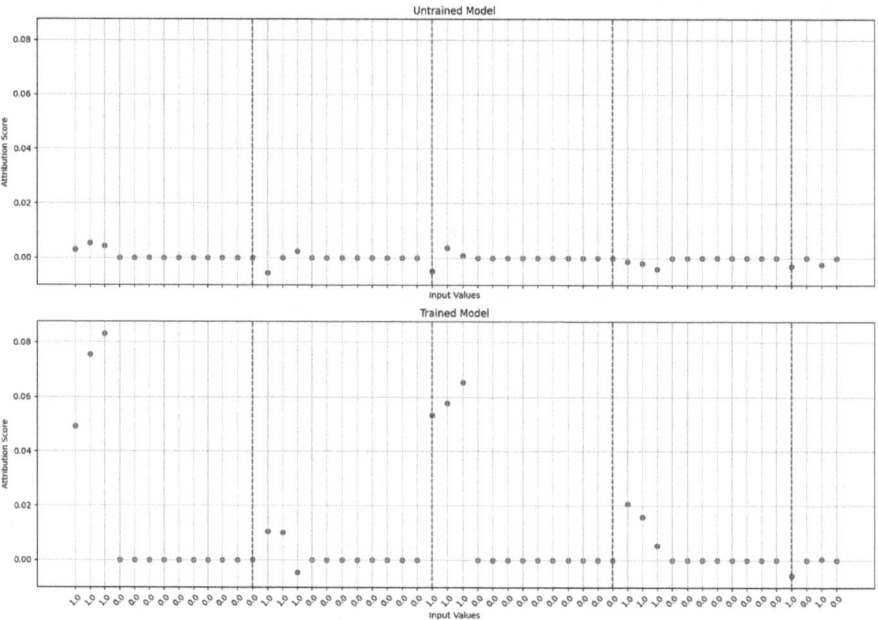

Fig. 12.2 Attribution scores for each input of the addition model for untrained (above) and trained (below) model. Red vertical lines indicate the positions for the numbers. Remember: the last four input features simply indicate that it is an addition problem

Classification report for binary classifier

	precision	recall	f1-score	support
0.0	0.90	0.86	0.88	154
1.0	0.70	0.77	0.74	65
accuracy			0.84	219
macro avg	0.80	0.82	0.81	219
weighted avg	0.84	0.84	0.84	219

Now that we have the binary classifier we can move on to calculate the integrated gradients. For complex LLMs such as BERT, the `captum` module provides `LayerIntegratedGradients`, which keep track of the gradients throughout the layers. As a first step, you initialize the the class `lig = LayerIntegratedGradien` `ts(...)`, and then apply it with the respective `input_ids` for a particular sentence (see Python code).

Calculate integrated gradients for BERT model

```
lig = LayerIntegratedGradients(
        evaluations_model,
        evaluations_model.bert.embeddings)

# Compute attributions using LayerIntegratedGradients
attributions_ig, delta =
        lig.attribute(
            inputs=input_ids,
            baselines=input_ids * 0,
            additional_forward_args=(attention_mask,),
            target=0,
            return_convergence_delta=True )
```

There is also a visualization class provided in the `captum` module: `VisualizationDataRecord`. You find more information on how to apply it in the online supplement. In Fig. 12.3 you see a visualization of sample sentences. Words that are highlighted in green contribute positively to the classification, whereas red highlighted words contribute negatively. As might be expected, evaluative words such as "good" or "bad" score positively for classifying a segment as evaluation. In fact, when calculating attributions for all tokens and averaging their attribution values we can create a list of the top 20 words that most positively contribute to classifying a segment as evaluation (see 12.1.4). Interestingly, negative words ("bad", "missing", "typical") score highest. This might relate to the issue that teachers tend to negatively judge teaching, which would be an important research topic given that positively and productively judging one's own and others teaching could contribute to a more positive image of the teaching profession.

Top 20 words with highest attribution values.

```
[('schwierig', 0.5768174216263907),
 ('keine', 0.5834361107807192),
 ('ansonsten', 0.5851991596999772),
 ('verloren', 0.5872418299226642),
 ('fast', 0.6489399595433871),
 ('kaum', 0.6511844824380291),
 ('nicht', 0.6672528365162796),
```

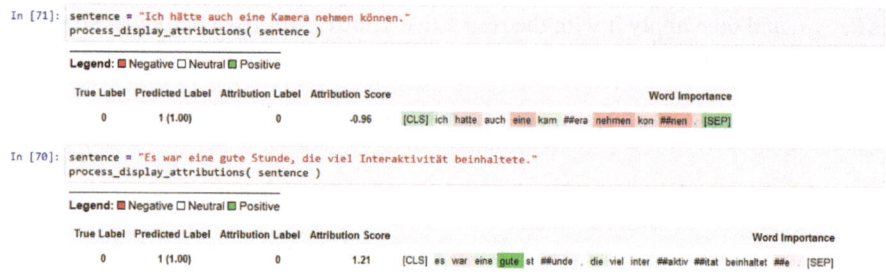

Fig. 12.3 Caption

```
('weniger', 0.668545931361491),
('positiv', 0.6770763150585619),
('erweist', 0.7085249840084905),
('will', 0.7228655488377653),
('erinnert', 0.749258007727709),
('schlechte', 0.7839921846245187),
('bereit', 0.8026147073352417),
('kein', 0.8189304757309708),
('geringer', 0.8370266931064063),
('mangelnde', 0.8586549366857567),
('fehlt', 0.9314228513614882),
('typisch', 0.9571113614605622),
('##spart', 0.9681103299410895)]
```

12.2 Summary

In this chapter, we applied supervised ML to text data: pre-service physics teachers' written reflections on teaching enactments. We showed how particularly LLMs could be leveraged to be fine-tuned to your particular data and achieve accurate classification performance. We also showed how you can make sense of model decisions, even though they come from rather complex LLMs. Unfortunately, the utilized inspection methods do not allow to derive a comprehensive understanding of the decision-making of the LLMs and they remain black boxes as of now.

12.3   Tasks

Comprehension

1. Explain the significance of supervised ML in classifying written reflections according to the reflection-supporting model.
2. What goals will you typically pursue with supervised ML, and what kinds of constructed-response items do you typically encounter in science education research?
3. What is your research focus, and what kind of model would be suitable for such a classification problem?
4. What is tokenization and why is it important in NLP?
5. *What are integrated gradients, and how and why are they used in the context of this chapter?

Application

1. Based on the Python code provided for training the BERT model, extend the evaluation code for using the pretrained reflection-supporting model in a function (`def function(...)`)) where `input_text` can be provided as an input parameter and the resulting category is predicted. (Remember, the text has to be in German, given that the model was trained in German. You could also use Google translate in Python the translate the language input into German as a possible extension.)
2. Implement a Python function using the BERT tokenizer to tokenize and print out a sample sentence.
3. *Create a complete script that trains a BERT-based classifier on a data set of science teachers' reflections (or a data set of your choosing) and evaluates its performance.
4. Propose a strategy to deal with unbalanced data in the context of classifying written reflections.
5. *Write Python code to visualize the integrated gradients for a specific input sentence classified by a BERT model.

12.4   Solutions

Comprehension

1. The outlined reflection-supporting model allows researchers to annotate written reflections, based on substantive theory on experiential learning and professional actions. The annotated data can then be used as training samples to train a classifier that automatically annotates this data. These annotations can spare

resources and enable instantaneous feedback for pre-service teachers. The outlined reflection-supporting model is rather generic as to be applicable across different contexts. More domain-specific models would require annotation/coding of content-related themes. For this task computational grounded theory and unsupervised ML approaches can be helpful, as outlined in Chaps. 13 and 15.

2. Typical goal include classification and regression. Input-output mappings could be language data (transformed into numerical representation) to categories (multilabel or multi-class classification), or numerical data with scores (regression). Typical constructed-response item formats would be essays, or short-text responses.

3. If your research focus were argumentation processes in science education, you could employ a shallow ML model such as random forest classifier to automatically and reliably assign categories to sentences (claim, evidence, or reasoning). Of course, segmentation plays a crucial role here, and also cross-segment correlations would be important to capture. More involved models would include LLMs such as BERT to improve classification accuracy or even generative LLMs to automate annotation.

4. Tokenization refers to the process of transforming a text string into elementary units that form a meaningful representation of the text.

5. Integrated Gradients are one way to help understand decision-making processes in ANNs. They can even be used with LLMs (which often include transformer-based ANNs). Integrated Gradients assign the input tokens/features a score which attributes to the importance of this input for the output. In the reflection example above we showed that tokens with high scores were also considered by us as important for certain categories in the reflection-supporting model.

Application

Please see online notebook.

References

de Vries, A. (2023). The growing energy footprint of artificial intelligence. *Joule, 7*(10), 2191–2194.

Dood, A., Winograd, B., Finkenstaedt-Quinn, S., Gere, A., & Shultz, G. (2022). Peerbert: Automated characterization of peer review comments across courses: Lak22, March 21–25, 2022, online, USA (pp. 492–499).

Gombert, S., Di Mitri, D., Karademir, O., Kubsch, M., Kolbe, H., Tautz, S., Grimm, A., Bohm, I., Neumann, K., & Drachsler, H. (2022). Coding energy knowledge in constructed responses with explainable nlp models. *Journal of Computer Assisted Learning*.

Korthagen, F. A., & Kessels, J. P. A. M. (1999). Linking theory and practice: Changing the pedagogy of teacher education. *Educational Researcher, 28*(4), 4–17.

Mathet, Y., Widlöcher, A., & Métivier, J.-P. (2015). The unified and holistic method gamma for inter-annotator agreement measure and alignment. *Computational Linguistics, 41*(3), 437–479.

Nowak, A., Kempin, M., Kulgemeyer, C., & Borowski, A. (2019). Reflexion von physikunterricht [reflection of physics lessons]. In C. Maurer (Ed.), *Naturwissenschaftliche Bildung als Grundlage für berufliche und gesellschaftliche Teilhabe: Jahrestagung in Kiel 2018* (p. 838). Regensburg: Gesellschaft für Didaktik der Chemie und Physik.

Ostendorff, M., Bourgonje, P., Berger, M., Moreno-Schneider, J., Rehm, G., & Gipp, B. (2019). Enriching bert with knowledge graph embeddings for document classification. arXiv:1909.08402v1.

Winograd, B. A., Dood, A. J., Moeller, R., Moon, A., Gere, A., & Shultz, G. (2021). Detecting high orders of cognitive complexity in students' reasoning in argumentative writing about ocean acidification: Lak21, April 12–16, 2021, Irvine, CA, USA (pp. 586–591).

Wulff, P., Buschhüter, D., Westphal, A., Mientus, L., Nowak, A., & Borowski, A. (2022). Bridging the gap between qualitative and quantitative assessment in science education research with machine learning — a case for pretrained language models-based clustering. *Journal of Science Education and Technology, 31*, 490–513.

Wulff, P., Westphal, A., Mientus, L., Nowak, A., & Borowski, A. (2023). Enhancing writing analytics in science education research with machine learning and natural language processing—Formative assessment of science and non-science preservice teachers' written reflections. *Frontiers in Education, 7*, 1–18. https://doi.org/10.3389/feduc.2022.1061461

Chapter 13
Unsupervised ML with Language Data

Peter Wulff, Marcus Kubsch, and Christina Krist

Abstract This chapter introduces a case study of how to apply unsupervised ML to science education-related language data. We will start with dimensionality reduction and hierarchical-agglomerative clustering. Later on, we will extend these analyses using LLMs and more involved clustering techniques.

13.1 Collecting Unstructured Language Data in a Science Education Research Context

As previously indicated, unsupervised ML typically seeks to find patterns in unlabelled data. As we also stressed, most data today must be considered unlabelled. In fact, it would be of great value to have algorithms that can extract meaningful patterns in these data sets or compress the data in ways that keeps as much meaningful information as possible. Unsupervised ML algorithms are well versed to do so. As with supervised ML algorithms, there is a vast variety of different unsupervised ML algorithms and the researcher has to decide for herself, which algorithm suits the problem at hand best. Unsupervised ML algorithms can be applied to all kinds of numerical data. However, particularly with images or language data, the researcher has to perform more involved pre-processing in order to transform the data at hand into a numerical representation.

Outline of the data set

Let's dive into a concrete example of a research context where unsupervised ML might come handy. In a collaborative project, we collected students' written

P. Wulff (✉)
Heidelberg University of Education, Heidelberg, Baden-Württemberg, Germany
e-mail: peter.wulff@ph-heidelberg.de

M. Kubsch
Freie Universität Berlin, Berlin, Germany

C. Krist
Graduate School of Education, Stanford University, Stanford, CA, USA

© The Author(s) 2025
P. Wulff et al. (eds.), *Applying Machine Learning in Science Education Research*,
Springer Texts in Education, https://doi.org/10.1007/978-3-031-74227-9_13

267

reflections on two short physics videos that showed a teacher inciting students to think about energy and force. The students were instructed to reflect upon this video, i.e., they wrote on what the observed situation was about, how they evaluate the situation, what alternative actions they would imagine, and what personal consequences they might draw based on the video for their own professional development (based on original research by Nowak et al. (2019)). Our research goal was to identify themes that the students' addressed in their written reflections. Data from 19 teachers was collected, with an average of 23 (SD=6) sentences, with an average of 501 (SD=119) words. Overall, we received some 441 sentences, which are all different except for one. As a matter of fact, one can see how complex language data is. Only 19 teachers produced a variety of sentences that are connected in some ways, however, different in others. For this reasonably-sized context, we could engage student workers and begin inductively (we have no specific hypotheses on what the teachers notice) to code the data through content analysis. However, given that we could have easily received data from 200 students, then content analysis becomes increasingly infeasible. After all, computers are versatile tools to systematically apply rules to large data sets and group this data. In particular, unsupervised ML offers a variety of methods for purposes such as dimensionality reduction and clustering.

Pre-processing the data

Let us start with transforming our data into a numerical representation. As a first step, we load the data set, i.e., the written reflections. Also, we load the `spaCy` library as we introduced it as a good choice for many NLP-related tasks. We can also calculate the overall vocabulary size (unique words), which is 833. Having loaded the relevant NLP models, we can then simplify our data, as, for example, punctuation or highly redundant words (so-called stopwords) could be discarded for further analyses. Please note that this might not be true for other research questions, given that stopwords such as "and" can have discourse-related functions. Moreover, we use the lemma of a word (asking → ask), as this reduces this vocabulary size and also might improve clustering results, as similar terms are more easily recognized by the clustering algorithm. All this is performed with the `spaCy` module in Python (see accompanying notebook).

Now we have a list of sentences (stored in the `texts_structured`) that we can use further. Our goal now is to group together sentences with a similar meaning/common theme or topic. This is trickier than one might expect. For example, the representation of the data plays a crucial role on what features the algorithm can pick up. The arguably simplest representation is a one-hot encoded vector for each sentence (see Chap. 3), indicating which words in the vocabulary are present in the sentence and which are not, which can be done with the help of the `scikit-learn` (`sklearn`) library in Python as outlined in the Python code.

Pyhton code: create term-document matrix

```
# restructure dataset into Sent[ words ]
sent_word_list = [ ' '.join([ w for w in s ])
                    for t in texts_structured for s in t ]

from sklearn.metrics.pairwise import cosine_similarity
from sklearn.feature_extraction.text import TfidfVectorizer,
                                            CountVectorizer

cv = CountVectorizer()
raw_count_df = cv.fit_transform( sent_word_list ).toarray()
```

The term-document matrix (Table 13.1) is a sparse matrix, i.e., many cells are zero. This is a rather inefficient way of storing information, however, it enables us to utilize it for clustering. Before clustering, a meaningful step here would be to find some form of dense representation for the sentences. Researchers used for similar applications (actually the co-occurrence matrix of words in a corpus) PCA to find robust, dense representations for words. We will start here by using the familiar SVD (that also powers PCA, see also Chap. 5) to compress the sparse term-document matrix into a much smaller matrix with 50 dimensions (it can be verified that 50 dimensions already capture most of the variation in this data set). From applying SVD we now received a reduced matrix (dimension 441 x 50). Essentially, we went down from 833 columns to "only" 50 columns at the expense of interpretability. It is often asked what these dimensions refer to, or what they mean. Quite unfortunately, the 50 columns do not refer to any specific word in the vocabulary, but rather are rotated axes that account for variance in the data. This is, by the way, also why many researchers consider SVD and PCA as a learning approach, because new axes are learned to represent the complex data. However, we can further reduce the matrix to only 2 dimensions, which is a common procedure to visually inspect the data in a scatterplot (which, in this case, does not yield much valuable information).

Table 13.1 Excerpt of term-document matrix. Rows refer to sentences, columns to terms in the vocabulary

	Ability	Able	Accompaniment	Accompany	Accord	Accordingly	Accurately	Achieve
0	0	0	0	0	0	0	0	0
1	0	0	0	0	0	0	0	0
2	0	0	0	0	0	0	0	0

13.2   Finding Clusters in the Data

We will first apply (see Python code) a rather simple (bottom-up) approach, namely hierarchical agglomerative clustering using Ward distance that (as many other of these approaches) proceeds as follows:

1. Initially, imagine there to be a forest including all data points (i.e., sentences) from the above reduced representations in which each sentence forms its own cluster.
2. One can then calculate a distance matrix (oftentimes Euclidean distance), where distances between any two sentences are calculated.
3. Most similar clusters are then merged (agglomerated) and updated distances are calculated, and clusters are merged according to their similarity again.
4. This algorithm proceeds until all clusters are merged into only one large cluster.

Python code: apply SVD to term-document matrix and clustering

```python
from sklearn.cluster import AgglomerativeClustering
from sklearn.decomposition import TruncatedSVD
import scipy.cluster.hierarchy as shc
import matplotlib.pyplot as plt

n_components = 50  # The number of dimensions you want
svd = TruncatedSVD(n_components, random_state=42)
X_reduced = svd.fit_transform(raw_count_df)

(...)

clustering_model = AgglomerativeClustering(
        n_clusters=num_clusters,
        metric='euclidean',
        linkage='ward')
clustering_model.fit(X_reduced)

(...)

plt.figure(figsize=(10, 7))
plt.title("Dendrogram")

clusters = shc.linkage(X_reduced,
            method='ward',
            metric="euclidean")
shc.dendrogram(Z=clusters)
plt.show()
```

This algorithm can be implemented in Python as outlined in the Python code, and results can be visualized in a stepwise dendrogram as equally outlined in this code chunk. The resulting dendrogram (see Fig. 13.1) indicates that there are groups of sentences that are closer related (more similar) than others. It is now the task of the researcher to use this dendrogram as well as representative documents, and interpret the clusters, e.g., by comparing them with each other. A horizontal line can be drawn through the dendrogram as a cutoff for clusters. The belonging leaves are included sentences within this cluster. We can then determine which words are used most often in the respective clusters. This will give us an idea of what the clusters actually are about. Some clusters are rather small with only one or three sentences in it.

To choose a number of topics, an approach as performed by Sherin (2013) can be utilized. The frequency distribution of topics for each different cut of topics is depicted. We can see in the beginning that there is a cluster with only 3 sentences (see accompanying notebook), which remains quite stable throughout the algorithm. Moreover, we can see that at certain points, large clusters disintegrate into smaller ones at certain points. We consider a number of 10 clusters (which yields the following counts of documents in topics: 47, 23, 86, 20, 56, 33, 71, 74, 28, and 3) as a reasonable first choice to do some interpretations and see what patterns can be found in the data. In particular, a number of 9 clusters had one large cluster of 107 documents, which breaks down into smaller topics in the following steps.

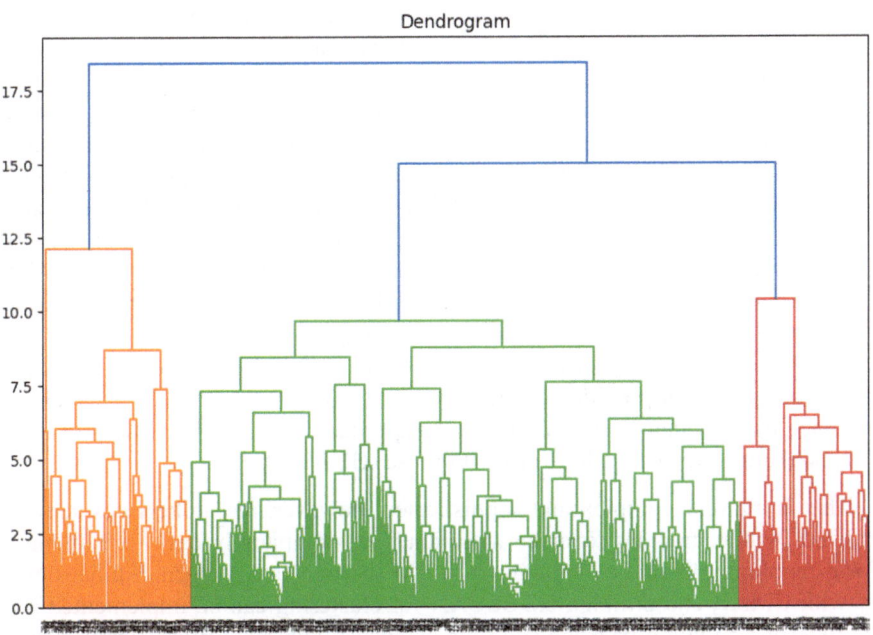

Fig. 13.1 Stepwise dendrogram of hierarchical agglomerative clustering approach

Finally, it is the task of the researcher to make sense of the resulting clusters. Interpretability also hinges on the preprocessing of the data. For example, if a bag-of-words assumption was introduced as in our case, word ordering effects play no role in determining the clusters which should be considered. Typically, most representative words for clusters and most representative documents (here: sentences) are displayed (Table 13.2). For example, cluster 0 with 47 documents seems to relate to more general statements with representative words such as: 'teacher', 'student', 'answer', 'research', and 'question'. Other clusters such as cluster 6 with 43 documents seem more content-related, with representative words such as: 'energy', 'conversion', 'student', 'lk', and 'concept'.

13.3 Utilizing LLMs and Advanced Clustering

The above mentioned hierarchical agglomerative clustering approach is comparably simple as related to analysis of language data. As such, there are many aspects of the analysis that we can further advance:

1. First, we saw that the representation of sentences in the term-document matrix is sparse and arguably rather poor. After all, the term-document matrix eliminates all information on word order (bag-of-words assumption). However, word order is certainly important when more nuances in scientific reasoning should be analyzed ("current causes voltage", and "voltage causes current" are semantically different).
2. Second, dimensionality reduction can be improved. Local and global structure in complex data sets are important (see also Chap. 3). Dimensionality reduction algorithms typically preserve either one of those two. For example, SVD and PCA can be categorized into the former, while t-SNE can be categorized into the latter. Algorithms such as Uniform Manifold Approximation and Projection (UMAP) can preserve (to some extent) both. UMAP preserves local structure like t-SNE while better keeping also global structure. It is thus a versatile algorithm to reduce complex data with the goal to preserve meaningful information.
3. Third, clustering methods can be improved as well. For example, density-based approaches might provide beneficial results for very complex data, such as language data. A density-based clustering approach is hierarchical density-based spatial clustering of applications with noise (HDBSCAN).

This is all very complex and sophisticated, however, researchers devised specialized Python modules to make these analyses comparably easy to implement. For example, a pipeline of applying an LLM to your data, reducing dimensionality with UMAP, and utilizing HDBSCAN for clustering can be implemented with the BERTopic Python package (find the extensive code with more elaboration on interpretation in the notebooks):

Table 13.2 Top 5 words, top 5 words with tf-idf scaling, and representative sample sentences for the extracted topics.

No. (N)	Top words (count)	Top words (c-tf-idf)	Sample sentences
0 (47)	Question, research, answer, student, teacher	Yes, xy, explain, expert, experimentation	n observed teaching unit question collect question parking lot clarify; observed classroom situation different guide question relate unit energy collect drive question board answer
1 (47)	Student, subject, experiment, knowledge, important	Activity, continually, right, engage, arguing	Possible visible student sit row high motivation average; student actively involve classroom activity
2 (13)	Term, energy, teacher, student, use	Colloquial, hint, undiscovered, remain, notion	Misconception false notion term remain undiscovered; inertia hint colloquial term mention
3 (74)	Question, student, answer, teacher, ask	Twice, classmate, single, improvement, exact	Room apparently specialized classroom equip smartboard; work classmate ask improvement
4 (23)	Object, question, motion, set, fast	Rise, gravitational, fast, call, object	Focus body set motion fast height rise; question high object rise
5 (65)	Class, teacher, video, evidence, student	Sparingly, sufficiently, negatively, overwhelming, smile	Think sparingly; evident smile way speak
6 (43)	Energy, conversion, student, lk, concept	Typical, stack, analogy, bring, distinction	Clarify distinction concept motion force energy highlight connection; energy conversion stack magnet discuss
7 (3)	Energy, kinetic, conversion, potential, motion	Velocity, equate, engagement, internalize, talk	Student apparently internalize engagement energy equate high kinetic energy high velocity teacher let explain cause motion conversion energy kinetic energy; teacher talk conversion energy general make student aware phenomenon potential energy kinetic energy magnetic energy kinetic energy actually phenomenon conversion energy
8 (19)	Energy, kinetic, form, student, convert	Elastic, thermal, yes, expert, string	Require discuss additional form energy elastic energy thermal energy; form energy conversion process address overall concept energy closed nature system overlook
9 (33)	Teacher, student, explanation, contribution, lesson	Table, nice, motivated, camera, willing	Teacher willing deviate plan lesson extent; teacher work table document camera
10 (18)	Correct, student, teacher, explanation, answer	Interject, expand, incomplete, imprecise, attempt	Teacher briefly interject correct word; opinion teacher correct change expand student response bit

(continued)

Table 13.2 (continued)

No. (N)	Top words (count)	Top words (c-tf-idf)	Sample sentences
11 (21)	Lesson, student, structure, energy, unit	Definitely, consideration, apart, planning, care	Reason slip tongue possibly superficial lesson preparation; need spontaneity improvisation possibly depend course multiple plan lesson
12 (1)	Energy, statement, student, physical, model	Storage, finding, appropriately, practice, solution	Teacher practice physical reasoning appropriately student take student statement provide formulation model solution supplement statement necessary technical term finding physical phenomenon energy storage distance potential energy magnetic energy etc
13 (24)	Explanation, student, physical, question, scientific	Name, integration, induce, equipment, recap	Integration experimental equipment recap support explanation; risk student develop inadequate understanding scientific explanation
14 (10)	Student, answer, teacher, energy, question	Accurately, significantly, guess, exclusively, rarely	Student answer precisely teacher summarize supplement answer accurately; classroom discussion result fully derive student teacher hardly anticipate content serve support student promote independent scientific explanation

Python code: Apply BERTopic

```python
from bertopic import BERTopic
from umap import UMAP

umap_model = UMAP(n_neighbors=9,
                  n_components=5,
                  min_dist=0.0,
                  metric='cosine')

topic_model = BERTopic(
            language="english",
            top_n_words=3,
            min_topic_size=4,
            embedding_model="all-MiniLM-L6-v2",
            umap_model=umap_model)

topic_model.visualize_documents(docs,
             reduced_embeddings=reduced_embeddings)
```

It can be verified that well interpretable topics can be identified with BERTopic (see Fig. 13.2, more visualizations and reasoning can be found in the accompanying notebook) without too much effort such as tuning hyperparameters. Physics-related topics separate from other, more general, topics (see Fig. 13.2). Moreover, similar sentences are grouped into the similar clusters. You then have many options with BERTopic to inspect the clusters. In Fig. 13.3 you can see association strengths of tokens in the sentence "The students note that there must always be another form of energy present for kinetic energy to be created" with all relevant topics that are partly existent in this sentence. You see that the token "energy" has particularly high association with different, energy-related topics. However, in this sentence the forms of energy are most central. Moreover, you can visualize the genesis of clusters (see Fig. 13.4). Note, for example, the joint origin of the "energy" clusters, topics 29, 10, 2, and 14.

The BERTopics module thus offers you valuable tools to better understand clustering decisions of these complex models. It is also important to remember that these models oftentimes are not sensitive to science education specific contexts. We therefore cannot expect the models to make meaningful classifications with respect to science education contents, but rather based on usage of terms in public parlance and co-occurrence statistics in the present texts.

Documents and Topics

Fig. 13.2 Parts of the scatterplot of the first two dimensions of BERTopic's embedded documents. Different colors refer to different topics, indicated also by the numbers

	The	students	note	that	there	must	always	be	another	form	of	energy	present	For	kinetic	energy	to	be	created
2_pendulum_energy_magnet								0.111	0.222	0.339	0.476	0.503	0.600	0.698	0.561	0.423	0.216		
5_structuring_would_students									0.101	0.301	0.458	0.616	0.617	0.417	0.364	0.207	0.104	0.104	
10_energy_system_kinetic								0.100	0.201	0.344	0.564	0.684	0.914	1.124	1.009	0.789	0.459		0.105
14_lk_energy_concludes			0.187	0.364	0.543	0.766	0.713	0.697	0.636	0.524	0.578	0.606	0.765	0.964	0.776	0.587	0.309		
18_have_everything_formulate															0.111	0.111	0.111		0.111
19_form_must_another	0.159	0.348	0.658	1.049	1.295	1.600	1.700	1.679	1.554	1.399	1.419	1.741	1.643	1.412	1.259	1.070	0.689	0.440	0.193
21_concept_about_had												0.101	0.101	0.101	0.101				
29_when_energy_force														0.101	0.101	0.101	0.101		
31_experiments_topic_on														0.143	0.310	0.310	0.310	0.167	

Fig. 13.3 Association strengths of tokens in a given sentence (top) and the respective Topics (left)

13.4 Applying Unsupervised ML to Cluster Language Data

In this chapter, we applied a shallow and deep-learning based clustering technique to unlabelled text data, which counts as unsupervised ML. Unsupervised ML for language data can reveal clusters of sentences that relate to each other. The relation is then defined by the feature selection process. For example, when one-hot encoding is employed to encode the language data, unsupervised ML algorithms will determine topics (i.e., clusters of sentences or words) based on the similarity in the words used in the sentences. A novel approach to encode language data is by utilizing LLM-based contextualized, dense embeddings. LLM-based embeddings can often account better for context and synonyms. Furthermore, using non-linear dimensionality reduction and clustering approaches such as UMAP and HDBSCAN can help to partly preserve local and global structure of your data. In our applied example, we saw that the BERTopic Python library yielded well interpretable clusters without any further necessity to fine-tune the LLM or the hyperparameters. LLM, UMAP,

Hierarchical Clustering

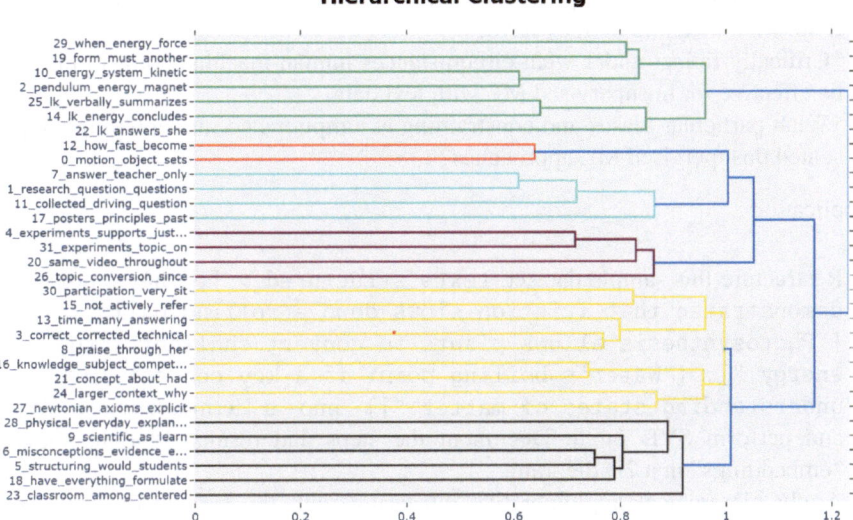

Fig. 13.4 Genesis of topics in the present clustering approach

and HDBSCAN have been utilized in conjunction to cluster language data in science education research. However, to get a full understanding for clustering decisions we are still lacking appropriate means in order to increase transparency.

13.5 Tasks

(Note: Some of the tasks and solutions were generated with ChatGPT, by prompting the LLM with the entire chapter's text and the instruction to generate comprehension and application tasks; difficult tasks are marked by *)

Comprehension

1. Summarize the process described in the text for transforming written reflections into a numerical representation.
2. Explain the significance of using SVD (Singular Value Decomposition) in reducing the dimensions of the term-document matrix.
3. Describe the hierarchical agglomerative clustering approach mentioned in the text. What steps are involved in this method?
4. *What are advantages of using LLM-based embeddings for clustering language data compared to traditional one-hot encoding?
5. Interpret the importance of visualizing clusters using dendrograms as described in the text. How does this visualization aid researchers?

6. Do the numbers of the topics in the agglomerative clustering have any meaning? If so, what meaning do they have?
7. *Critically reflect under what circumstances human-machine collaboration can be effective for unsupervised ML with text data.
8. *What particular biases and constraining assumptions go into applying the presented unsupervised ML approaches?

Application

1. Restructure this sample dataset: `texts_structured = [["The experiment demonstrated that friction slows down a rolling ball."], ["Photosynthesis allows plants to convert sunlight into energy."], ["Water's boiling point is a key concept in understanding states of matter."]]` into a term-document matrix and perform SVD on it. Document the steps and results. Plot the reduced "embeddings" in a 2D diagram.
2. Apply hierarchical agglomerative clustering with the sample dataset: "biology_wikipedia_articles_sample.csv", found in Google Drive folder for chapter 13, /data/Wikipedia_science_data/. Visualize the resulting dendrogram and give some reasoning for how you interpret one or more clusters formed.
3. *Using the `BERTopic` Python package, cluster a collection of text documents (data set: "biology_wikipedia_articles_sample.csv", found in Google Drive folder for chapter 13, /data/Wikipedia_science_data/). Provide a detailed analysis of the clusters identified, including representative words and sample sentences.
4. *Compare the outputs of the previous two tasks (using traditional methods versus advanced methods, e.g., `BERTopic`). Discuss the differences in cluster quality and interpretability.
5. Develop a simple NLP pipeline in Python to preprocess text data (remove stopwords, lemmatize, and vectorize). Use this pipeline on a dataset of your choice and explain the preprocessing steps and their significance.

13.6 Solutions

Comprehension

1. Transforming written reflections into a numerical representation can be performed in multiple ways. A common way is to use one-hot encoding, where each word is represented as a column in a matrix (term-document matrix) and the rows refer to documents, which can be sentences, responses, or other meaningful structures.
2. SVD calculates dimensions (eigenvectors) of maximum variance in a data set. This is a commonly used technique to reduce the dimensionality of the original data. Oftentimes, such dimensionality reduction can be sensibly used in ML analyses

to retrieve essential features in a data set (feature extraction). These features can then be used as compressed, new features in the further ML pipeline.

3. The hierarchical agglomerative clustering starts with calculating each pairwise distances (e.g., Euclidean distance or Manhatten distance) between the data points. Closed clusters (each data point is a cluster at first) are merged and the distance matrix is calculated once again. These steps proceed until eventually only one cluster exists. The resulting distances and merges can be represented visually in a dendrogram.

4. LLM-based embeddings, also called contextualized embeddings, capture the context in which a word is used. I.e., the embedding vector depends on the context. This is advantageous, given that word meaning is a function of context (think of using "force" in the context "may the force be with you," "they forced them to do this," or "the gravitational force pulls objects.").

5. A dendrogram provides you some insight into what documents evolved from a similar branch, or when certain clusters form. These information can then be utilized to make sense of the resulting clusters or even determine a sensible number of clusters.

6. The agglomerative (hierarchical) clustering will by design produce as many clusters as there are data points. Hence, the researcher has to critically examine what the clusters actually mean. This can be done in multiple ways, many of which are presented in this textbook. First, determining the number of clusters is challenging and hinges on assumptions such as what the individual dimensions refer to. Agglomerative (hierarchical) clustering does not produce a number which determines the number of clusters. One meaningful way to determine a number of clusters was through observing the distribution of clusters. In many research contexts it is probably not meaningful to have clusters with only one document in them. Moreover, a cluster with nearly all documents in it is also not meaningful. Hence, a good balance of clusters would be desirable. Moreover, interpreting the clusters is a next challenging task where the human researcher's substantive domain knowledge is required. The researcher could concurrently analyze the documents that belong to one particular cluster, e.g., by comparing frequent words in the sentences.

7. Human-machine collaboration is a crucial part in any unsupervised ML project. ML and NLP techniques might help you find patterns, however, deciding on the algorithms, the data format and representation, and interpretations are key tasks for human researchers that cannot be outsourced.

- Hierarchical agglomerative clustering: In this approach, we represented the data through a term-document matrix and used SVD do compress the data. We already mentioned that one-hot encoding cancels information on word ordering which is disadvantageous when nuanced analysis of language data is required. Moreover, SVD is a linear dimensionality reduction procedure and if your data set is non-linear, this might be inappropriate.
- BERTopics: This LLM-based approach hinges on the quality of the pre-trained LLM that is utilized, as well as on the dimensionality reduction procedure (here:

UMAP), and the clustering approach (here: HDBSCAN). The pre-trained LLM might be spoiled by biased training data. We reviewed that the training data is in fact crucial for quality of the LLM. UMAP and HDBSCAN introduce a range of hyperparameters that have to be set. For detailed instructions and information on these hyperparameters, see: Tschisgale et al. (2023).

Application

See notebooks online.

References

Nowak, A., Kempin, M., Kulgemeyer, C., & Borowski, A. (2019). Reflexion von physikunterricht [reflection of physics lessons]. In C. Maurer (Ed.), *Naturwissenschaftliche Bildung als Grundlage für berufliche und gesellschaftliche Teilhabe: Jahrestagung in Kiel 2018* (p. 838). Regensburg: Gesellschaft für Didaktik der Chemie und Physik.

Sherin, B. (2013). A computational study of commonsense science: An exploration in the automated analysis of clinical interview data. *Journal of the Learning Sciences, 22*(4), 600–638.

Tschisgale, P., Wulff, P., & Kubsch, M. (2023). Integrating artificial intelligence-based methods into qualitative research in physics education research: A case for computational grounded theory. *Physical Review Physics Education Research, 19*(020123), 1–24.

Chapter 14
Unsupervised ML with Text Data

Kevin Hall and Christina Krist

Abstract This chapter builds on the techniques introduced in the previous Chaps. 12 and 13. Specifically, we will demonstrate how unsupervised pattern recognition approaches can be applied to text data to answer a research question. Because preprocessing and "knowing" your data are especially important when using unsupervised approaches with text, we will introduce additional techniques with an emphasis on exploratory data analysis tools such as token frequency analysis, basic text analytics, and n-gram analysis to explore large text-based datasets before using this information in support of the application of unsupervised natural language processing (NLP) techniques.

14.1 Basics of Natural Language Processing

As stated in previous chapters, working with language as data comes with a range of challenges based on the complex intricacies of human language and the difficulties faced by computers when working with that language. Words have multiple and sometimes conflicting meanings that require contextual information to decode properly. While the mature human brain does this naturally (using context cues, etc.), computers make a lot of mistakes. They are tripped up by lexical, syntactic, or semantic ambiguity. For example, the statement "Flying planes can be dangerous" represents a sentence that exhibits numerous ambiguities. A first ambiguity is lexical whether the term "flying" represents a verb or an adjective, or if the word "plane" is referring to an airplane or alternatively the geometrical sense of it being a "flat surface". A second ambiguity is syntactic. Is the sentence suggesting that the act of flying a plane (i.e., piloting) is a dangerous one? Or does it mean that planes that are flying, rather than grounded, can be dangerous? A third ambiguity is a semantic one in which these other ambiguities lead to confusion about where the danger itself

K. Hall (✉)
College of Education, University of Illinois Urbana-Champaign, Champaign, IL, USA
e-mail: knhall@illinois.edu

C. Krist
Graduate School of Education, Stanford University, Stanford, CA, USA

P. Wulff et al. (eds.), *Applying Machine Learning in Science Education Research*, Springer Texts in Education, https://doi.org/10.1007/978-3-031-74227-9_14

as well as where it stems from. Is it the physical planes in flight themselves or is it specifically talking about the act of flying planes? These forms of ambiguity become compounded when additional aspects of human language are introduced, such as idioms, slang, dialects, or other variations.

These word-level distinctions matter because, as described in Chap. 12, the dataset will be tokenized into smaller pieces that the computer will use as the unit of analysis. However, unlike in Chap. 12, when we are planning to utilize unsupervised approaches to analyzing text-based data, carefully considering how data can and should be tokenized and how it can be put back together in order for the human analyst to make sense of it is an essential consideration throughout the analytic process—including in the pre-processing phase!

As such, in this chapter, we will discuss how the three important stages (Fig. 14.1) that have been introduced so far—pre-processing, feature extraction, and pattern exploration and interpretation—are overlapping and non-linear when one's goal is an unsupervised analysis of text-based data. Specifically, we emphasize how decisions made in each "stage" must be informed by the theoretical and empirical goals of the other stages. We will give examples of how the same techniques can be used for multiple "stages" and why opposing decisions might be made within each stage, given your specific research goals.

14.1.1   Text Pre-processing

Text pre-processing includes cleaning and standardizing the text in order to lower the background "noise" that is intrinsic to large text-based datasets. Common steps in this process include ***lowercasing*** (converting all characters in the data corpus to lowercase) and ***removal of punctuation and special characters***. Both of these steps lead to the standardization of words which helps with the removal of duplicates (e.g., "large" vs. "Large" vs. "large." vs. "large;" will now all be treated as the same token: "large").

In addition, although often treated as a standard part of preprocessing (as done in Chap. 12), processes such as tokenization, lemmatization, stemming, and stop-word removal involve decisions that should be carefully considered and theoretically informed when engaging in these steps when planning on utilizing unsupervised techniques. This is because, depending on one's research question, different decisions about each may be more or less relevant–or even essential–when exploring patterns in the data.

- **Tokenization**: The process of converting the unstructured characters of a text corpus into discrete structured elements with clear properties that allow for easier processing with a computer. For example, the prior sentence when read by a computer would show up as a string of 175 characters. Keeping text as individual characters can become computationally intensive when trying to figure out

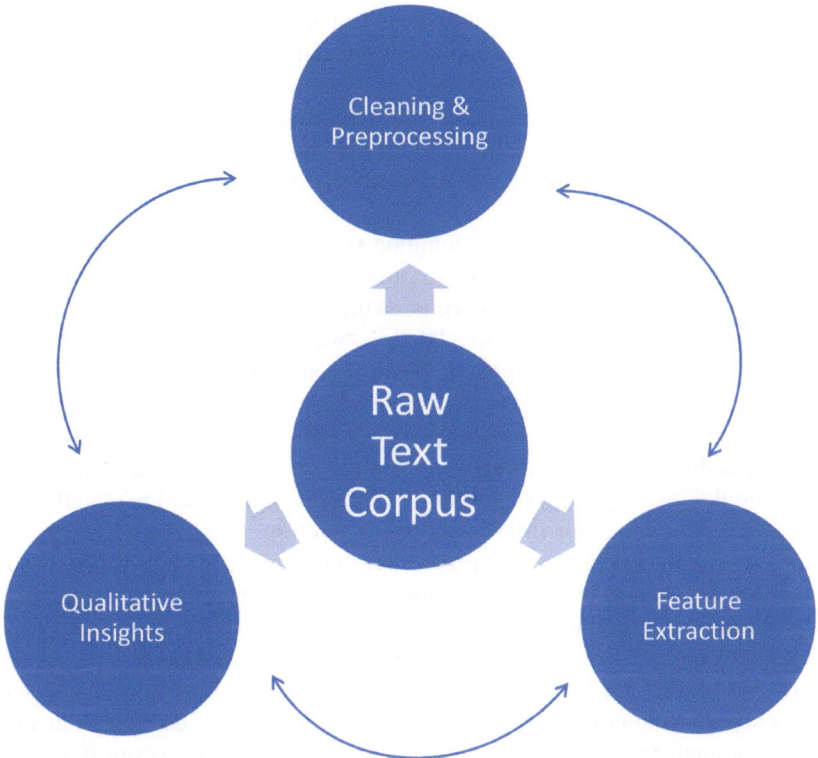

Fig. 14.1 Non-linearity of early stages of working with text data for unsupervised pattern recognition

context and meaning. Strings of text can instead be tokenized into various recognizable chunks, such as sentences and words. Sentence tokenization breaks down paragraphs into their constituent sentences. Word tokens are formed from decomposing sentences into their constituent words. NLTK has a unique tokenizer called Text Objects that adds an additional wrapper of metadata, allowing for advanced analysis features.

- **Stemming**: The process of converting words back into their normalized root form by using hard and fast predetermined rules. Stemming uses a brute force approach by trimming suffixes ("-ing", "ly", "-es", "-s", or "-er"). This serves to convert the words "walking", "walked", "walks" into one stem: "walk". This brute force approach allows for quick processing with much lower use of computational resources. However, stemming is done devoid of context. Depending on the stemmer used, this can lead to issues of over- or under-stemming where unrelated words are converted to the same stem or vice versa. For example, "university", "universe", and "universal" would all be stemmed to "univers."

- **Lemmatization**: Similar to stemming in that it aims to reduce words to their normalized base words so that duplicates are eliminated, lemmatization takes a more sophisticated approach by considering context. It does this by analyzing each word's part of speech and comparing the words to a lexical database. Depending on the lemmatizing tool used, the word "better" is more likely to be reduced to its contextual lemma of "good" instead of possibly being over-stemmed and converted to the word "bet".
- **Stopword removal**: Eliminating common words that are unlikely to contribute to the analysis of the data corpus. A common practice is to remove the 50–100 most common words in the data corpus, as these tend to include words like "the", "and", "as", "be", "because", "no", "their", "was", "should", "won't", etc. However, sometimes words that appear in the most common words list are essential for the research question at hand. For example, if my question was about differences in pedagogical agency (i.e., how responsibility for pedagogical decision-making is indexed differently) in various versions of curriculum materials, modal verbs like "should" vs. "could" vs. "might" vs. "must" contribute significantly to differences in that indexing (Martin, 2016). Therefore, these modal verbs should NOT be included in the list of stopwords, even if they are quite common. Another important decision in a science education context is one about technical vocabulary. While standard practice is to leave these words in, given their relative rarity, they also become strong signals. In other words, all documents containing the word "molecule" are likely to be analyzed as more similar to each other than to documents containing the word "equilibrium." This is a desirable outcome if sorting or analyzing based on specific content is the goal. However, if a research question is attempting to understand a construct that is cross-cutting or to analyze how it compares across content areas (Rosenberg and Krist, 2021), then including content-specific words as part of the stoplist (i.e., removing them from the dataset) may lead to cleaner identification of the patterns of interest.

14.1.2 Feature Extraction

Similar to preprocessing, feature extraction serves to improve computational accuracy by selecting relevant components, or removing irrelevant components, from the dataset, which will ideally lead to better results and more informative analysis. However, feature extraction is different in that the goal is selecting only those aspects of the data corpus that are useful for a particular task. These important features will differ depending on the goals that we have when working with the data and might change at different times throughout the process.

Feature extraction techniques generally involve various ways of quantifying or otherwise processing text-based data into a format that can then be processed by machine learning algorithms. These techniques typically involve converting words into vectors or matrices. Converting text into a bag of words is the most fundamental version of feature extraction. This involves representing a document or set of text as a frequency distribution of those words, without attending to order or context. There are other more sophisticated versions of this as well, such as Term Frequency-Inverse Document Frequency (TF-IDF), which provides a statistical measure of how often a term appears in a given document relative to the number of documents containing the word, or n-grams, which provide frequencies of word sequences of n length. These techniques vary greatly in terms of both complexity and purpose. Our rule of thumb is that simpler is often better, and we recommend starting with that assumption. More complex techniques can be leveraged as needed, but with more complexity comes more computational obfuscation—which can hinder, rather than support, the kind of qualitative exploration that unsupervised uses of ML with text data tend to be oriented towards.

Feature extraction techniques can also be leveraged for necessary (human) processing of a dataset. For example, one could utilize part-of-speech tagging to run a search for all proper nouns found within the data corpus to anonymize personal data found throughout a transcription of an interview or classroom discussion. However, here is where decisions about pre-processing become critical. If all words in the data corpus have been lowercased during preprocessing, then it might now be more difficult to find and anonymize someone whose last name is "Brown" or "White," for example. In this scenario it might be best to do an initial round of feature extraction before blindly running all preprocessing options.

When aiming to do unsupervised analyses with text data, feature extraction can also function as a way of beginning to familiarize oneself with the dataset. In this way, feature extraction can also be used as a form of initial data exploration. For example, you could compare word frequencies of a subset of key terms (e.g., the 50 most frequent words; a self-constructed dictionary of key terms relevant to your research topic) across multiple documents. Or you could again utilize part-of-speech tagging to look at differences between amount of pronoun use vs. proper noun use across various documents, or to extract all the questions in a document set. However, once again, pre-processing decisions really matter in terms of the specific words or punctuation marks that were removed. We hope these examples encourage you, even as a beginning user, to ask questions and think carefully about the implications of any step or process that is described as a "standard" part of NLP!

We will also pause here to point out even more overlaps and nonlinearities between "stages" present in these examples. The feature extraction examples described above could also be utilized as pattern exploration. This is actually the first stage of computational grounded theory introduced in Chap. 3. In examining documents from the first and second wave feminist movements in Chicago vs. New York, Nelson (2020) used parts-of-speech tagging and identified notable differences between abstract vs. proper noun use that seemed to vary by location (Chicago vs. New York) but not by time. This then motivated her deeper qualitative content analysis of the documents, including guiding her selection of which subset of the 200+ documents to analyze in depth. This example highlights how it is the purpose and goal of using a specific technique, rather than the technique itself, that matters.

14.1.3 The Unsupervised ML Workflow

Finally, as we have hinted at above, unsupervised ML with text data follows the same general workflow introduced in Chapter Pattern Recognition—Unsupervised Machine Learning: (1) pattern recognition, (2) qualitative pattern interpretation, (3) pattern validation. Again, all of the techniques described in previous chapters can be applied to these stages. Rather than reiterating these techniques here, the remainder of this chapter will focus on how to go about the pre-processing and feature extraction techniques relevant to working with text data, as well as some additional tools for getting to know your data in ways that matter for the kinds of empirically and theoretically guided decision-making that are needed for utilizing unsupervised techniques with a text-based data corpus.

14.2 Pre-processing and Feature Extraction with *General Science Quarterly/Science Education* Editorials

14.2.1 Introduction to the Data Corpus

To introduce and practice using various pre-processing and feature extraction techniques in service of unsupervised (and qualitative-leaning) explorations of text-base data, we will utilize two datasets generated from the journal *Science Education* (originally named *General Science Quarterly*).

The first data corpus that we will be using includes the editorials from volumes I and II of *General Science Quarterly (GSQ)* published between 1916 and 1918. *GSQ*, now *Science Education (SE)*, has long been a seminal journal for the publication of academic research connected specifically to science education. In fact the very first article published in *GSQ* was *Method in Science Teaching* written by the well-known philosopher and educational scholar John Dewey. These original early copies are easily found on the web and have been made available in digitized form thanks to Google's OCR work on older undigitized documents and publications. We will use this data corpus to do some preliminary natural language processing, as well as to gain some initial familiarity with the data. Specifically, we are wondering: *What was topical to research in science education in 1916?* We will draw on any insights gained to aid our decisions about the usage of unsupervised learning NLP techniques in the second dataset. This second dataset consists of two corpora: one of some early-period editorials published in *GSQ/SE* and another of later-period editorials, published in the 2000s. We will use this broader corpus to look for historic trends in science education's stated priorities. For both datasets, we will use Python's Natural Language Toolkit (NLTK).

14.2.2 Pre-processing the Documents for Data Corpus 1

For this original data corpus (GSQ volume I), we have already converted the 1916 editorials to an easy-to-use text document—a text file. Using text files reduces issues related to special characters and other formatting problems. If any documents in your text corpus are not in text file format, they should be converted before we start processing. While it is possible to start with other document types such as DOC, CSV, Excel, or JSON files, working with text files is easier to begin with.

14.2.3 Setting Up the Environment

For this chapter we will be using Python's Natural Language Toolkit (NLTK) and several support libraries:

1. Natural Language Toolkit (NLTK): Library containing a wide range of tools for natural language processing
2. Matplotlib: Library for creating data visualizations
3. NumPy: Library for performing complex mathematical operations as well as data manipulation.
4. Pandas: Another library for data analysis and manipulation.

The following code will load the packages that we will use throughout this chapter. You will only need to install these packages once unless you decide to change the version of Python that you were using when you installed the packages. The code that we will utilize throughout this chapter can be downloaded and found within the Jupyter Notebook that it is linked in the resources section. This notebook can be opened and interacted with using the free source-code editor Visual Studio Code after simply installing the Jupyter Notebook support extensions. Alternatively, feel free to interact with the Python code in whichever Integrated Development Editor (IDE) you are comfortable with.

Python code snippet

```
# load required Python packages
pip install nltk
pip install matplotlib
pip install numpy
pip install pandas
```

14.2.4   Loading Your Text File into a Data Corpus

Next, we will import the first data corpus so that you can begin to "read" it and get to know it in the NLTK environment. The following code will import the text file and allow you to view it to ensure that it has loaded properly.

Python code snippet

```python
# Define the path to the text file
filename = 'editorials\editorialGSQ1916.txt'

try:
    # Open the text file for reading with UTF-8 encoding and
    replace any errors in character encoding with
    open(filename, 'rt', encoding='utf-8', errors='replace') as
    editorialGSQ1916Text:
        # Read the entire contents of the file into a variable
        raw_GSQ1916Text = editorialGSQ1916Text.read()

    # Print the contents of the file
    print(raw_GSQ1916Text[0:249])

except FileNotFoundError:
    # Handle the case where the file is not found
    print(f"Error: The file '{filename}' was not found. Please
    check the file path.")

except IOError:
    # Handle other input/output errors
    print(f"Error: An error occurred while reading the file
    '{filename}'.")
```

As this script executes it will locate and read the contents of the text file named editorialGSQ1916.txt at the path listed, after which it will copy this entire text into a variable named raw_GSQ1916Text before printing the the first 250 characters of raw_GSQ1916Text so that you can do some cursory checking to see that the process worked. Alternatively, if it does not find that specific file or is unable to open it, you will be notified in the form of an error message. If this process does not work, the most likely error is that the path to the text is incorrect in some way and that the filename variable needs to be fixed. The path convention used for the filename variable shows the relative path from the Python script to the text file. Currently, this script will only work if the text file is located in a subfolder named 'editorials' next to the Python script or Jupyter Notebook that you are running. If the text file was not located in a subfolder then the code could be changed to read:

```
filename = 'editorialGSQ1916.txt'
```

Alternatively, an absolute path could instead be used that could point to any location on your computer or network. For example:

```
filename = 'C:\GeneralScienceQuarterlyResearch\editorials\
editorialGSQ1916.txt'
```

While it may make sense to use absolute paths throughout this initial process as you practice, it is not a great idea over the long run. Using absolute paths limits the flexibility of the scripts that you will be using due to issues that will occur if someone else tries to use your code on their machine. You end up with broken links that will then need to be fixed, whereas you will not face this same issue with relative links. We will be using relative paths as well as trying our best to put files in folders moving forward.

Another consideration to begin thinking about is naming conventions for the different text corpora that will be created and interacted with throughout this process. At this point, we have only two, but that is already double the number that we started with, and we will be adding three more during our next step. What do we mean by this? Well, we started with a text file called `editorialGSQ1916.txt` whose entire contents were copied into a variable called `raw_GSQ1916Text`. In doing so, we moved the contents of the text file from the hard drive to the system's memory (RAM) where it can be accessed and manipulated faster and more efficiently. These two formats are identical other than the fact that if you were to restart your computer only `editorialGSQ1916.txt` would exist on reboot.

14.2.5 Converting the Data Corpus into Tokens

Now that we have the text corpus loaded into our system memory as a named variable, we can interact with it in numerous ways that we could not when it was solely a text file. In fact `raw_GSQ1916Text` is only useful to us as an intermediate stage as part of the tokenizing of our data corpus. We separated this stage in this chapter so that you could get a better understanding of tokenization. As described above, tokenization serves to break down or help to normalize our large stream of text into words, sentences, paragraphs, or other meaningful textual elements which are then individually called tokens. Converting streams of text into tokens allows the algorithms used to better understand the context and meaning throughout the text. For example, if the text is tokenized into words, then the computer knows that it is dealing with a word and does not have to "figure this out" from a string of letters.

These tokens also allow for an easier path to extracting the specific features that we want from the data corpus. We will largely be working with word tokens and Text objects for our cursory analysis, but sentence tokens will become important when we begin our focus on looking for patterns and themes in the entire *Science Education* Editorial text corpus later on.

These next lines of code will convert `raw_GSQ1916Text` into three different variables. One contains **word tokens**, another contains **sentence tokens**, and the last one will be made up of NLTK-modified tokens called **text objects**.

Python code snippet

```python
# Importing necessary libraries
import nltk
from nltk.tokenize import word_tokenize, sent_tokenize
nltk.download('punkt')

# relative file path
filename = 'editorials/editorialGSQ1916.txt'

try:
    # 1. Open and read the text file
    with open(filename, 'rt', encoding='utf-8',
    errors='replace') as file:
        raw_GSQ1916Text = file.read()

    # 2. Tokenize the raw text into words
    GSQ1916_wordTokens = word_tokenize(raw_GSQ1916Text)

    # 3. Tokenize the raw text into sentences
    GSQ1916_sentTokens = sent_tokenize(raw_GSQ1916Text)

    # 4. Convert word our tokens into NLTK Text objects
    GSQ1916_wordTextObjects = nltk.Text(GSQ1916_wordTokens)

except FileNotFoundError:
    print(f"Error: The file '{filename}' was not found. Check
    the file path.")
except Exception as e:
    print(f"An error occurred: {e}")
```

Running this code will result in the generation of three more data corpora all stemming from our raw text. The text will still be identical in each corpus, the big difference being the number of tokens in each as well as what each token represents:

```
GSQ1916_wordTokens = Our original raw text converted into word
    tokens
GSQ1916_sentTokens = Our original raw text converted into
    sentence tokens
GSQ1916_wordTextObjects = Our original raw text converted into
    NLTK Text object tokens
```

14.2.6 The Need for Normalization During the Pre-processing Phase

Run the print functions to look inside each container. This code will print the first two tokens inside each container:

Python code snippet

```
print("Here are the first 2 tokens in number 0:
    ",raw_GSQ1916Text[0:2])
print("Here are the first 2 tokens in number 1:
    ",GSQ1916_wordTokens[0:2])
print("Here are the first 2 tokens in number 2:
    ",GSQ1916_sentTokens[0:2])
print("Here are the first 2 tokens in number 3:
    ",GSQ1916_wordTextObjects[0:2])
```

You should see the following output:

```
Here are the first 2 tokens in number 0:
    GE
Here are the first 2 tokens in number 1:
    ['GENERAL', 'SCIENCE']
Here are the first 2 tokens in number 2:
    ['GENERAL SCIENCE QUARTERLY\nEditorials\nGeneral Science
    at the National Education Association In New York.', 'The
    four special sessions on four different days were devoted
    respectively to chemistry, physics, biology, and science.']
```

```
Here are the first 2 tokens in number 3:
    ['GENERAL', 'SCIENCE']
```

Based on these outputs, we can intuit some of the differences between these token types. The first two characters in the raw file are "G" and "E", which we can see are the first two letters in GENERAL. In this untokenized raw text, we can see that the computer will have to process every letter individually to discern meaning, a task that has incredible computational overhead even for humans. When we get to our tokenized file we can see that each list starts and ends with brackets [ ], and each token is separated by a comma. Looking at the word tokens and text object tokens we see that the first two tokens are indeed words: "GENERAL" AND "SCIENCE". And the first two tokens in the sentence tokens file contain similar separation characters, but much more text:

```
1.'GENERAL SCIENCE QUARTERLY\nEditorials\nGeneral Science at
the National Education Association In New York.',
2.'The four special sessions on four different days were
devoted respectively to chemistry, physics, biology, and
science.'
```

You might also notice \n showing up in the first sentence and possibly wondering if that is a typo contained in the text itself. It is not; rather, it is an artifact of sentence tokenization. There are various tokenizers within NLTK. These tokenizers differ in the features and heuristics that they use to identify the boundaries that they are looking for. For example, words can be tokenized based on whitespace, punctuation, special characters, regular expression, or other features and heuristics. Sentence boundaries for tokenization can be discerned using punctuation marks, capitalization, and so on. Each boundary selection choice brings with it its own affordances and constraints. One such constraint is seen here with the \n indicating that a new line began in the text file, basically denoting a paragraph break.

The following text and formatting is actually what appears at the beginning of the first editorial section:

```
GENERAL SCIENCE QUARTERLY
Editorials
General Science at the National Education Association In New York.

The four special sessions on four different days were devoted
respectively to chemistry, physics, biology, and science. One
paper at least on general science was on the program...
```

In this case, the tokenizer we used started at the first word and kept moving forward until it made it to the period found at the end of New York, at which point it tokenized that whole section as one sentence.

If you are ever uncertain as to what type of token you are dealing with, using the Python type and len functions we can also take a quick structural look:

Python code snippet

```
print("Number 0 is a: ",type(raw_GSQ1916Text), "It contains:
    ",len(raw_GSQ1916Text), "tokens")
print("Number 1 is a: ",type(GSQ1916_wordTokens), "It contains:
    ",len(GSQ1916_wordTokens), "tokens")
print("Number 2 is a: ",type(GSQ1916_sentTokens), "It contains:
    ",len(GSQ1916_sentTokens), "tokens")
print("Number 3 is a: ",type(GSQ1916_wordTextObjects), "It
    contains: ",len(GSQ1916_wordTextObjects), "tokens")
```

You should get an output similar to the following:

```
Number 0 is a:  <class 'str'> It contains:  11300 tokens
Number 1 is a:  <class 'list'> It contains:  2073 tokens
Number 2 is a:  <class 'list'> It contains:  91 tokens
Number 3 is a:  <class 'nltk.text.Text'> It contains:  2073
    tokens
```

Here we can see that word and sentence tokens are converted from a raw sentence string to a list of either words or sentences. Word tokens and NLTK Text objects are identical in token count, but show up as different types of tokens. We can also use this for a quick data point, seeing that the editorials found in the 1916 volume of GSQ contained 2073 words divided into 91 sentences.

So far we have run some cursory commands to look at the outputs of the tokenization preprocessing. The next step of preprocessing is the normalization of our resulting word tokens to remove multiple versions of our words. To begin this step, we will run a word frequency distribution query that will output a list of the most frequently appearing words along with their frequencies.

There are numerous ways to query word frequencies using Python. In this case, we will use the text tokenized as Text objects to generate a list of the top 26 most common words found in the GSQ1916 editorial corpus.

Python code snippet

```
#Run the NLTK Frequency distribution method using our Text
    objects
FdGSQ1916 = nltk.FreqDist(GSQ1916_wordTextObjects)
```

```
print("Most common words in GSQ1916 editorial:
    ",FdGSQ1916.most_common(26))
```

You should see the following output:

```
Most common words in GSQ1916 editorial: [('the', 106),
(',', 103), ('.', 91), ('of', 79), ('science', 58),
('to', 54), ('in', 44), ('and', 42), ('general', 31),
('a', 31), ('is', 30), ('teachers', 25), ('for', 18),
('be', 18), ('some', 18), ('on', 17), ('was', 15),
('school', 15), ('The', 14), ('high', 14), ('it', 13),
('are', 13), ('which', 12), ('that', 12), ('year', 12),
('General', 10)]
```

This check helps us to notice the issues that occur based on a lack of normalization of words in our corpus. Unsurprisingly, the most frequent word found in the dataset is "the". We also see frequencies for punctuation marks, which were tokenized and counted as words. In addition, note that the word "general" (9th most frequent word) is tallied separately from "General" (26th most frequent word). Remember that any difference in word case leads the computer to believe that these words are distinct from each other. As such, while we have tokenized our text, we have not normalized the resulting tokens and are therefore still left with quite a bit of background noise.

As mentioned before, there are numerous ways to tokenize the text and there is no perfect sequence of tokenizing. The process will be specific to the type of research that you are planning to conduct with the text corpus. In all likelihood, you will need to make multiple sets of tokenized text and each will need to be preprocessed in their own specific ways. It is possible that you want to do the preprocessing iteratively or all in one go depending once again on what you are searching for in your data (Fig. 14.2). For the exploration you will be doing for the rest of this chapter, we will tokenize and normalize our raw text corpus using a script that does it all in one go. The code snippet below performs multiple steps to first tokenize and then normalize our text corpus. These steps include:

1. Tokenization: Creating sentence and word tokens, as well as NLTK Text objects.
2. Lowercasing: normalizes all characters in the text corpus to their lowercase version.
3. Remove Punctuation: Punctuation marks are removed from each token. This is done by translating each string using a translation table that removes punctuation marks.
4. Retain Alphanumeric characters: characters that are not purely alphabetic are filtered out to keep only words containing alphabetic characters.
5. Remove Stopwords: Finally, common stopwords like 'and', 'the', etc., are removed from the list of tokens. The NLTK library provides a list of such stopwords for the English language.

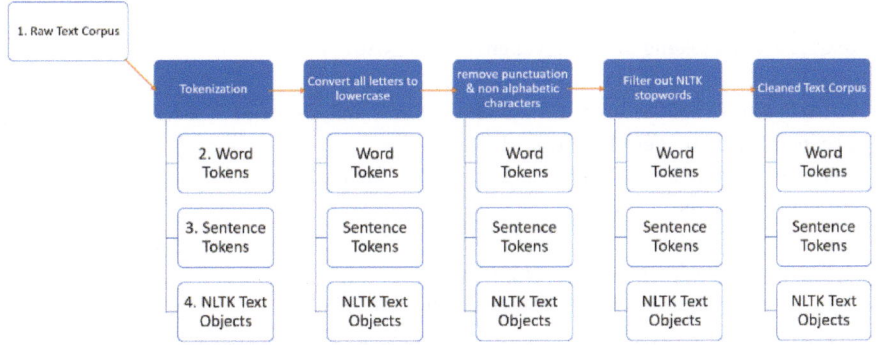

Fig. 14.2 Graphical representation of tokenization and normalization steps in NLTK's combined all in one pre-processor

The resulting list will contain tokens that are all lowercase, clear of punctuation, alphanumeric, and are not stopwords. Use the following code to run this all-in-one script:

Python code snippet

```python
#Combined All in one Pre-processor (Tokenization/Normalization)

import os
import nltk
import string
from nltk.corpus import stopwords
from nltk.tokenize import sent_tokenize, word_tokenize
from nltk.text import Text

nltk.download('punkt')
nltk.download('stopwords')

filename = 'editorials/editorialGSQ1916.txt'

try:
    with open(filename, 'rt', encoding='utf-8', errors='replace')
    as file:
        raw_GSQ1916Text = file.read()
except FileNotFoundError:
    print(f"Error: The file '{filename}' was not found. Check the
    file path.")
    exit()
```

```python
# Tokenize the raw text into sentences
GSQ1916_sentTokens = sent_tokenize(raw_GSQ1916Text)

# Initialize containers for cleaned data and removed elements
cleaned_GSQ1916_sentTokens = []
removed_elements = {'punctuation': [], 'non_alpha': [],
    'stop_words': []}

# Load stopwords once for efficiency
stop_words = set(stopwords.words('english'))

for sentence in GSQ1916_sentTokens:
    words = word_tokenize(sentence.lower())
    cleaned_words = []
    for word in words:
        if word.isalpha() and word not in stop_words:
            cleaned_words.append(word)
        else:
            if not word.isalpha():
                removed_elements['non_alpha'].append(word)
            if word in stop_words:
                removed_elements['stop_words'].append(word)
            if any(char in string.punctuation for char in word):
                removed_elements['punctuation'].append(word)

    cleaned_sentence = ' '.join(cleaned_words)
    cleaned_GSQ1916_sentTokens.append(cleaned_sentence)

output_filename = 'editorials/preProcessed_GSQ1916Text.txt'
os.makedirs(os.path.dirname(output_filename), exist_ok=True)
with open(output_filename, 'w', encoding='utf-8') as file:
    for sentence in cleaned_GSQ1916_sentTokens:
        file.write(sentence + '\n')

# Now, read back the cleaned text for further processing
with open(output_filename, 'rt', encoding='utf-8') as file:
    cleaned_text = file.read()


# Tokenize the cleaned text into sentences and words
preProcessed_GSQ1916_wordTokens = word_tokenize(cleaned_text)
preProcessed_GSQ1916_textObjects = Text(cleaned_word_tokens)

# Output results for verification
print(f"Cleaned text saved to {output_filename}")
```

```
print(f"Number of cleaned word tokens:
    {len(preProcessed_GSQ1916_wordTokens)}")
```

```
#Run the NLTK Frequency distribution method using our Text
    objects
FdGSQ1916 = nltk.FreqDist(preProcessed_GSQ1916_textObjects)
print("Most common words in Pre-Processed GSQ1916 editorial:
    ",FdGSQ1916.most_common(26))
```

You should now see the 26 most common word tokens and their counts:

```
Most common words in Pre-Processed GSQ1916 editorial:
[('science', 72), ('general', 45), ('teachers', 28),
('school', 20), ('high', 18), ('year', 12), ('one', 9),
('teaching', 9), ('course', 8), ('junior', 8), ('new', 7),
('special', 7), ('club', 7), ('physics', 6), ('two', 6),
('college', 6), ('much', 6), ('pupils', 6), ('different', 5),
('subject', 5), ('given', 5), ('discussion', 5), ('best', 5),
('find', 5), ('five', 5), ('per', 5)]
```

Compare this output to the previous word frequency output. What are the differences that you notice?

You should see that all occurrences of "general" have been normalized to all fall into one bucket, giving us a better overall count of common words in our corpus. We also see some reshuffling of the list as stop words and punctuation have been removed from the corpus.

To be clear, this is not an ideal corpus by any stretch of the imagination. There is probably some level of data loss that occurs with each one of these preprocessing steps. Going through all possible ways of preprocessing is out of the scope of this book chapter and much of it will have to be glossed over due to space constraints. It should also be noted that we have made a variable container for each step of preprocessing done (i.e., one tokenized by sentences; one tokenized by Text objects), and we can continue to use any or all of them for analysis. This will provide flexibility and allow us to do some exploration look for interesting themes or patterns before settling on one processed version of our text corpus.

14.2.7   Extracting Features from Our Corpora

Before moving on to our data analysis phase we will also produce two more text corpora, creating each by extracting a specific feature from our text corpus. The two feature extraction techniques we will use here are extracting n-grams and part-of-speech (POS) tagging.

Extracting n-grams: The n-gram extraction will result in a compilation of words that are found together in pairs (bigrams), or the three words most frequently found together (trigrams). We will use the frequency and distribution of n-grams later to explore patterns and trends across the hundred-plus years of *Science Education* editorials. During our upcoming EDA, we will analyze the frequency and the distribution of our extracted n-grams. Use the following code to extract n-grams:

Python code snippet

```python
import nltk
from nltk.tokenize import word_tokenize
from nltk.util import ngrams
from collections import Counter

tokens = preProcessed_GSQ1916_wordTokens

# Generate bigrams
bigrams = list(ngrams(tokens, 2))

# Generate trigrams
trigrams = list(ngrams(tokens, 3))

# Count and print the frequency of the top 10 bigrams
bigram_frequency = Counter(bigrams)
print("Bigram Frequency:", bigram_frequency.most_common())

# Count and print the frequency of the top 10 trigrams
trigram_frequency = Counter(trigrams)
print("Trigram Frequency:", trigram_frequency.most_common())
```

You should see the following output. What do you notice? What does it make you wonder?

```
Bigram Frequency: [(('general', 'science'), 41),
(('high', 'school'), 14), (('science', 'teachers'), 13),
```

```
(('junior', 'high'), 8), (('per', 'week'), 4),
(('new', 'york'), 3), (('senior', 'high'), 3),
(('science', 'club'), 3), (('year', 'junior'), 3),
(('science', 'required'), 3)]

Trigram Frequency: [(('junior', 'high', 'school'), 4),
(('senior', 'high', 'school'), 3),
(('general', 'science', 'club'), 3),
(('year', 'junior', 'high'), 3),
(('general', 'science', 'required'), 3),
(('general', 'science', 'quarterly'), 2),
(('teaching', 'general', 'science'), 2),
(('general', 'science', 'main'), 2),
(('science', 'teachers', 'associations'), 2),
(('attention', 'given', 'general'), 2)]
```

Even a cursory look over this output gives us a general idea of some of the topics discussed in the 1916 editorials. For example, if you did not know the nature of the publication, it should start to become clear to you that this journal is based in the field of education. The group that these editorials seem to talk most about are "science teachers", it also seems to have an educational leaning towards 7–12th grade students. It also seems to be overwhelmingly focused on the discussion of "general science", which of course makes sense based on the fact that at this point the journal was named General Science Quarterly. In less than ten years the journal will transition away from this title towards Science Education. This first set of data points from this n-gram extraction captures this moment in time and will be used as a comparison as we look at the evolution of the journal over time.

Part-of-Speech (POS) tagging: POS tagging looks at all words found in the text corpus and assigns parts of speech to them based on features and heuristics. The resulting list can then be used to perform other NLP tasks and analyses. Once again, there are numerous POS taggers accessible through Python; NLTK defaults to the Penn Treebank tagset (example link https://www.cis.upenn.edu/~bies/manuals/tagguide.pdf). Each tag represents a different part of speech. In this case:

- JJ = adjective
- NN = Noun, singular or uncountable nouns
- NNS = Noun, plural

Use the following script to carry out POS tagging on our data corpus:

Python code snippet

```
import nltk
from nltk.tokenize import word_tokenize
from nltk import pos_tag
```

```
# Ensure you've downloaded the necessary NLTK data
nltk.download('punkt')
nltk.download('averaged_perceptron_tagger')


try:
    # Read the cleaned text file
    with open(filename, 'r', encoding='utf-8') as file:
        cleaned_text = file.read()

    # Apply POS tagging to the word tokens
    preProcessed_GSQ1916_pos_tags =
        pos_tag(preProcessed_GSQ1916_wordTokens)

    # Print the POS tags for the first few tokens as a sample
    print("First 10 POS tags:")
    for token, tag in preProcessed_GSQ1916_pos_tags[:10]:
        # Adjust the slice for more or fewer samples
        print(f"{token} - {tag}")

except FileNotFoundError:
    print(f"File {filename} not found. Please check the path.")
```

You should see the following output:

```
First 10 POS tags:
general - JJ, science - NN, quarterly - JJ, editorials - NNS,
general - JJ, science - NN, national - JJ, education - NN,
association - NN, new - JJ
```

In the output, we can see that each word has been connected to some tag and that tags can be repeated.

14.3   Exploratory Data Analysis Using a Larger Data Corpus

We are now able to start doing some exploratory data analysis using our cleaned word tokens and textObjects as the corpus. At this early stage in our analysis, it is sometimes difficult to ascertain what might be interesting about our data corpus. This is where carefully utilized unsupervised analyses can help to support the (qualitative) analyst in doing good qualitative research: observing patterns, noting themes, and making decisions about when and how a human should dive in more deeply.

In the interest of diving into some exploratory data analysis, we will now be looking at more than just the one set of editorials from 1916. For this exploration, we will do a comparative analysis of an early (1916, 1917, 1926, and 1929) vs. a late set (2000, 2002, 2005, and 2016) of published editorials from Science Education. We have (somewhat arbitrarily) decided to segment the corpus by period (editorials from the first and last four volumes of the journal) to make it manageable to take a cursory look for patterns over time in how the themes and content of editorials about science education have evolved over 100+ years.

An important note about this text corpus: editorials were not written for every volume of this journal, and even then they differ greatly in format, size, or who wrote the editorial (e.g., editor-in-chief vs. section editor). There are also sometimes guest editorials as well as swaths of volumes that do not contain editorials at all. This is important to keep in mind as we move forward because this affects our choice of editorials to analyze for this exercise.

For this exploration, we have already converted the individual editorials into clean text files, which were then merged based on the time they were published, resulting in two different text corpora to run our analysis. These text files were then pre-processed and had features extracted following the methods discussed in the previous section.

In the following subsections, we will provide guidance for conducting a range of analyses to begin to give us a glimpse into the data. However, we will leave it up to the reader to observe, ask questions, and begin to make sense of these patterns. Again, as this is exploratory, the goal is to look for patterns that you, as the human analyst, find interesting or worth exploring further, based on your own theoretically-informed questions and curiosities. This is a cyclical and iterative process, and we recommend utilizing these NLTK-based tools in conjunction with other qualitative analysis tools like analytic memos and jottings to keep track of your observations and questions along the way.

As you try out each of the following analyses, keep in mind the broad research question that motivated us to compile this dataset in the first place: How have the expressed priorities of the journal Science Education (previously General Science Quarterly) evolved over time? We were also interested specifically in how priorities related to diversity and equity have entered into editorial priorities. If you have a more specific topical focus, feel free to jot that down and keep an eye out for indicators related to that topical focus as you go along as well.

With these corpora, we will conduct a set of analyses that are all versions of **token frequency analysis**. Looking at these various frequencies and related descriptive data will offer us potential analytic insights about the richness of the vocabulary used, trends in highest frequency word usage, domain specific word usage, differences in editor writing styles, trends and patterns in science education, and changes in areas such as readability or complexity of words used.

We will introduce a description and the input code for each technique. We will also provide a sample output text in order to show the structure and syntax of what this code should be producing for you. However, keep in mind that the substance of that output (e.g., the specific words or counts) will likely differ from what you see, depending on the corpus you are working with. The code for pre-processing and comparing the new corpora is found in the Jupyter notebook.

14.3.1 Average Sentence Length

Average Sentence Length uses the total number of words and total number of sentences in a corpus to calculate exactly what it says: the average sentence length. While the equation is very basic and straightforward it provides information that can be used to infer, for example, how complex sentences are on average throughout a given text corpus. Because we will be comparing by decade, this can also be useful to identify potential indicators for patterns like changes in writing styles and stylistic preferences when looking at how editorials have been written over time.

Python code snippet

```python
print("Early File:")
print("Number of sentences:", len(cleaned_early_sentences))
print("Number of word tokens:", len(early_word_tokens))
# Calculate the average sentence length
average_sentence_length_early = len(early_word_tokens) /
    len(cleaned_early_sentences)
print("Average sentence length for early is:",
    average_sentence_length_early)

print("\nLate File:")
print("Number of sentences:", len(cleaned_late_sentences))
print("Number of word tokens:", len(late_word_tokens))
# Calculate the average sentence length
average_sentence_length_late = len(late_word_tokens) /
    len(cleaned_late_sentences)
print("Average sentence length for late is:",
    average_sentence_length_late)
```

Here is an example of what your output structure should look like:

```
Number of sentences: 284
Number of word tokens: 3144
Average sentence length for early is: 11.070422535211268
```

14.3.2 Word Frequency Distributions

Word frequency distributions are what we ran initially when looking at the most common words found in a text corpus. While simple at first glance, this metric can also point to qualities such as author's style and tone, giving us the ability to again infer trends and patterns over time such as main topics, themes, or subjects discussed in a text corpus. Word frequency distributions can also be used in applications such as topic modeling.

Python code snippet

```
#Top 25 most common words with their counts
fdEarly = nltk.FreqDist(early_text_objects)
print("Most common words in early file:",fdEarly.most_common(25))

fdLate = nltk.FreqDist(late_text_objects)
print("Most common words in late file:",fdLate.most_common(25))
```

Here is an example of what your output structure should look like:

```
Most common words in early file: [('science', 199),
('general', 108), ('school', 67), ('teachers', 53),
('high', 49), ('one', 32), ('education', 28), ('new', 28),
('course', 25), ('college', 24), ('year', 22), ('university',
('elementary', 20), ('work', 19), ('teaching', 17),
('courses', 16), ('schools', 16), ('special', 15),
('junior', 15), ('may', 15), ('physics', 14), ('club', 14),
('pupils', 14), ('michigan', 14), ('free', 13)]
```

14.3.3 Lexical Diversity

Lexical diversity quantifies the variety of unique words found in a document. It produces a numerical measure that indicates how diverse the vocabulary is that is used in a text. Broadly speaking, scores of 0.8–1 are considered extremely high and difficult to maintain in typical communicative texts. Scores of 0.4–0.79 are considered moderate to high; most high-quality texts fall in this range. Scores of 0–0.39 are considered low lexical diversity and tend to suggest highly specialized or technical texts (e.g., instruction manuals) or texts aimed at young readers. While this measure is sensitive to text length (longer texts have more opportunities to repeat words), comparing lexical diversity scores over time can allow for quantitative comparison that might suggest potential changes in how the usage of academic language has changed over time. This might indicate changes in, for example, editorial stylistic preferences, changes in editors, or changes in readership expectation.

Python code snippet

```
lexical_diversity_early = len(set(early_word_tokens)) /
    len(early_word_tokens)
lexical_diversity_late = len(set(late_word_tokens)) /
    len(late_word_tokens)

print("Lexical diversity for early file:", lexical_diversity_early)
print("Lexical diversity for late file:", lexical_diversity_late)
```

Here is an example of what your output structure should look like:

```
Lexical diversity for early file: 0.3880407124681934
Lexical diversity for late file: 0.43950617283950616
```

14.3.4   Word Length Distribution

Again, a fairly straightforward measure that can provide insight into how long, on average, words are in a given corpus.

Python code snippet

```
# Average word length
avg_word_length_early = sum(len(word) for word in early_word_tokens)
    / len(early_word_tokens)
avg_word_length_late = sum(len(word) for word in late_word_tokens)
    / len(late_word_tokens)

print("Average word length for early file:", avg_word_length_early)
print("Average word length for late file:", avg_word_length_late)
```

Here is an example of what your output structure should look like:

```
Average word length for early file: 6.74236641221374
Average word length for late file: 7.292901234567902
```

14.3.5   Unique Words

Unique words can be used to identify the frequency of words that appear only once in a given corpus. We can also print a list of these word tokens. Looking at unique words between or across text corpora can allow us to look for the appearances and disappearances of specialized educational terminology over time.

To find the frequency (number) of unique words, use the following code:

Python code snippet

```python
unique_words_early = set(early_word_tokens)
unique_words_late = set(late_word_tokens)

print("Number of unique words in early file:", len(unique_words_early))
print("Number of unique words in late file:", len(unique_words_late))
```

Here is an example of what your output structure should look like:

```
Number of unique words in early file: 1220
Number of unique words in late file: 1424
```

To generate a list of words that are unique to that corpus compared to the others, use the following code:

Python code snippet

```python
# Convert word tokens to sets for set operations
unique_words_early = set(early_word_tokens)
unique_words_late = set(late_word_tokens)

# Find words that are in the early set but not in the late set
exclusive_to_early = unique_words_early - unique_words_late

# Find words that are in the late set but not in the early set
exclusive_to_late = unique_words_late - unique_words_early

print("Words exclusive to early set:", exclusive_to_early)
print("\nWords exclusive to late set:", exclusive_to_late)
```

Here is an example of what your output structure should look like:

```
Words exclusive to early set: {'small', 'electric', 'sought',
'subjects', 'judgment', 'termed', 'kalamazoo', 'nine', 'china',
'lecture', 'divided', 'pretentious', 'bound', 'periodical',
'soon', 'richard', 'discoursed', 'lines', 'usage', 'contribute',
'needs', 'suitability', 'providence', 'said', 'encouraged',
'albany', 'whose', 'columbia', 'hotel', 'provoked', 'advised',
'miller', 'title', 'webb', 'situation', 'ably', 'stratification',
'market', 'subscribers', 'es', 'continued', 'suggestive',
'craig', 'numerous', 'enough', 'forty', 'richardson', 'full',
'expound', 'ohio', 'souled', 'carrying', 'seldom'
```

```
Words exclusive to late set: {'community', 'consistently',
'hold', 'gap', 'ongoing', 'compromise', 'inequities',
'sophisticated', 'visibility', 'assert', 'unified', 'assess',
'rudolph', 'middle', 'level', 'unrealized', 'production',
'robb', 'interaction', 'observing', 'criteria', 'observed',
'ends', 'relations', 'agree', 'reading', 'retort', 'intent',
'disability', 'round', 'advocates', 'glancing',
'quantification', 'produce', 'kong', 'symposium',
'understanding', 'inherent', 'able', 'evaluate', 'proposals',
'upshot', 'engagement', 'target', 'superficial', 'tantalizing',
'judge', 'plenary', 'exercise', 'proficiency', 'mobility',
'fr{\o}yland', 'impossible', 'regarding'
```

14.3.6 N-grams

N-grams point out recurring word combinations found throughout the text corpus. As described earlier, bigrams (e.g., New York; science teachers; high school) and trigrams (e.g., gold standard research; junior high school) are the most commonly used. In our earlier explorations of the 1916 corpus, we saw how bigrams conveyed a lot of information about the contents of the text corpus.

To generate an ordered list of the most common bigrams, use the following code:

Python code snippet

```
# display frequency of highest 25 bigrams
finder = nltk.collocations.BigramCollocationFinder.from_words
    (early_text_objects)
finder.ngram_fd.tabulate(50)
finder2 = nltk.collocations.BigramCollocationFinder.from_words
```

```
(late_text_objects)
finder2.ngram_fd.tabulate(50)
```

Here is an example of what your output structure should look like:

Early List
('general', 'science') ('high', 'school') ('science', 'teachers')
('junior', 'high') ('new', 'york') ('teachers', 'college')
('elementary', 'science') ('science', 'education')
('high', 'schools') ('science', 'courses') ('senior', 'high')
('science', 'club') ('special', 'science') ('board', 'education')
('school', 'education') ('science', 'general') ('per', 'week')
('school', 'science') ('sent', 'free') ('normal', 'school')
('university', 'michigan') ('college', 'columbia')
('columbia', 'university') ('science', 'one')
('physics', 'chemistry') ('special', 'sciences')
('science', 'teaching') ('science', 'course') ('ninth', 'grade')
('specialized', 'sciences') ('introductory', 'science')
('state', 'board') ('science', 'work') ('education', 'university')
('editorial', 'board') ('york', 'city') ('science', 'quarterly')
('science', 'taught') ('science', 'association') ('year', 'junior')
('school', 'general') ('science', 'required') ('required', 'subject')
('two', 'periods') ('periods', 'per') ('science', 'elementary')
('seventh', 'eighth') ('domestic', 'science') ('club', 'new')
('new', 'england')

Late List
('science', 'education') ('gold', 'standard') ('good', 'research')
('science', 'teaching') ('educational', 'research')
('test', 'scores') ('teaching', 'learning') ('science', 'educators')
('john', 'dewey') ('standard', 'research') ('review', 'process')
('education', 'researchers') ('scientific', 'literacy')
('student', 'learning') ('research', 'science')
('general', 'science') ('manuscript', 'central')
('scholars', 'science') ('education', 'research') ('aims', 'science')
('policy', 'practice') ('learning', 'outcomes')
('nancy', 'brickhouse') ('nature', 'science') ('school', 'science')
('plenary', 'articles') ('science', 'quarterly') ('united', 'states')
('latest', 'issue') ('across', 'globe') ('research', 'programs')
('lead', 'good') ('educational', 'sciences') ('improve', 'quality')
('quality', 'education') ('educational', 'aims')
('potential', 'influencing') ('nclb', 'gold') ('value', 'science')
('everyday', 'lives') ('science', 'curriculum')

```
('education', 'programs') ('educational', 'researchers')
('way', 'getting') ('getting', 'process') ('years', 'science')
('learning', 'science') ('multicultural', 'science')
('modern', 'science') ('stanley', 'brickhouse')
```

To generate an ordered list of the most common trigrams, use the following code:

Python code snippet

```
# display frequency of highest 25 trigrams
finder = nltk.collocations.TrigramCollocationFinder.from_words
    (early_text_objects)
finder.ngram_fd.tabulate(25)

finder2 = nltk.collocations.TrigramCollocationFinder.from_words
    (late_text_objects)
finder2.ngram_fd.tabulate(25)
```

Here is an example of what your output structure should look like:

```
Early File
('junior', 'high', 'school')    ('senior', 'high', 'school')
('general', 'science', 'club') ('teachers', 'college', 'columbia')
('college', 'columbia', 'university')
('science', 'general', 'science')    ('high', 'school', 'science')
('school', 'education', 'university')   ('new', 'york', 'city')
('general', 'science', 'quarterly') ('general', 'science', 'general')
('general', 'science', 'teachers')   ('general', 'science', 'course')
('junior', 'high', 'schools')    ('year', 'junior', 'high')
('high', 'school', 'general')    ('school', 'general', 'science')
('general', 'science', 'required')  ('periods', 'per', 'week')
('general', 'science', 'elementary')    ('science', 'club', 'new')
('club', 'new', 'england')  ('normal', 'school', 'salem')
('michigan', 'schoolmasters', 'club')
('state', 'board', 'education')

Late File
('science', 'education', 'researchers')
('gold', 'standard', 'research')    ('aims', 'science', 'education')
('science', 'teaching', 'learning')
('general', 'science', 'quarterly')
('scholars', 'science', 'education')    ('lead', 'good', 'research')
```

```
('improve', 'quality', 'education')
('educational', 'aims', 'science') ('nclb', 'gold', 'standard')
('science', 'education', 'research')    ('way', 'getting', 'process')
('years', 'science', 'education') ('teaching', 'learning', 'science')
('multicultural', 'science', 'education')
('issue', 'science', 'education') ('began', 'general', 'science')
('science', 'quarterly', 'article') ('quarterly', 'article', 'john')
('article', 'john', 'dewey')
('journal', 'credible', 'significant')
('science', 'education', 'continue') ('research', 'lead', 'good')
('research', 'science', 'education') ('child', 'left', 'behind')
```

14.3.7 Concordance

Concordance is an NLTK Text object method that also looks for word distribution, but specifically searches for words found before and after a specific word of choice. Concordance allows you to find out how words are used contextually throughout the corpus. This can be particularly powerful when looking at trends over time. For example, in the sample below we search for the all the contextual occurrences of the word "science" in two separate corpora. What trends or patterns do you see between the two? What might those changes be pointing to in terms of shifts over time?

Python code snippet

```
print(early_text_objects.concordance("science"))
print(late_text_objects.concordance("science"))
```

Here is an example of what your output structure should look like:

```
Early
Displaying 25 of 199 matches:
general science quarterly editorials general scienc
cience quarterly editorials general science national educati
pectively chemistry physics biology science one paper least
ogy science one paper least general science program sessions
s many subjects sidetracked general science prominence subje
discussion provoked subject general science general science
ked subject general science general science one live issue i
 authors discoursed bearing general science later courses ph
s chemistry general feeling general science taught prepare f
```

```
erce plan two years course required science advocated replac
 advocated replace one year general science widely adopted r
ion special sciences must done make science general final sc
ust done make science general final science session dewey ga
on dewey gave us masterpiece method science teaching general
ece method science teaching general science main theme enthu
```

```
Late
Displaying 25 of 144 matches:
editorial talking issues many years science education provid
ant ideas related teaching learning science perhaps livelies
multiculturalism debate began pages science education derek
son elegant rationale multicultural science education sensit
e cultural context hodson described science education would
 standard cultural framework modern science rationale came w
eliefs experiences including rubric science would fact viole
 rubric science would fact violence science science universa
science would fact violence science science universal hence
tandard universalist account nature science flawed nature sc
ccount nature science flawed nature science fact reflect mul
 good strongly rejected idea school science curricula expand
```

Whew—you have made it to the end of our descriptive exploration! We hope you now have a rich set of jottings and memos giving you ideas for more sophisticated analyses.

At this point, based on the insights you have garnered, you can choose to utilize any of the ML techniques described in the previous chapters on unsupervised machine learning, such as Topic Modeling, K-Means Clustering, or Latent Derelicht Analysis (LDA). We encourage you to refer back to those chapters and revise the code provided as needed to conduct the analyses that best support your desired analyses.

14.4 Tasks

Comprehension

1. Discuss the importance of pre-processing in text data analysis and how it affects the outcomes of unsupervised learning models. What are issues that may occur if the pre-processing stage is skipped?
2. Describe the challenges associated with lexical ambiguity in natural language processing and how they affect text analysis.
3. How does feature extraction impact the subsequent stages of text analysis in unsupervised learning?

4. What are the primary challenges mentioned in the chapter when applying unsupervised learning to text data?
5. What is the role of stopword removal in preprocessing text data, and how might it differ depending on the research question?

Application

1. Much of the code in this chapter has been kept as simple as possible to avoid obscuring what is happening in the background. For instance, we use paths to single files in a folder, which ensures we always know what the subsequent variable names are. However, processing 100 files with this current code would be incredibly cumbersome. This is an ideal situation to use an LLM such as Chat-GPT to extend the capabilities of the existing code. Formulate a prompt using the current code as a base, asking ChatGPT to generate code that can automate the pre-processing of a folder of text or CSV files. Additionally, you will need to discuss with ChatGPT how to modify the subsequent notebook cells to work with the new variable names.
2. Develop a workflow using NLTK to pre-process, extract features, and conduct unsupervised analysis on a corpus of your own choosing.
3. Take that text corpus and apply different pre-processing techniques such as tokenization, stemming, and stopword removal. Compare the outputs and discuss how each technique affects the analysis of the text.
4. Implement an n-gram analysis of the dataset. Identify the most frequent bi-grams and tri-grams and interpret what their frequency might suggest about the dataset's content.
5. Construct a system using unsupervised learning to automatically generate summaries of the dataset, by identifying key themes and patterns in the text.

References

Martin, J. (2016). The grammar of agency: Studying possibilities for student agency in science classroom discourse. *Learning, Culture and Social Interaction, 10*, 40–49.
Rosenberg, J. M., & Krist, C. (2021). Combining machine learning and qualitative methods to elaborate students' ideas about the generality of their model-based explanations. *Journal of Science Education and Technology, 30*, 255–267.

Chapter 15
Triangulating Computational and Qualitative Methods to Measure Scientific Uncertainty

Joshua M. Rosenberg, Hadi Bhidya, and Cody Pritchard

Abstract This chapter outlines steps to analyze a complex construct of interest to science education researchers in a very commonly used digital media platform, YouTube, particularly popular science education-related videos. The construct of interest is uncertainty—established as important but challenging for teachers and researchers alike to recognize and understand as many definitions and operationalizations of uncertainty as it relates to learning science exist. To study uncertainty, transcripts of videos are created using Python and the Python packages pytube and Whisper, and a two-step triangulation approach that combines a computational (a dictionary-based text analysis) and qualitative approach. In the text analysis step, transcripts of videos are searched for key uncertainty-related terms using the statistical software R. Next, qualitative coding of the transcripts is carried out, with the output from the first step as a support for the task of developing an initial set of codes for the types of uncertainty present in the science education videos.The proposed chapter contributes to the book by providing a practical guide for researchers interested in studying complex constructs using an approach that merges some of the benefits of quantitative and qualitative approaches. Python and R code are provided to support researchers to replicate and draw on the analysis carried out.

15.1 Introduction

Educational stakeholders are often most interested in supporting students in developing competencies that are complex, like an identity as someone who can do science (Brickhouse and Potter, 2001) or students' conceptual understanding of scientific

J. M. Rosenberg (✉) · H. Bhidya · C. Pritchard
University of Tennessee, Knoxville, USA
e-mail: jrosenb8@utk.edu

J. M. Rosenberg
513 Claxton, 1122 Volunteer Blvd., Knoxville, TN 37996, USA

H. Bhidya · C. Pritchard
1122 Volunteer Blvd., Knoxville, TN 37996, USA

313

P. Wulff et al. (eds.), *Applying Machine Learning in Science Education Research*,
Springer Texts in Education, https://doi.org/10.1007/978-3-031-74227-9_15

mechanisms or processes (Krist et al., 2019). A key challenge pertinent to support-ing students to develop such competencies is measuring or assessing them, which is concomitantly complex. One way to approach the complexity of measuring complex educational outcomes is to triangulate methodological approaches. In this chapter, we pursue such a triangulation approach.

Our triangulation strategy for measuring complex constructs differs from more conventional approaches in that it involves the combination not of quantitative and qualitative methods, but instead from combining *computational* and qualitative meth-ods in a manner inspired by (Nelson, 2020) in the *computational grounded theory* three-step approach. Such an approach has been used in some prior science education research (Rosenberg and Krist, 2021). Here, we loosely follow this approach's first two steps of *computationally exploring* a large and complex data source (the first of the three computational grounded theory steps) and then *qualitatively analyzing* the data in light of the output of the first step (the second computational grounded theory step). We do so with a focus on the construct of uncertainty, a central element of science teaching and learning (Manz and Suárez, 2018).

The goal of this chapter is to demonstrate how we employ a triangulation approach. Our approach loosely follows Nelson (2020) computational grounded the-ory approach to specifically measure a complex construct (i.e., the use of uncertainty within an informal science learning environment (Youtube)). While this chapter specifically measures how uncertainty is used with science learning videos on YouTube, a similar approach can be applied to measuring other scientifically-related complex constructs in other online educational spaces (e.g., Khan Academy and Coursera), with the exception of YouTube-specific data mining techniques discussed within the chapter.

In addition to showing how we can carry out a triangulation approach to measure a complex construct, we have a secondary aim: showing how a relevant data source can serve as the basis for a new kind of learning analytics- or educational data science-inspired investigation of uncertainty as it pertains to science teaching and learning. Namely, we show how we can efficiently collect data from a large corpus of science education videos, using the speech (and transcripts of speech that we can create) to play out and test the ideas about the nature of uncertainty that have been explored primarily in classroom contexts, but not in the context of educational media.

Before describing our data collection and data analysis method, we first briefly review some of the prior research on uncertainty and science education in the next section.

15.2   Prior Research on Uncertainty and Science Education

Uncertainty is ubiquitous in conversations among scientists Kirch (2008, 2010) and is a fundamental driver for continued research and the refinement of knowledge Allchin (2012). The *Framework for K-12 Science Standards* states that "Scientific knowledge is a particular kind of knowledge with its own sources, justifications, ways of dealing with uncertainties, and agreed-on levels of certainty" (National Research Council, 2012, p. 251). Despite its presence and importance in scientific inquiry and learning, there is little consensus on how uncertainty is conceived in specific domains.

In the scientific community, uncertainty is communicated alongside its relationship with probability and often refers to the level of variability that exists in relationship to a scientific model or hypothesis. However, uncertainty is not a concept exclusive to the scientific community, but rather is a fact and condition present in our daily lives Pollack (2003). Jordan and McDaniel (2014) define uncertainty as a cognitive feeling or "an individual's subjective experience of doubting, being unsure, or wondering about how the future will unfold, what the present means, or how to interpret the past" (p. 492). (Kahneman and Tversky, 1982) discuss how variants of psychological uncertainty, which may not follow rules of formal scientific inquiry, may be correlated with expressions within our natural language. The discourses surrounding our understanding of uncertainty are complex enough that some researchers have suggested discipline-specific typologies of uncertainty because data types and scientific models vary so widely based on one's domain Bateman et al. (2022).

To determine how "uncertainty" is conveyed through language, we employ (Kirch, 2010) typology of uncertainty as a starting point for our exploration of uncertainty in educational videos. Kirch (2010) define uncertainty as both a psychological experience and a *mathematical object*. Kirsch writes, "uncertainty refers to a psychological condition of being in doubt (e.g., I am uncertain about something or someone...). It also refers to a statistical (or mathematical) object (e.g., a statistical estimation of uncertainty)" (Kirch, 2011, p. 57). As a psychological condition, uncertainty can be procedural, sociocultural, epistemological, and ontological. As a statistical object, it refers to measurement, sampling, repeatability, and predictive value Kirch (2011). We use this framework later in our analysis, focusing on psychological uncertainty as an initial step.

Thus, our aim in this analysis is to try to understand the nature of psychological uncertainty as it is expressed in educational videos in the science domain. To do so, we triangulate methodological approaches, using both an automated, computational approach, and a qualitative approach.

15.3 Data Analysis Overview

This is structured into four distinct sections. The first section—Data Analysis Step #1—guides you through extracting audio streams from YouTube playlist URLs using 'Pytube'. In the second section—Data Analysis Step #2—we focus on converting these audio files into transcriptions, which can be in the form of SRT or plain text files.

Each section is designed to be comprehensive, providing step-by-step instructions to ensure a smooth and effective learning experience.

We then proceed to two analytic steps, one computational (Data Analysis Step 3) and one qualitative (Data Analysis Step 4).

15.4 Data Analysis Step #1: Programmatically Accessing YouTube Video Data

The primary aim of this and the second step is to develop a method that converts YouTube playlists into transcriptions of all videos within those playlists. For these steps, we employ Python. If you haven't used python before, the easiest way to get started in our view is to download the Anaconda distribution, a particular version of Python combined with some already-included python libraries and tools for using Python: https://www.anaconda.com/.

We will be using two Python libraries: 'pytube' for downloading videos and 'whisper' for converting audio to text.

To replicate this process, the prerequisites are minimal. You'll need a computer capable of running Python and some understanding of how Python works. For beginners, assistance from AI tools like ChatGPT can also be invaluable in navigating the learning curve.

15.4.1 Converting Playlists to Audio

Pytube is a highly capable Python library, freely available for interacting with YouTube via URL links. Its primary function is to download audio and video streams, captions, and search results. In this first part of our tutorial, we will focus on using Pytube to retrieve audio streams from YouTube playlists.

Before diving into the code, it's essential to install Pytube. This can be done by running the following command in your terminal or command prompt:

Python code snippet: installing pytube

```
pip install pytube
```

Additionally, familiarizing yourself with Pytube's documentation (available at https://pytube.io/en/latest/) is highly recommended. It provides a wealth of information and will enhance your understanding of the library's capabilities.

This segment of the tutorial will guide you through creating a program that downloads and saves the audio streams of a YouTube playlist's videos into a specified folder using Pytube.

Start by importing the Playlist class from Pytube and then create a Playlist object by passing the URL of the playlist—here, for a playlist from the YouTube channel Veritasium on the physics concept of intertia—as a string:

Python code snippet: importing the playlist class and reading in a playlist URL

```
from pytube import Playlist
playlist_url = 'https://www.youtube.com/playlist?list=PLAB27A3C12C31E663'
playlist = Playlist(playlist_url)
```

Iterate through each video in the playlist using a for loop. The loop will handle two main tasks: extracting the highest resolution audio stream and downloading it to a specified path. This ensures all videos in a playlist are stored in a single folder, aiding in organization and future analysis. We did this for around one dozen playlists from two channels: Crash Course (https://www.youtube.com/user/crashcourse) and Veritasium (https://www.youtube.com/channel/UCHnyfMqiRRG1u-2MsSQLbXA).

Python code snippet: iterating through each video

```
for video in playlist.videos:
    try:
        # Get the highest resolution audio stream, the first

        audio_stream = video.streams.filter(only_audio=True).first()

        # Download the audio stream and save it to the path below
        # For each playlist downloaded, change the folder it goes to

        print(f'Downloading: {video.title}')
        audio_stream.download(output_path=
        '/Users/actualuser/Desktop/path/to/save/veritasium/Inertia')

    except Exception as e:
        print(f"Error downloading {video.title}: {str(e)}")
```

By following these steps, you will be able to download audio streams from YouTube playlists, setting the stage for the next part of our tutorial where these audio files will be transcribed. We did this for a total of 282 videos, 146 from Crash Course and 126 from Veritasium.

15.5 Data Analysis Step #2: Creating Transcripts Using an Automatic Speech Recognition Tool

Whisper, developed by OpenAI (the same organization behind ChatGPT), is an open-source speech recognition model designed to handle a variety of tasks involving audio files, including automatic speech recognition. Some early work suggests that it may be better than other automatic speech recognition tools, especially when the audio is not wholly clear (Palaguachi et al., 2023).

In this part of our project, we'll use Whisper to convert the audio files we've gathered into subtitle—transcript—files.

To work with Whisper, you need to install its Python library. This can be done using the command:

Python code snippet: installing the Python package Whisper

```
pip install -U openai-whisper
```

Additionally, you'll need the Command Line tool 'ffmpeg', essential for handling multimedia files. Detailed instructions and additional information about Whisper, including its GitHub repository, can be found at https://github.com/openai/whisper. Note that the other required libraries for this part of the project are already included in the standard Python installation, so there's no need for additional downloads.

Whisper, while robust, does have some known limitations. Specifically, it may struggle with transcribing videos longer than 10 minutes and could inaccurately insert words like "oks" and "yeahs" in the transcript. If you encounter these issues or want to learn more about potential solutions and workarounds, visit the discussion at https://github.com/openai/whisper/discussions/679. This page offers valuable insights and community-driven advice on addressing these challenges.

This program utilizes Whisper to transcribe audio streams into subtitle files from the folder created using Pytube.

Begin by importing the necessary libraries. Whisper for transcribing audio, os and pathlib for handling file paths, and get_writer from Whisper utils for creating subtitle files.

Python code snippet: importing Whisper and other libraries

```
import whisper
from whisper.utils import get_writer
import os
import pathlib
```

Define the folder containing your audio files and convert it into a path object for easier handling. This should be the same path the files in the above step were saved to. Then, generate a list of all files in the specified directory, filtering only audio files (in this case, .mp4 files).

Python code snippet: installing the Python package Whisper and listing the audio files

```
directory_path = r"path_to_files"
directory_path = pathlib.Path(directory_path)

all_files = os.listdir(directory_path);
all_mp4 = [audio for audio in all_files if audio.endswith(".mp4")]
```

Next, load the Whisper model with a suitable model size. The choice of model ('small', 'medium', and 'large') impacts the speed and accuracy of transcription, with larger models potentially being more accurate, but also taking considerably longer. For this tutorial, we use the 'small' model, though for research uses, the additional time may be worth the investment. A counter is used to track progress.

Python code snippet: loading the Whisper model

```
model = whisper.load_model("small")
```

Python code snippet: iterating over each audio file

```
i = 0
for mp4s in all_mp4:

    # Convert the current .mp4 file to a string as a parameter

    mp4s_path = directory_path / mp4s
    result = model.transcribe((str(mp4s_path)), fp16=False)

    # whisper.utils get_writer will output the text with timestamps
    srt_writer = get_writer("srt", directory_path)
    srt_writer(result, (str(mp4s_path)))

    # printing out the name of the file just written
    name = str(mp4s)
```

```
print(name)
i += 1
print(i)
```

The transcript files should be saved to the same folder as the audio files. These end in the extension `.srt`.

The result was transcripts for each of the 272 videos we accessed in the last stpe. On average, each video has 8 minutes, 12 seconds of speech, for a total of 36,780 utterances, or around 1.54 days worth of speech (36.96 hours, or 2,217 minutes). In other words, a fairly large collection of transcribed speech data, motivating the computational approach described next.

15.6   Data Analysis Step #3: Computational Analysis

Let's pick up where we left off from the first two steps, with one big difference – we'll be using R, a statistical software and programming language used in other chapters in this book—instead of Python.

You can find instructions on downloading R and RStudio here, in a chapter in the book *Data Science in Education Using R* (Estrellado et al., 2020): https://datascienceineducation.com/c05.

We'll assume some basic knowledge of R here.

First, let's load the tidyverse library—a set of R packages that work together for common analytic tasks.

```
library(tidyverse)
```

Then, let's find the `.srt` files we created in the last step and then read them in, saving them to the object l. We have now read in the transcripts! They should look like this (here is the first one, accessed by indexing the first list item—the first ten rows):

R code snippet and output: Reading in the transcripts

```
file_paths <- list.files(path = ".",
                        pattern = "\\.srt$",
                        recursive = TRUE,
                        full.names = TRUE)

l <- map(file_paths, read_lines)

l[[1]] %>%
  head(10)
```

```
## [1] "1"
## [2] "00:00:00,000 --> 00:00:07,160"
## [3] "Hey folks, Phil Plait here, and for the past few episodes"

## [4] ""
## [5] "2"
## [6] "00:00:07,160 --> 00:00:11,880"
## [7] "know about the structure, history, and evolution of the universe"
## [8] ""
## [9] "3"
## [10] "00:00:11,880 --> 00:00:14,160"
```

Now, we have our transcripts loaded, and our data ready. The next step is a big one that we'll introduce primarily through comments—this is code to create a manual *function* that we will use to read each of the transcript files:

R code snippet: A function to process transcripts

```
process_transcripts <- function(d) {

  my_nrow <- length(d) # this is to find out how long each transcript file is

  d %>%
    as_tibble() %>%
    rename(X1 = 1) %>% # to make this easier to type
    # create different values for each of the rows of the transcript
    mutate(id = rep(c("i", "time", "transcript", "blank"), my_nrow/4),
           index = rep(1:(my_nrow/4), times = 4) %>% sort()) %>%
    spread(id, X1) %>% # change the data from long to wide format
    select(-i) %>%
    # process the time stamps
    separate(time, into = c("start", "end"), sep = "-->") %>%
    # trim the time stamps so they are easier to read and use
    mutate(start = str_trim(start),
           end = str_trim(end))
}
```

Whereas with Python we used a "for loop", in R, for loops are less common than *apply* functions. These two approaches share a commonality: they are both used for iteration. Given the kinds of data R is chiefly intended to work with and how R as a programming language most efficiently works, apply functions are generally the better way to go. Here, we will use the map() function that is a part of the tidyverse package (specifically, the purrr package).

R code snippet: Iterating to create a single data frame with processed transcripts

```
ll <- l %>%
  # here, we use the apply function; possibly is used to handle errors
  map(possibly(process_transcripts, NULL))

# this removes any NULL list items that resulted from errors
ll <- compact(ll)

# this adds an index for the rows associated with each transcript

ll <- imap(ll, ~ mutate(.x, group = .y))

bound_rows <- ll %>%
  map_df(~.) # this changes the list of data frames into a single data frame
```

We are getting close to ready for analyses. Next, we process the transcript to create several variables that will be useful for our analysis. Most important among these is two variables we have created:

For these purposes, we employ Kirch's (2011) typology of uncertainty as both a **psychological experience** and a **mathematical object**. Kirsch writes, "uncertainty refers to a psychological condition of being in doubt (e.g., I am uncertain about something or someone...). It also refers to a statistical (or mathematical) object (e.g., a statistical estimation of uncertainty)" (Kirch, 2011, p. 57). As a psychological condition, uncertainty can be procedural, sociocultural, epistemological, and ontological. As noted earlier, we focus on psychological uncertainty as a starting point and illustration.

Our computational approach is a dictionary-based approach (see Nelson et al., 2021 for a definition). This approach is a relatively straightforward text-analysis technique—it involves searching for key words in text. The dictionary is provided by the analayst, but we can use R to conduct the search automatically. We acknowledge that more complex approaches could be helpful, but we chose this approach given our aim of triangulating evidence—qualitative analyses can complement this approach by providing context and depth to what the computational approach reveals.

Our dictionary corresponding to our conception of psychological uncertainty follows.

R code snippet: Defining dictionaries

```
psychological_uncertainty <- c(
  "unsure", "not sure", "maybe", "kind of", "sort of", "don't know",
  "doubt", "doubtful", "no clue", "unclear", "confused", "confusing",
  "hesitant", "don't get", "don't understand", "ambivalent",
```

```
        "can't decide", "questioning", "question", "wondering", "wonder",
        "weird", "strange", "odd", "weirded out", "puzzled", "puzzling",
        "don't get it", "weird feeling", "weirdly", "skeptical", "skeptic",
        "guessing", "guess", "vague", "ambiguous", "indefinite",
        "uncertain", "iffy", "on the fence", "mixed up", "unsure what to do"
)
```

First, let's process the transcripts a bit further to create some useful variables and select columns.

R code snippet and output:

```
out <- bound_rows %>%
  mutate(start = str_sub(start, 1, 8),
         end = str_sub(end, 1, 8)) %>%
  mutate(start = chron::chron(times = start),
         end = chron::chron(times = end)) %>%
  mutate(duration = end - start) %>%
  select(index, start, end, duration, everything())
```

Next, we can apply these lists.

R code snippet: Conducting the dictionary-based analysis

```
# function to count words from a dictionary in a text
count_words <- function(text, dictionary) {
  sum(str_count(text, paste0("\\b", dictionary, "\\b")))
}

# apply the function to each row of your dataframe
out <- out %>%
  mutate(transcript = tolower(transcript)) %>%
  mutate(
    count_psychological_uncertainty = map_dbl(transcript,
    ~count_words(.x, psychological_uncertainty))
  )
```

The result of this step is the following data frame (represented through the use of an R function that summarizes data frames) below. We can see that the data frame contains over 36,000 rows, one for each utterance in the video. We can also see counts of psychological uncertainty for each utterance.

R code snippet: Resulting transcript

```
> d %>% glimpse()
Rows: 36,780
Columns: 12
$ group                             <dbl> 1, 1, 1, 1, 1, 1, 1, 1, 1, 1...
$ channel                           <chr> "crash course", "crash cours...
$ playlist                          <chr> "astronomy", "astronomy", "a...
$ video                             <chr> "audio_A Brief History of th...
$ start                             <time> 00:00:00, 00:00:07, 00:00:1...
$ end                               <time> 00:00:07, 00:00:11, 00:00:1...
$ duration                          <time> 00:00:07, 00:00:04, 00:00:0...
$ blank                             <lgl> NA, NA, NA, NA, NA, NA, NA, ...
$ transcript                        <chr> "hey folks, phil plait here,...
$ count_psychological_uncertainty   <dbl> 0, 0, 0, 0, 0, 0, 0, 0, 0, 0...
$ path                              <chr> "./crash course/astronomy/au...
```

We can briefly explore the prevalence of psychological uncertainty with the following R code, which shows us that 747 utterances contain one word from our psychological uncertainty dictionary, and 36 utterances contain two.

R code snippet: Exploring the frequency of words associated with psychological uncertainty

```
> d %>% count(count_psychological_uncertainty)
# A tibble: 3 x 2
  count_psychological_uncertainty       n
                            <dbl> <int>
1                               0 35997
2                               1   747
3                               2    36
```

We are now ready to proceed to the qualitative analysis phase.

15.7 Data Analysis Step #4: Qualitative Analysis

In this step, we conduct a qualitative analysis in two phases.

15.7.1 Inspecting the Utterances with the Most Uncertainty Detected

First, we inspect the utterances with the most uncertainty detected for psychological uncertainty. We do so by arranging the above data frame in descending order based simply on the *count* (or frequency) of the number of uncertainty-related words in our dictionary detected.

For psychological uncertainty, we examined the 50 utterances with the most uncertainty-related words by reading the utterances and considering them in light of our definition of psychological uncertainty: "a psychological condition of being in doubt" (Kirch, 2011, p. 57), which includes procedural, sociocultural, epistemological, and ontological elements.

The most common forms of psychological uncertainty were procedural and epistemological.

Epistemological uncertainty was fairly common, evidenced by around one-third of the utterances with the greatest amounts of psychological uncertainty detected. Examples are as follows.

- "We don't know what kind of atmospheres these planets will have or what they're composed." (Crash Course—Astronomy)
- "like maybe that animal is hard to find in the wild, or maybe it can't be kept in captivity." (Crash Course—Zoology)

Procedural uncertainty was also fairly common, present in around one-half of the utterances.

- "Well, we don't know what we don't know." (Crash Course—Biology)
- "i guess maybe about that far?" (Veritasium—Misconceptions)

We also saw a degree of measurement or statistical uncertainty, even in these utterances that the computational analysis suggested were psychological in nature; these were relative uncommon:

- "maybe the observation is wrong, or maybe we're misinterpreting it." (Crash Course—Astronomy)
- "although the numbers are a little bit uncertain, something like a third to half of all stars"

As this is a tutorial, let us consider that a more systematic qualitative analysis indeed suggested that there are three forms of uncertainty that are common in science education videos: epistemological, psychological, and measurement. Were we to expand on this analysis, we would likely want to qualitatively investigate other videos to understand whether forms of uncertainty—and to substantially deepen our analysis of the utterances above by understanding their context in the videos and how truly common (or not) they are across the entire set of transcript data. For now, our purpose was to demonstrate how a finer-grained qualitative approach could complement the automated computational approach we carried out in the last step.

15.8 Findings and Discussion

In this chapter, we sought to demonstrate how a triangulating approach that combines computational and qualitative methods could be used to measure uncertainty. We played out this approach at the same that that we developed a data set that could be suitable for answering it—a collection of 272 science education-related YouTube videos. We showed four data analysis steps:

1. **Downloading YouTube Videos:** The study utilized Python libraries, namely 'pytube' and 'whisper', to extract and transcribe audio streams from selected YouTube playlists. This process efficiently converted a substantial corpus of science education-related videos into a format suitable for text-based analysis.
2. **Transcription Using Whisper:** The Whisper tool was employed to transcribe the audio files into textual data. This transformation was crucial in standardizing diverse video content into a uniform textual format, primed for computational analysis.
3. **Computational Text Analysis:** Applying a computational approach with the use of R, the study focused on a dictionary-based text analysis to identify the presence and frequency of terms related to scientific uncertainty in the video transcripts. This quantitative analysis provided an overarching perspective of the manifestation of uncertainty in the video content.
4. **Qualitative Analysis:** Complementing the computational analysis, a qualitative examination of the context and nuances surrounding uncertainty-related terms in the videos was conducted. This approach offered a deeper and more nuanced understanding of the nature and presentation of scientific uncertainty in the educational content.

The findings from this tutorial suggest that the representation of scientific uncertainty in educational videos is predominantly characterized by procedural and epistemological uncertainties. We note that the intent of working through these four steps was to illustrate how to access and create transcripts of YouTube videos and to demonstrate a triangulation approach. Of course, more systematic inquiry would be necessary to substantiate this finding. Here, we showed the very first stages of doing so, setting the stage for the establishment of the reliability and validity of a measure that we could use to answer substantive questions about the nature of uncertainty in educational videos. Later, such an approach could help us to better understand the role of uncertainty in science teaching and learning within and beyond classroom settings. We also note that the methodology delineated here can be adapted for analyzing other complex constructs across various digital platforms, thereby expanding the research scope within the field of educational technology and digital media analysis.

15.9 Tasks

Comprehension Tasks

1. Explain the two main steps in the triangulation approach used in this study to analyze uncertainty in science education videos.
2. Summarize why uncertainty was chosen as a focal construct for this analysis.
3. Describe the different use cases of python and R in the example.
4. Summarize Kirch's (2011) typology of uncertainty as presented in the text.
5. Explain what a dictionary-based text analysis approach is and how it was used in this analysis.

Application Tasks

1. Design a dictionary-based approach to analyze another complex construct in science education (e.g., scientific reasoning). Provide a list of ten or more key terms you would include in your dictionary.
2. Outline a research plan to apply the triangulation method described in this chapter to study the representation of a different scientific concept (e.g., evolution or climate change) in online educational resources.
3. Propose modifications to the computational analysis step that could potentially improve the detection of uncertainty in the video transcripts.
4. Develop a coding scheme for a more detailed qualitative analysis of uncertainty in science education videos, based on the findings from this study and your own insights.
5. Create a plan to validate the findings from this triangulation approach, including suggestions for additional data sources or methods that could be used to crosscheck the results.

Link to analytic code (python and R) in the OSF: https://osf.io/v2x7j/.

References

Allchin, D. (2012). Teaching the nature of science through scientific errors. *Science Education, 96*(5), 904–926.

Bateman, K. M., Wilson, C. G., Williams, R. T., Tikoff, B., & Shipley, T. F. (2022). Explicit instruction of scientific uncertainty in an undergraduate geoscience field-based course. *Science & Education, 31*, 1541–1566.

Brickhouse, N. W., & Potter, J. T. (2001). Young woman's scientific identity formation in an urban context. *Journal of Research in Science Teaching, 38*, 965–980.

Jordan, M., & McDaniel, R. R. (2014). Managing uncertainty during collaborative problem solving in elementary school teams: The role of peer influence in robotics engineering activity. *Journal of the Learning Sciences, 23*(4), 490–536. https://doi.org/10.1080/10508406.2014.896254

Kahneman, D., & Tversky, A. (1982). Variants of uncertainty. *Cognition, 11*(2), 143–157.

Kirch, S. A. (2011). Understanding scientific uncertainty as a teaching and learning goal. In B. Fraser, K. Tobin, & C. McRobbie (Eds.), *Second international handbook of science education, volume 24 of Springer international handbooks of education*. Dordrecht: Springer.

Kirch, S. A. (Ed.). (2008). *A comparative science study: Uncertainty in the laboratory and in the science education classroom*. New York.

Kirch, S. A. (2010). Identifying and resolving uncertainty as a mediated action in science: A comparative analysis of cultural tools used by scientists and elementary science students at work. *Science Education, 94*, 308–335.

Krist, C., Schwarz, C. V., & Reiser, B. J. (2019). Identifying essential epistemic heuristics for guiding mechanistic reasoning in science learning. *Journal of the Learning Sciences, 28*(2), 160–205.

Manz, E., & Suárez, E. (2018). Supporting teachers to negotiate uncertainty for science, students, and teaching. *Science Education, 102*(4), 771–795.

National Research Council (2012). A framework for K-12 science education: Practices, crosscutting concepts, and core ideas. National Academy of Sciences.

Nelson, L. K. (2020). Computational grounded theory: A methodological framework. *Sociological Methods & Research, 49*(1), 3–42.

Nelson, L. K., Burk, D., Knudsen, M., & McCall, L. (2021). The future of coding: A comparison of hand-coding and three types of computer-assisted text analysis methods. *Sociological Methods & Research, 50*(1), 202–237.

Palaguachi, C., Cox, E., Rosenberg, J., Dyer, E., & Krist, C. (2023). Automatic speech recognition (asr) in noisy classrooms: Evaluating the usefulness of three popular asr tools. In *Learning sciences graduate student conference 2023*.

Pollack, H. N. (2003). *Uncertain science. . .Uncertain world*. New York: Cambridge University Press.

Rosenberg, J. M., & Krist, C. (2021). Combining machine learning and qualitative methods to elaborate students' ideas about the generality of their model-based explanations. *Journal of Science Education and Technology, 30*, 255–267.

Part III
Future Directions

Chapter 16
Risks and Ethical Considerations in the Context of Machine Learning Research in Science Education

Cynthia M. D'Angelo

Abstract Besides the tremendous potentials of machine learning (ML) methods, many ethical challenges such as biased datasets with regard to gender or race have to be considered. In this chapter, a conclusive reflection on the particular challenges in science education research based on the case studies and prior research will be outlined. Paths to address these challenges will finish this chapter.

16.1 Bias and Ethics and Equity

Hopefully you are committed to minimizing biases in your research and interrogating your process in order to help achieve this goal. It may not be possible to remove all biases, but the more that you can engage in the reflection and strategies necessary to minimize biases, the more your work will be able to address issues of equity and justice in science education.

There are lots of potential biases to consider: race/ethnicity, language/linguistics, gender, disability, and socio-economic status. While that is a long list of biases to consider and interrogate in your work, the more you can address these biases, the stronger your work will be and more able to address the true diversity of experiences that students and teachers bring into a science classroom or learning environment.

It's not magic. It's math. It's important to remember that as you use these advanced techniques. They do not magically get rid of the biases in our society by doing complicated math. The biases come from the humans that are creating these ML methods and models and from the data being fed into them, all of which reflect the biases of our society. These ML techniques have been created to do specific things by humans. The more you can understand these motivations and the designed use cases of these different approaches, the more you will be able to understand the inherent trade-offs when making decisions about whether or not to use ML techniques and which one is most appropriate for your purposes and research situation.

C. M. D'Angelo (✉)
1310 S. Sixth St., Champaign, IL 61820, USA
e-mail: cdangelo@illinois.edu

© The Author(s) 2025

P. Wulff et al. (eds.), *Applying Machine Learning in Science Education Research*,
Springer Texts in Education, https://doi.org/10.1007/978-3-031-74227-9_16

Part of this equity-focused approach is the need to be intentional about what you mean by equity. What or who are you designing for? Who are you centering? Who is being marginalized in this process? What kinds of questions are you asking and how are these questions privileging certain ways of being in a science class or teaching science? What does it mean for a ML model to be "accurate"? Who is it accurate for? Under what circumstances and contexts is it accurate? Is your model only accurate for students who fall in the most common categories that you are looking at or is it more inclusive of students that typically fall outside those majority categories? With ML techniques, you need to think carefully about low-occurrence categories or situations and the students/teachers that fall into them. It is much more difficult to accurately model these low-occurrence events, so if these kinds of events are something you are interested in, you might want to consider different methods or modifications that will allow more of a focus on these events. If you are working with text data and natural language processing types of approaches that might help with auto-grading or evaluating short answers in science there are many issues to be aware of. For short science answers specifically, it's really complicated to do well. If you just want to check for some keywords or simple constructions, that's not too difficult, but it's a simple approach and will give you limited information about the science concepts. It also privileges native English speakers, especially those who are particularly good at school English. You need to also ensure that you're not just checking to see if someone knows science vocabulary, but actually understands the concepts behind those words. That is much harder. More recent advances in ML are improving this type of task, but it is always important to look into what kinds of science answers were being used to train these more advanced models and who is represented (and not) in those data sets.

If you are working with video or image data and are using vision-based ML approaches you have another set of challenges. For instance, if you're trying to extract human skeletons to look at, you need to think about students with disabilities and why those skeletons (and the resulting analysis) might not be accurate/fair/appropriate for certain students and contexts. This challenge is not just for visible physical disabilities, but also for students that exhibit neurodivergent behaviors and how that might show up when tracking a person's movement. Again, with vision-based ML, it is essential to understand about the images (and labels) that have been used to train the models, as these historically have been not representative of the diversity of our student populations.

16.2 Purposes and Trade-Offs

When considering whether or which ML techniques you should use with your research, you need to think about your research questions and what you are trying to achieve with your research and for whom. In order to address the challenges of these kinds of approaches you will have to make decisions that involve trade-offs with different approaches. Part of this process is to think carefully about why you are using ML and what your goals with it are. Would other kinds of analyses or methods be better? Sometimes there is a tendency to want to use the more "advanced" or newer techniques just because they are more advanced or newer. But that doesn't mean that they are better for answering your particular research questions with your data set and context. What are the trade-offs with different kinds of uses of ML or algorithms? Is ML the right tool for accomplishing your goals or would another approach be more appropriate?

One potential pitfall is to choose ML techniques in order to be more efficient with your research. Prioritizing efficiency can lead to problems—you are always making a trade-off when choosing one approach over another. You might want to ask yourself why you are prioritizing efficiency. Why is it important that this process be efficient? Is it because you don't have sufficient resources to do this a different way? Is that a good enough reason to risk the many potential issues with a ML approach?

If you are using ML to predict outcomes for students or teachers, there are additional questions to consider. What is the goal of prediction for your study? Will the predictions end up coming back to the students or the teachers? How might a prediction about their future behavior or learning affect them? Care needs to be given if you are going to reveal these predictions to students or teachers, making sure to message effectively about their ability to change. It also means that you need to be even more sure that there are not major biases or errors in your model and analysis. It raises the stakes considerably for your analysis and you should be even more intentional in your design and reflection on your data set and approach.

Are these data about people learning (i.e., something that is in progress) or is it assessment data to evaluate learning that has happened? There are different considerations to make about your models depending on the kind of data you have. The stakes are higher for assessment or evaluative models, so it's more important to consider the ways in which your model might be biased towards certain groups or certain kinds of outcomes. These types of data are also typically missing context to a larger extent than in-progress learning data are. The stakes are lower for process learning data, but also it's important to consider the nature of learning (or teaching) and wanting to perhaps reach different kinds of conclusions or produce different kinds of models with this type of data.

16.3   Data Characteristics

The plan for collecting the data (including the structure of it, the modality of it, the levels of it, the conditions under which it was collected) to be used in your machine learning approach is the most important part of the process. The more you know about your data and why and how it was collected in the way that it was, the better able you will be to make careful and considered decisions about the construction of your model (including important pre-processing steps).

The modality of the data can also prioritize certain people and/or certain ways of demonstrating knowledge and skills. So it's important to consider these questions at the beginning of your research, when you are planning your data collection strategies, not just toward the end when you're doing analysis.

Is this (the output) going to narrow avenues for students or expand them? There are lots of different ways students can show up and demonstrate their knowledge. What do you do with any outliers? What do outliers even mean in this situation? Sometimes outliers are just ignored or even deleted. But, in a lot of science education research contexts, these outliers are students. Would it make sense to ignore a student? Trying to fit people into boxes or categories just to put them into those boxes or categories may not be a good use of this kind of technology/tool, even if that is what it is especially good at.

There is also the issue of context in your data set. How much information do you have about the contexts under which the data were collected? How much information should you have? How do different contextual factors change within your data set and how are you taking that into account in your model?

Missing data is another important aspect of your data set to consider when using ML techniques. There are different kinds of missing data. Are the data missing because a student was absent that day? Or is it incomplete because a student wasn't able to finish an assignment? Do you, as the researcher, know why a student didn't finish something? Could there have been a fire alarm in the building or a medical issue or did they run out of time or did they not know the answer? As much as possible it's important to know about why the data are missing—what that incompleteness means—as different kinds of missing data need to be handled differently in how you process and interpret your data.

16.4   Privacy, Transparency, and Agency

Privacy is an important issue to consider not just when reporting your results but throughout the whole process, including the plan for your data collection. One way to help protect your participants' privacy is to not collect more data than what you actually need to answer your research questions. That includes meta-data that could potentially identify or expose your participant if the data were stolen. Some advanced ML techniques are able to identify individuals, even if names aren't included, because of the amount and detail of the data collected. So, taking care to protect and anonymize data as early as possible can be crucial to protecting the privacy of your participants.

One way to mitigate some issues related to privacy is to allow your research participants more agency in the data collection process. For instance, you might build in ways for participants to opt-out temporarily of data collection when they don't want certain things about them being collected. Giving participants more agency by allowing them to choose how and when their data is being collected and providing them more context and information about how their data is going to be used by you and your team should be an important part of regular ethical practice.

You also need to think about what kinds of services you are accessing when using different ML approaches. For instance, some software might require you to upload your data to their servers in order to use the algorithms. This type of access is typically disallowed by the guidelines of most ethics boards (e.g., Institutional Review Boards)—mostly due to the risk of putting your potentially sensitive education-related data on the servers of a company. Local solutions (that is, solutions that reside on your local computer or on a server that you or your institution control) are occasionally more difficult to implement, but are much safer for your participants.

16.5 Paths to Address These Challenges

There is no one central path to address these challenges. The challenges themselves, as outlined above, are myriad and depend on many factors unique to your data set, research questions, and science education setting. But a set of questions can help with finding the right path for you to wrangle the challenges that you face. The sections above contain many questions that you can ask of yourself as a researcher, your dataset, and your models in order to help minimize the challenges of using these kinds of techniques.

Data and algorithm auditing can be an important strategy to help mitigate some of the risks of using these techniques. This involves scheduling time as part of your project to intentionally investigate your data and the algorithms used and models produced for biases. You can proactively look for instances of different kinds of bias in your data. If you find them, you can then make changes to whichever part of the process you find the bias. Additionally, being transparent about this process and reporting it along with your more typical results can help others see the limitations and caveats with your results (that are true with all findings, but are not always disclosed). This can also help you and others be more intentional in your next data collection plan to help minimize these biases in the future.

Part of the solution to address these challenges with the risks of ML approaches is to use these tools conscientiously, understanding the risks, and only when willing to mitigate the challenges and be responsible. Continuously educating yourself, students, and research partners on how these algorithms work and how they could impact their lives (or learning or teaching) remains an important element of conducting ethical ML research.

Chapter 17
Future Directions

Christina Krist

In many ways, it is a fool's errand to attempt to write a "future directions" chapter given the rapid pace at which AI is developing. Case in point: when we began writing this book, ChatGPT did not exist, and now, less than a year later, the conversation about the role of generative AI in education has taken up significant space in both formal and informal conversations. We have no doubt that technological advances are going to continue at a rapid pace. So, in light of this, rather than predict or project future applications of machine learning or AI more broadly in science education (see Zhai's (2023) concluding chapter if you are looking for these types of insights), we would like to offer our perspective on our responsibilities and obligations as we continue to engage with the methodological advances in the field.

In our view, more than simply naming ethical principles, our responsibility is to think carefully, critically, and creatively about whether and how to take up–and influence the direction of–those advances. We view our role as science education researchers as standing in the "bridge space" between technological advancements and educational theory and practice. In this role, we offer three touchpoints for shaping how we make decisions about taking up ML and AI in our future directions as a field.

17.1 Considering How to Include Human-in-the-Loop, and When to Utilize Which Tools, as Informed by the (Science) Education Literature

It has (hopefully) become clear that human-in-the-loop approaches to utilizing ML or AI lead to not only more effective predictions (e.g., Wang et al., 2023), but also leave

C. Krist (✉)
Graduate School of Education, Stanford University, Stanford, CA, USA
e-mail: stinakrist@stanford.edu

© The Author(s) 2025
P. Wulff et al. (eds.), *Applying Machine Learning in Science Education Research*,
Springer Texts in Education, https://doi.org/10.1007/978-3-031-74227-9_17

space for careful, theoretically informed consideration of which decisions humans versus computers should be making and why (Kubsch et al., 2023). Importantly, there are more than likely not universal principles for which decisions should be made by whom and when.

In evaluating who should be making these decisions, we suggest asking and attempting to answer two key questions. One: to what problem is the AI- or ML-based tool the solution? And two: what do we know from the science education literature about that problem—what works, and what conditions are important to make it work? The combination of these questions should lead researchers to carefully consider whether and how automation might undo part of "what works" in an attempt to address the focal problem through efficiency-oriented means.

Early versions of teacher dashboards or automated assessment systems suffered from this universal pivot to efficiency. In part this is due to an underspecification of the design problem: "teachers don't have enough time" is not precise enough to carefully define what should and should not be automated about the multitude of tasks and subtasks involved in good teaching. Recent advances in both of these areas have done a more careful job with this specification, targeting the integration of AI-based tools in ways that keep teachers in the driver's seat in terms of making the important professional decisions involved in tasks that they are aiming to support with the AI-based tools. For instance, Steinert et al. (2023) developed LEAP, a novel platform using large language models to provide feedback to students. An important feature of this platform is that it involves teachers pre-prompting and assigning tasks to the large language model, customized to fit their instructional needs. They also present examples of prompt designs that lead to feedback focused on a wide range of outcomes, including sense-making, elaboration, self-explanation, and metacognition and motivation. In another recent example, He et al. (in press) present a four-part framework for supporting teachers in using AI-based knowledge-in-use assessments as part of their teaching practices. The four stages of this framework are (a) engaging with the AI system via professional learning support; (b) evaluating the automatically generated assessment reports; (c) considering AI-suggested instructional strategies; and (d) determining instructional decisions and actions. Rather than a framework for how to design the AI tool itself, this framework aims to facilitate its meaningful integration into the work of teaching.

Another key area ripe for advancement is the integration of additional types of data (e.g., physiological data) to improve the accuracy and real-time efficiency of ML systems that aim to anticipate student learning needs and provide individualized instructional supports. For example, Bertolini et al. (2021) showed that ML could identify patterns in students' neurocognitive data when learning about DNA replication while watching a video or engaging with a VR simulation that predicted their post-session content scores more accurately (85 percent vs. 55 percent) than algorithms trained on other types of data, and could act on those predictions more quickly, enabling better real-time alerts for students in need of additional support. The paper shows the potential of machine learning for tapping into data that hold deep information about student learning but are too complex to be analyzed using traditional methods. Drawing on neural imaging data the paper also demonstrates how this combination of data accrued as students learned and how machine learning can be utilized to predict learning outcomes in a less invasive way.

However, applications focused on individual assessment or feedback tend to perpetuate individualized, and often cognitively-centric, theories and models of learning. These are not the only types of learning goals we have—and in fact, the current vision for science education within the US as well as elsewhere emphasizing students' participation in science and engineering practices as the means for building science knowledge is explicitly grounded in sociocultural theories of learning (e.g., NRC, 2012). So, while schooling systems tend to be dominated by logics and systems emphasizing individual achievement and are therefore more "friendly" towards AI-based interventions that align with and appear to ease the burdens associated with enacting these logics, as science education researchers we should be loudly challenging this emphasis as the locus for innovation that could be brought about by AI-based tools.

Breideband et al. (2023) have explicitly begun to do this via the development of CoBi, a multi-party AI tool that focuses on relational aspects of collaboration. By visualizing multi-party student talk, CoBi supports students to reflect on four dimensions of their collaborative processes: respect, equity, community, and thinking. Importantly, this tool was developed through intense collaboration between human-computer interaction (HCI) scholars, learning scientists, and team science researchers.

Another example of a base of knowledge in science education that further challenges the individualized default orientation of AI-based tools is the growing body of work on the role of epistemic affect in students' sense-making as they participate in science and engineering practices (e.g., Jaber and Hammer, 2016) as well as developing teachers' epistemic empathy to support responsive teaching (Jaber, 2021; Jaber et al., 2022). Distinct from the literature on the importance of collaboration more broadly, this body of scholarship emphasizes the importance of teachers and students developing an awareness of how it feels to engage in the uncertainties of scientific knowledge-building and to make instructional decisions rooted in that awareness. Thus, a key facet of teacher expertise involves developing this sense of epistemic empathy; and a key goal for student learning is to develop this sense of epistemic empathy as students engage with one another. In many ways, tools that emphasize individualized models of learning are antithetical to these teacher and student learning goals. Notably, the PISA 2025 Science Framework (https://pisa-framework.oecd.org/science-2025/) emphasizes similar goals at the international level. It includes attitudes and dispositions toward science as well as environmental awareness, concern, and agency as target assessment outcomes for science education.

We see (at least for now) two potential ways of engaging with constructs such as epistemic empathy that we know are central for science teaching and learning but are in conflict with the underlying ideologies of AI. One is to take the approach that Breideband et al. (2023) have taken: to explicitly center those literature-based goals, and to harness a multidisciplinary team of scholars to creatively develop AI-based tools as part of systems that support those goals. A second is less sexy, but equally important: to explicitly draw boundaries around interactions and goals that AI should not touch. Arguing that in-person, individualized interactions between teachers and small groups of students, unmediated by digital tools, is the most powerful means of developing and fostering something like epistemic empathy is just as important of an empirical outcome as showing that the integration of an AI-based tool may improve some other learning- or teaching-related outcome (with a similar caveat to the one that we have articulated above: just like AI applications originally over-specified the problem of "teachers don't have time," it is all too easy if you find yourself hesitant about, suspicious of, or downright opposed to AI to offer gestalt scorched-earth declarations that AI should never be used. This is a similarly unhelp-ful argument, and we encourage a similar level of theoretical precision in articulating when, why, and how AI is not helpful in particular learning context, for particular learning goals).

The studies cited above help us to expand and articulate some different dimen-sions of consideration when thinking about what keeping a teacher in the so-to-speak professional drivers seat means: considering which decisions they should be making, which aspects of a task can be automated in ways that support this decision-making, and critically evaluating what might be lost (including time, given the investment required for learning to navigate a new system). These decisions include design deci-sions (e.g., which types of questions or prompts do I want to use for assessment and feedback, and why; He et al., in press; Küchemann et al., 2023); decisions grounded in teachers' knowledge of individual students (e.g., Steinert, personal communication); and decisions about which learning goals to center alongside careful articulation of how AI can be integrated (or not) to support those goals (e.g., Breideband et al., 2023).

17.2   Remaining Vigilant for Unintended Consequences and Unanticipated Impacts

A second touchpoint is the encouragement to consider both short-term and long-term consequences of the use of AI in education. We think about this as something like the intersection of the call to focus on ethics in the use of AI (which has been elaborated elsewhere in this book) and the call to focus on ethics in science (e.g., genetics research or the introduction of new species into an ecosystem; Hammer, personal communication). This kind of ethical thinking requires considering both short-term and long-term consequences. Some of these short-term consequences are already obvious, such as large language models being based on datasets that include social (racial, gendered, ableist) biases and therefore perpetuate these social biases as they are used–often amplifying them in ways that surpass expectations (Straw and Wu, 2022; Tanksley, 2022). As additional types of data, such as physiological data, are brought into the development of AI models, additional expertise in medical ethics and psychology should be brought in to help anticipate (or study) both intended and unintended short-term consequences for students and teachers.

The longer-term consequences are much more difficult to predict, but are perhaps more important to attempt to anticipate. What might be the long-term consequences–for teachers, for learners, for society's view of the role of teachers and schools, for students' affective and social engagement in schooling–of how we introduce AI into education? For example, Dennett put forth the argument that generative AI has the potential to "counterfeit people," ultimately spoiling a sort of societal contract where you can with reasonable certainty assume what other societal members know (common sense) and do (2023). Be that as it may, fake videos of politicians or celebrities are just recent (as of early 2024) examples of how generative AI might interfere with interpersonal affairs.

We offer three actionable strategies for how to concretely maintain this vigilance throughout a research project. First, attempting to even begin to anticipate these short- and long-term consequences requires deep technical expertise. In other words— please do not become an armchair ethicist about AI without at least attempting to gain a rudimentary understanding of the computer science involved! We hope that this book is a helpful starting point in developing a baseline of computational literacy for the science education community.

Second, in many ways, AI is just the next technological advancement. Looking to past integrations of technological tools in education can help us see past the shiny (or scary) newness of AI and the tech-bro discourses of "disruption" and instead take a measured, perhaps more principled approach to how we should proceed. For example, the introduction of television was ushered into education with similar promises of "revolutionizing" and "democratizing" education–of making the quality of schooling not dependent on the quality of teachers, but instead making the same type of programming available to all students. Similarly, the introduction of graphing calculators in the 1990s was both celebrated and met with fierce skepticism and hand-wringing. These, and other, integrations are well-documented and we can learn from them. What was the hype, and how was it framed? What was the panic? When things were tried, what worked well? What promises never panned out? What were the unintended or unanticipated consequences of those tools, and how does their historical introduction continue to shape our educational landscape? Although AI feels new, history can be our teacher.

And finally, a third strategy is to conceptualize the process of learning as an impossibly complex system. This means that attempting to anticipate unintended consequences requires holding broad and complex models of, and goals for, learning. Drawing from the examples introduced above, one needs to hold goals of learning individual content knowledge and self-regulation of that knowledge, col-laboration, collective sense-making, and epistemic empathy together. How does an influx of resources to support self-regulated learning, for example, impact a group's capacity for collective sense-making, both positively and negatively? In addition to considering these questions, empirically documenting and reporting these kinds of systems-level impacts is a crucial focus for research, moving beyond simply demonstrating a single positive impact of an AI-based tool on a single outcome in isolation.

17.3 Centering Student and Teacher Well-Being Over Technocratic Priorities to Set Educational Outcomes

The final touchpoint that we offer is one requiring attention to the ideological bases underpinning the rhetoric surrounding science education goals. While the NGSS-based vision does promote interaction and co-construction of ideas, the NGSS standards themselves are still guided by neoliberal vision of science: a utilitarian approach to understanding the natural world for technological development and economic benefit (Morales-Doyle et al., 2019; Weinstein, 2017). Without intentional efforts otherwise (Carlone et al., 2016; Strong et al., 2016), NGSS-aligned instructional approaches do little to explicitly counter the dominant technocratic role of science in society (Lemke, 1990; Sharma and Alvey, 2021). Funding agencies around the world similarly tend to prioritize these technocratic discourses and aims, making it relatively likely for AI-focused research to be funded but incredibly challenging to carry it out in ways that then go on to challenge these aims. And similarly, neoliberal schooling pressures feed into individualistic and abstracted measures of success which often come at the expense of attending to holistic student well-being (Dadvand and Cuervo, 2020).

In other words: at every level, the education and educational research systems are set up to promote the individualized, utilitarian, andro-centric, military-industrial, and technocratic ideologies that drive dominant developments and uses of AI. Attempting to push back on these, at any level, will be met with resistance.

And yet, we take the stance that attending to holistic student well-being and promoting human thriving–including but more importantly going beyond economic thriving–is at the heart of what it means to support learning. And so, again, as scholars who are shaping the national and international knowledge bases and priorities about how people learn and teach, we are obligated to continually re-center these ideologies and goals, both in theory and in practice. Similar to how attempting to anticipate unintentional consequences requires technical expertise, recognizing, and then choosing how to counter, these ideological underpinnings requires expertise in critical scholarship within and beyond the science education literature. We see this again as highlighting the need for interdisciplinary teams and organizational structures within those teams that elevate and integrate this expertise throughout the technical design and research processes.

In closing, we hope that this chapter–and this book as a whole!–will empower you to develop new skills, expand your expertise, and to position yourself as a science education leader in the bridge space between our contemporary technological advances and a vision for science education theory and practice that contributes to human flourishing.

References

Breideband, T., Bush, J., Chandler, C., Chang, M., Dickler, R., Foltz, P., Ganesh, A., Lieber, R., Penuel, W. R., Reitman, J. G., Weatherley, J., & D'Mello, S. (2023). The community builder (cobi): Helping students to develop better small group collaborative learning skills: Cscw '23 companion (pp. 376–380).

Carlone, H. B., Benavides, A., Huffling, L. D., Matthews, C. E., Journell, W., & Tomasek, T. (2016). Field ecology: A modest, but imaginable, contestation of neoliberal science education. *Mind, Culture, and Activity, 23*(3), 199–211.

Dadvand, B., & Cuervo, H. (2020). Pedagogies of care in performative schools. *Discourse: Studies in the Cultural Politics of Education, 41*(1), 139–152.

Dennett, D. C. (2023). The problem with counterfeit people. *The Atlantic.*

He, P., Shin, N., Zhai, X., & Krajcik, J. S. (in press). A design framework for integrating artificial intelligence to support teachers' timely use of knowledge-in-use assessment. In X. Zhai, & J. S. Krajcik (Eds.), *Uses of artificial intelligence in STEM education.* Oxford University Press.

Jaber, L. Z. (2021). He got a glimpse of the joys of understanding - the role of epistemic empathy in teacher learning. *Journal of the Learning Sciences, 30*(3), 433–465.

Jaber, L. Z., Dini, V., & Hammer, D. (2022). "well that's how the kids feel!"–epistemic empathy as a driver of responsive teaching. *Journal of Research in Science Teaching, 59*(2), 223–251.

Jaber, L. Z., & Hammer, D. (2016). Engaging in science: A feeling for the discipline. *Journal of the Learning Sciences, 25*(2), 156–202.

Kubsch, M., Sorge, S., & Wulff, P. (2023). Emotionen beim reflektieren in der lehrkräftebildung. In L. Mientus, C. Klempin, & A. Nowak (Eds.), *Reflexion in der Lehrkräftebildung: Empirisch - Phasenübergreifend - Interdisziplinär* (pp. 261–270). Potsdam: Universitätsverlag Potsdam.

Küchemann, S., Steinert, S., Revenga, N., Schweinberger, M., Dinc, Y., Avila, K. E., & Kuhn, J. (2023). Physics task development of prospective physics teachers using chatgpt. arXiv.

Lemke, J. L. (1990). *Talking science: Language, learning, and values.* Language and educational processes. Norwood, NJ: Ablex Publ.

Morales-Doyle, D., Childress Price, T., & Chappell, M. J. (2019). Chemicals are contaminants too: Teaching appreciation and critique of science in the era of next generation science standards (ngss). *Science Education, 103*(6), 1347–1366.

National Research Council. (2012). A framework for K-12 science education: Practices, crosscutting concepts, and core ideas. National Academy of Sciences

Sharma, A., & Alvey, E. M. (2021). The undercurrents of neoliberal ethics in science curricula: A critical appraisal. *Ethics and Education, 16*(1), 122–136.

Steinert, S., Avila, K. E., Ruzika, S., Kuhn, J., & Küchemann, S. (2023). Harnessing large language models to enhance self-regulated learning via formative feedback. arXiv.

Straw, I., & Wu, H. (2022). Investigating for bias in healthcare algorithms: A sex-stratified analysis of supervised machine learning models in liver disease prediction. *BMJ Health & Care Informatics, 29*(1).

Strong, L., Adams, J. D., Bellino, M. E., Pieroni, P., Stoops, J., & Das, A. (2016). Against neoliberal enclosure: Using a critical transdisciplinary approach in science teaching and learning. *Mind, Culture, and Activity, 23*(3), 225–236.

Tanksley, T. (2022). Race, education and #blacklivesmatter: How online transformational resistance shapes the offline experiences of black college-age women. *Urban Education*, page 004208592210929.

Wang, H., Fu, T., Du, Y., Gao, W., Huang, K., Liu, Z., Chandak, P., Liu, S., van Katwyk, P., Deac, A., Anandkumar, A., Bergen, K., Gomes, C. P., Ho, S., Kohli, P., Lasenby, J., Leskovec, J., Liu,

T.-Y., Manrai, A., ... Zitnik, M. (2023). Scientific discovery in the age of artificial intelligence. *Nature, 620*(7972), 47–60.

Weinstein, M. (2017). Ngss, disposability, and the ambivalence of science in/under neoliberalism. *Cultural Studies of Science Education, 12*(4), 821–834.

Zhai, X. (2023). The Role of AI and Technologies in Science Learning Progression: Commentary for Section IV. In Handbook of Research on Science Learning Progressions (pp. 499-512). Routledge.

Chapter 18
Conclusions

Marcus Kubsch, Christina Krist, and Peter Wulff

You have come a long way. Congratulations! On you journey into the world of machine learning in science education research you traveled along the coasts of technical fundamentals of ML and methodological considerations that guide the use of ML. Later, you ventured into the wild and took your first steps on the sometimes slippery slopes of the case studies. Finally, you took a walk through the park of ethical and procedural principles that may guide your future ML journey and where offered a vista on the emerging topics in the field.

We hope that this journey made a lasting (and pleasant) impression on you and has left you a skillful navigator for your future ML journeys. We quoted Richard Feynman in the beginning with "What I cannot create, I cannot understand." Take this with a grain of salt: We showed you some important decisions in applying ML and how to implement them. Obviously, we build on shoulders of (thousands of) giants— those who created the open-source software and packages that we utilized. It cannot be our responsibility to create such packages. Computer scientists are much better equipped to build efficient and versatile packages and software. However, we should maintain some degree of control over important steps in the workflows of applying ML. We believe that open-source software and basic programming as displayed in this textbooks can be of great value to maintain this control, and thus develop a better understanding of the validity with which you can make scientific claims in your research.

M. Kubsch (✉)
Freie Universität Berlin, Berlin, Germany
e-mail: m.kubsch@fu-berlin.de

C. Krist
Graduate School of Education, Stanford University, Stanford, CA, USA

P. Wulff
Heidelberg University of Education, Heidelberg, Baden-Württemberg, Germany

© The Author(s) 2025   347
P. Wulff et al. (eds.), *Applying Machine Learning in Science Education Research*,
Springer Texts in Education, https://doi.org/10.1007/978-3-031-74227-9_18

Moreover, we hope that beyond technical skills, you took away a set of principles that allow you to make informed decisions on when, how, and why to use ML in science education research. Prime among these, is to never forget that ML—however exciting and powerful the possibilities it provides are—is just *one* tool among many in our methodological toolkit and not every tool fits every task. Sometimes ML may allow you to solve intricate problems with ease and provide great value in terms of advancing both—our knowledge about science teaching and learning and science teaching and learning itself. At other times, ML will not help you solve your problems and the consequences of applying it may actually counteract the goal of advancing science teaching and learning. As science education researchers it is our moral obligation to apply ML responsibly ourselves and hold others in our community accountable to the same standard.

Let us also emphasize that such a short book can only provide you with rather general conceptual basics of ML and with a handful of applications with different ML algorithms. We point again to the excellent references (listed in Chap. 1) that you might be better equipped after reading this book to delve into. We also only marginally touched upon image data. This is supposed to be a feature, and not a bug. Other scholars are better equipped to provide examples of ML-based image processing and analysis. We consider language and numeric data to be a good starting point, given that it is considered rather straightforward to analyse (as many science education researchers already did), and common to more traditional analyses.

In this sense, we hope to see you leveraging the power of ML to do great things—from answering your research questions to building the tools, apps, and learning environments that help students and teachers thrive.

Peter, Marcus & Stina.

Notes

Chapter 1

- Complexity of learning and teaching processes, see: Koopmans and Stamovlasis (2016)
- Complexity of constructs such as competencies, or intelligence, see: Marcus (2019), McClelland (1973), Harris et al. (2019)
- Motivational determinants of decision processes, see: Eccles and Wigfield (2020)
- Interconnections with knowledge, see: Reinhold et al. (1999); Chunking in experts, see: Simon and Gilmartin (1973), Berliner (001), Ericsson (2003), van Es and Sherin (2002)
- Digitally enhanced learning environments, see: Kubsch et al. (2022)
- Rethinking the science education research methodological toolset, see: Fleener (2016), Brunton and Kutz (2019)
- Applications of AI, see: Domingos (2015), Du Sautoy (2020), Mitchell (2020), Bishop and Bishop (2024)
- ML as inductive, data-driven problem solving, see: Rauf (2021), Hastie et al. (2008)
- Automated feedback for students, see: Graesser (2016)
- Reproducibility crisis in ML, see: Ball (2023), Kapoor and Narayanan (2023)
- Propagation of bias with ML, see: Christian (2021)
- Intransparency of complex ML models, see: Cheuk (2021)
- Hallucination of LLMs in simple tasks, see: Metz (2023); Some suggested to call hallucination "confabulation" as human memory at times is struck by similar pitfalls, see: https://www.youtube.com/watch?v=N1TEjTeQeg0, last access: May 2024
- Knowledge-in-use assessments, see: Harris et al. (2019)
- Importance of feedback for learning, see: Hattie and Timperley (2007)
- Resources required for individualized instruction, see: Chen et al. (2020), and Gay (2015)

© The Editor(s) (if applicable) and The Author(s) 2025   349
P. Wulff et al. (eds.), *Applying Machine Learning in Science Education Research*,
Springer Texts in Education, https://doi.org/10.1007/978-3-031-74227-9

Chapter 2

- Historical roots and multi-disciplinarity of AI, see Russell and Norvig (1995);
- For an interesting discussion of differences in problem solving capabilities between machines and humans, see Aaronson (2011).
- Implicit assumptions in argumentation, see: Walton (2008)
- Differentiation of knowledge-engineering approaches (Cyc) and connectionism, see: Domingos (2015).
- It is said that ELIZA's originator was somewhat irritated by the suggestion to develop computational therapists based on such simple rule-based programs and became a critic of society, see: https://de.wikipedia.org/wiki/ELIZA, last access: Oct 2023.
- Story of early ANN training, see: Sutton and Barto (2015).
- DeepMind's AlphaGo startled the community by counter-intuitive moves, see: Tegmark (2018); and for AlphaStar, see: Vincent (2019)
- KataGo tracked weaknesses, see: https://www.engadget.com/human-convincin gly-beats-ai-at-go-with-help-from-a-bot-100903836.html?guce_referrer=aHR0 cHM6Ly9kdWNrZHVja2dvLmNvbS88&guce_referrer_sig=AQAAAAspAcZ-DP k0z5wKIZGsOgn4HD_g3X1kmrEaYuzr4jayhO-JKfUuyqiu-JqBX_hIzkL8imH KT4oIVstg49HJmZPoqYLiDJ6zJlQwzHzhYnn9pz6AlsgsSdk62beJJYfCwt5x01 -cSu609WIpdXCBNEk9qpCsp1-zdqLNYZgH76K4\&guccounter=2, last access: Dec 2023.
- Recognizing hand-written digits with precision, see: Hinton et al. (2006), Géron (2017)
- Moravec's paradox, see: Zador (2019)
- Coining of the term "deep learning", see: Géron (2017); Not only the scientific community is interested: companies provide the small Caribbean island Anguilla a fortune because it owns the ".ai" internet domain, outcompeting Pacific island Tuvalu which owns the ".tv" domain.
- See important applications of deep learning technologies regarding language in Jurafsky and Martin (2014), Manning (2022)
- Use of ML for air-pollution prediction, see: Bodnar et al. (2024)
- LLMs and prompting, see: Vaswani et al. (2017), Devlin et al. (2018), Brown et al. (2020)
- LLMs as sources of collective opinion, see: Crokidakis et al. (2023)
- Limitations of generative LLMs in domain-specific tasks, see: Gregorcic and Pendrill (2023)
- Advanced LLMs become more performant in domains such as physics education, see: West (2023)
- The term "stochastic parrots" for LLMs was introduced by Bender et al. (2021)
- Issues of "true understanding" in LLMs, see: Mitchell (2023); researchers devised a novel test, the ARCathon (Abstraction and Reasoning Corpus) to better test capabilities of LLMs, see: https://lab42.global/arcathon/ (last access: June 2024)
- Original research on protein folding, see: Jumper et al. (2021)
- ML for Covid-19 research, see: Keshavarzi Arshadi et al. (2020)

- Issue of providing explicit instruction for problems, see: LeCun et al. (1998)
- AI and ML for magnetic confinement in nuclear fusion, see: Degrave et al. (2022)
- AI and ML should complement first-principles based approaches to scientific research, see: Brunton and Kutz (2019)
- Research related to explanding scientific understanding with AI and ML: Krenn et al. (2022)
- Definition and involved disciplines in ML, see: Mitchell (1997). In the list provided in the main text, physics is missing (Marsland, 2015). In fact, statistical physics has contributed tremendously to the inceptions of machine learning, such as in Hopfield networks or simulation of random processes, etc.
- ML as deviating from traditional programming paradigms, see: Chollet (2018)
- Definition of ML: For a comprehensive list of tasks T, performance P, and experience E, see Goodfellow et al. (2016).
- Risks of generalization in inductive learning, see: Vapnik (1996) and Valiant (1984)
- Growth of AI and ML: The exponential growth in many technological domains is captured in the "Law of Accelerating Return" (Kurzweil, 2004), and Moore's law captures the relationship between doubling of transistors on an integrated circuits doubles every two years or so, depending on source, see https://en.wikipedia.org/wiki/Moore%27s_law, last access: Dec 2023. Reasons for AI growth are outlined (among others) here: https://www.diamandis.com/blog/scaling-abundance-series-26, last access: Dec 2023.
- Surprising trainability of ML models where number of parameters compared to training samples is large, see: Domingos (2015)
- Training models in ImageNet competition, see: Krizhevsky et al. (2012)
- Theoretical arguments that buttress the capabilities of ML models, see: Engel and den van Broeck (2001), Giraud-Carrier and Provost (2005)
- Universal approximation theorem, see: Hornik et al. (1989) and Goodfellow et al. (2016); Note that this does not assure that appropriate parameter weights are found in training with the training data, and this also does not account for extrapolation, see here: https://en.wikipedia.org/wiki/Universal_approximation_theorem, last access Nov 2023.
- ML algorithm vs. ML model, see: Marsland (2015)
- Issue of explainability versus prediction in science, see: Domingos (2015), Raz et al. (2024); Problems of model mis-specification, see: Vansteelandt (2021)
- Differentiation of forms of ML, see: Hastie et al. (2008), Zhai et al. (2020); For further learning approaches, such as online-learning, batch learning, and instance- and model-based learning see Géron (2017).
- Supervised ML as common approach, see: Chollet (2018), Zhai et al. (2020); cross-validation approaches, see: Chollet (2018); typical tasks, see: Rauf (2021), Bishop (2006), Goodfellow et al. (2016); a comprehensive review of model classes for supervised ML in Hastie et al. (2008); Linear models in science education, see: Theobald et al. (2019); kernel-based methods, see: Smola and Schölkopf (2003), Hilbert et al. (2021); tree-based methods, see: Hilbert et al. (2021) and Kotsiantis (2007); ANNs, see: Hilbert et al. (2021); Other (but similar) differentiations of model classes, see: Kotsiantis (2007);

- Unsupervised ML, tasks, see: Bishop (2006), Carleo et al. (2019), Khanum et al. (2015), Rauf (2021); Re-use pretrained representations, see: Khanum et al. (2015); Latent semantic analysis, see: Deerwester et al. (1990), Hofmann (2001); Tuning hyper-parameters in unsupervised ML, see: Campello et al. (2013); Latent Dirichlet allocation, see: Blei et al. (2003) and (in science education) Odden et al. (2020) (they used the intruder words in the topics); Model classes, see: https://en.wikipedia.org/wiki/Unsupervised_learning (last access Dec 2023); Capture behavior of physical systems with encoders, see: Cranmer et al. (2020); Markov Chain Monte Carlo sampling, see: Bishop (2006);
- Next token prediction and active inference in humans, see: Pezzulo et al. (2021), Clark (2023)
- Discussion about learning meaningful representations, see: Li et al. (2023), Hewitt and Manning (2019), Patel and Pavlick (2022), and Mitchell et al. (2023).
- Semi-supervised learning, see: Brunton and Kutz (2019); Zhu and Goldberg (2009); Ice-cake metaphor: https://www.youtube.com/watch?v=Ount2Y4qxQo&t=1072s (timestamp: 20:40, last access Dec 2023); application of semi-supervised ML, see: Zhu and Goldberg (2009)
- Reinforcement learning, see: Bishop (2006)
- Metalearning, see: Brazdil et al. (2022); applications in science education research, see: Carpenter et al. (2020), Wulff et al. (2022)
- Combine physics laws with ML algorithms, see: Liu and Tegmark (2021), Udrescu and Tegmark (2020), Karniadakis et al. (2021)
- Reproducibility crisis in ML, see: Kapoor and Narayanan (2022)
- Foundation models, see: Bommasani et al. (2022)
- Estimates for required training samples, see: Mitchell (2020)
- Call for data sharing, see: Hey et al. (2009)
- Use LLMs for educational data augmentation, see: Kieser et al. (2023), Fang et al. (2023)
- Issue with Amazon models, see: https://incidentdatabase.ai/cite/37/#r2461 (last access: Dec 2023)
- Researchers currently do not address issues of missing data, see: Kapoor and Narayanan (2022)
- Vision model to detect water birds picks up on water as irrelevant feature, see: Christian (2021)
- Imbalanced data as problem in ML research, see: Kapoor and Narayanan (2022); an insightful example of how to deal with it in physics education research, see: ?.
- Regularization as means to control for bias-variance trade-off, see: Goodfellow et al. (2016)
- Hypertension study, see: Ye et al. (2018)
- Data set contamination as a problem in LLMs, see: Li et al. (2023)
- Sampling bias, see: Bone et al. (2015)
- Importance of understanding influencing factors for phenomena, see: Pearl (2021)
- XAI, see: Molnar (2022); Best explanation of a simple model is the model itself Lundberg and Lee (2017); methods for XAI, see: Molnar (2022) and Wickramas-

inghe et al. (2021); additive feature explanation techniques, see: Lundberg and Lee (2017)

- Application of SHAP values in science education, see: Martin et al. (2023)
- Integrated gradients, see: Sundararajan et al. (2017); application in science education, see: Wulff et al. (2022a)
- Statistical causal modeling, see: Adlakha and Kuo (2023), Kapoor and Narayanan (2023); ML predominantly engaged with predictive modelling, see: Malik (2020); identification of conflating effects, see: Adlakha and Kuo (2023)
- X-ray identification declined with different scanners, see: Liang et al. (2021); problem of identifying tanks, see: Domingos (2015) and Yang et al. (2022); tricking ML models with adversarial examples, see: Christian (2021); clever prompting LLMs, see: https://github.com/giuven95/chatgpt-failures (last access: Dec 2023)
- Bias and stereotypes, see: Mehrabi et al. (2022); gender bias in LLMs, see: Caliskan et al. (2017), Brown et al. (2020), Borchers et al. (2022); 90 percent of nurses were female in 2008, see: Ulrich (2010); natural science and engineering dominated by men, see: Handelsman et al. (2005); modern forms of colonialism and AI, see: https://www.chathamhouse.org/publications/the-world-today/2023-10/why-ai-must-be-decolonized-fulfill-its-true-potential (last access: Dec 2023), and ?
- In Silicon Valley, the term "the GPU poor" refers to agents that have less specific compute power, indicating the novel resource categories for innovation and power
- "OpenAI used Kenyan workers on less than $2 per hour to make ChatGPT less toxic", see: https://time.com/6247678/openai-chatgpt-kenya-workers/ (last access: June 2024)
- AI Act by EU, see: https://artificialintelligenceact.eu/ (last access: May 2024)
- Alignment problem in AI, see: Christian (2021)
- Policies of OpenAI for tackling with hazardous information, see: OpenAI (2023)
- It took an AI less than 6 hours to invent over 40,000 potentially lethal molecules, many among them similar to the most potent nerve agent VX, see: https://www.theverge.com/2022/3/17/22983197/ai-new-possible-chemical-weapons-generative-models-vx (last access: June 2024)
- Doubts about easily discernable human values, see: Krauss (2023)
- Unfortunate incentives for teachers, see: Thomas and Uminsky (2022)
- Capabilities of LLMs, see: Wolfram (2023), Gregorcic and Pendrill (2023); performance of ChatGPT as "thoroughly remarkable" (Wolfram, 2023); AGI and ASI, see: Chalmers (2010); Human brain versus machine in terms of performance, see: Bostrom (2017)
- Brittleness of LLM performance, see: Lenat and Marcus (2023), Marcus (2019), Mitchell and Krakauer (2023)
- Power requirements of the human brain, see: Krauss (2023); Energy use for training and using GPT-4, see: de Vries (2023)
- Fossil fuels in supply chains and electricity demands in global tech sector, see: https://www.greenpeace.org/eastasia/press/7698/microsoft-google-reliant-on-fossil-fuels-despite-100-renewable-energy-pledges-study/, last access: May 2024.
- Problem of privacy leakage of personally identifiable information, see ?.
- Positivist science, see: Bortz and Döring (2002).

Chapter 4

- Data in educational research, see: Baig et al. (2020)
- Size of data in the world: "If data had mass, the earth would be a black hole" (Marsland, 2015, p.1)
- Importance of numerical approaches for science, see: Wolfram (2002); importance of data-driven discovery, see: Pontzen (2023)
- ML as data-driven discovery, see: Prince (2023)
- Importance and examples of inductive discovery approaches in science, see: Rothchild (2006)
- Expected data size in the world, see: https://www.diamandis.com/blog/scaling-abundance-series-26 (last access Dec 2023)
- V's for Big Data, see: Da Xu and Duan (2019)
- Fish killed in pond as complex systems, see: Grotzer and Shane Tutwiler (2014)
- Complex systems modelling, see: Bar-Yam (1997), Solé et al. (2019)
- Computers as interacting transistors, see: Bar-Yam (1997); The brain is better equipped to detect large-scale patterns and symmetries, see: Bar-Yam (1997), and Aaronson (2011)
- Attributes of complex systems, see: Koopmans and Stamovlasis (2016) and Bar-Yam (1997); Complex systems are characterized by high-dimensional state spaces, see: Brunton and Kutz (2019); Phases of complex systems, see: Sole (2011)
- Emergence, see: Wei et al. (2022); Emergent capabilities in LLMs, see: Wei et al. (2022)
- Phase changes, see: Sole (2011)
- Also the brain has been posited to optimally function at states of criticality and near phase transitions (Chialvo, 2010)
- Hypothesis spaces, see: Nisbet et al. (2009)
- Assumption of linear effects in linear models, see: Kantz and Schreiber (2003); Inability of linear models to capture non-linear relationships, see: Nisbet et al. (2009) and Brunton and Kutz (2019)
- Complex systems in education sciences, see: Sawyer (2002), and emergentism, see: Sawyer (2002)
- Notably, there are attempts to derive a "theory of everyone" based on few underlying laws: (1) the law of energy, (2) the law of innovations and efficiency, the low of cooperation, and the law of evolution (Muthukrishna, 2023). Though, applying those to understand and improve teaching and learning is still non-trivial
- Intra- and interperson phenomena and complex systems, see: Hilpert and Marchand (2018)
- Applications of complex systems to educational research problems, see: Stamovlasis and Koopmans (2014), Thelen and Smith (1996), and Patriarcha et al. (2020)
- Some even suggested that the properties of complexity in language map to complexity in cognition (see: https://www.preposterousuniverse.com/podcast/2024/01/01/260-ricard-sole-on-the-space-of-cognitions/, last access: Jan 2024),

because language and cognition are intricately related to each other (Boroditsky, 2011; Deacon, 1997)

- Identify g-factor (as dimensionality reduction) in IQ research, see: Gottfredson (1998); reduce personality traits, see: Asendorpf (2004)
- Dimensionality reduction of data sets, see (Bishop, 2006, p. 559) for an illustrative example for images that can be reduced from 10,000 dimensional data space to three degrees of freedom (Bishop, 2006); Note that data space and state space are used interchangeably in this context. State space is a term originating from complex system's theory, whereas data space is a term more used in ML research
- Complexity reduction is possible, see: Bar-Yam (1997)
- Problem solving, see: Dunbar (1998); Insight problems as emergent phenomena, see: Koopmans and Stamovlasis (2016); Davis and Sumara (2006); Process models for problem solving, see: Friege (2001), and Docktor et al. (2016)
- Differentiation of Get the data, Explore the data, and Prepare the data, see: Géron (2017)
- Constructed-response items, see: Chung et al. (2003), and Nehm and Härtig (2012)
- Unreasonable assumption of normality, see: Taleb (2020), Nunnally (1978), and Micceri (1989)
- Language as a medium for sense making and communication, see: Halliday (1978); Language as comprised of units, see: Nowak et al. (2002)
- Complex system's perspectives on language, see: Montemurro and Zanette (2010), Zanette (2014)
- Subject-verb-object ordering, see: Alvarez-Lacalle et al. (2006)
- Language is noise, unsegmented, and ambiguous, see: Jurafsky and Martin (2014)
- Word meaning is a function of context, see: Evans (2006); Multi-facetedness of meaning, see: McNamara et al. (1996)
- Variety of patterns in language, see: Batterink and Paller (2016)
- Data exploration and visualization as science, see: https://en.wikipedia.org/wiki/Exploratory_data_analysis (last access: Oct 2023); You can find valuable resources for further reading in Gelman (2004).
- Train and test data are generated by a probability distribution, see: Goodfellow et al. (2016)
- Access to conceptual physics problem, see: Tschisgale et al. (2023), and Wulff (2023)
- Genome with short- and long-range patterns, see: Lyubchenko et al. (2002); Local and global structures in networks, see: Steyvers and Tenenbaum (2005)
- ANNs find effective representations of data, see: Goodfellow et al. (2016)

Chapter 5

- Workflows for supervised ML, see: Kotsiantis (2007) and Chollet (2018)
- Complexity of constructed responses in language, see: Meurers (2012)
- Perceptron as hydrogen atom, see: Engel and den van Broeck (2001)
- Representations are formed in ANN layers, see: Chollet (2018)
- Importance of loss function, see: Russell and Ermon (2016), and Chollet (2018); Penalities in loss function, see: Bishop (2006), and Russell and Ermon (2016)
- ANNs as directed acyclic graphs, see: Chollet (2018)
- Data transformation in ANNs as tensor multiplications, see: Chollet (2018)
- Jaron Lanier (see this video: https://www.youtube.com/watch?v=caepEUi2IZ4, last access Feb 2024) offers the metaphors of deep learning as learning a forest of blended towers, where generating enables accessing different towers and combining them to produce outputs. The notion of "stochastic parrots" holds true, however, as Lanier puts it, the magic comes from combination of towers. Generative AI can generate virtual towers, but it cannot surpass existing towers by far, which might be linked to missing genuine creativity of these architectures.
- ANNs as universal function approximators, see: Hornik et al. (1989), and Prince (2023)
- Depth-efficiency of ANNs, see: Prince (2023)
- Non-linear separable problems are harder to learn, see: Elizondo (2006)
- Empirical risk and loss function, see: Wang et al. (2022)
- Details on many different loss function can be found in reference textbooks (see Chap. 1), as well as in Wang et al. (2022)
- Stochastic gradient descent, see: https://optimization.cbe.cornell.edu/index.php?title=Adam (last access: Dec 2023)
- Connectionists' master algorithm, see: Domingos (2015), Bishop (2006)
- See an overview of different ANN architectures in: Brunton and Kutz (2019), p. 222
- Slowing down weight loss in ANNs, see: Kirkpatrick et al. (2017)
- Sleep and ANNs, see: Tadros et al. (2022)
- Foundation models, see: Vilalta and Meskhi (2022), Bommasani et al. (2022)
- AI winter, see: Bishop (2006)
- GOFAI and symbolic AI, see: Chollet (2018)
- For a concise overview of human brain and learning, see: Marsland (2015) or Dehaene (2021).
- ANNs with billions of parameters should be able to learn in principle, see: Giraud-Carrier and Provost (2005)
- More is typically better in ANNs, see: Kaplan et al. (2020)
- Assumptions constrain the hypothesis space, see: Géron (2017)

Chapter 6

- Dominance of supervised ML in science education, see: Wang (2016)
- Predominance of unlabelled data, see: Marsland (2015)
- Generative LLMs such as ChatGPT as lossy compressions of data ("ChatGPT as a blurry jpeg of all the text on the Web"), see: https://www.newyorker.com/tech/annals-of-technology/chatgpt-is-a-blurry-jpeg-of-the-web.
- Lower dimensionality of text, see: Deerwester et al. (1990)
- Patterns in images, see: Flicker (2022)
- Capturing an image with three latent dimensions, see: Marsland (2015)
- Mathematical details of some clustering techniques, see: Kriegel et al. (2011), and Müllner (2011); Density-based clustering, see: Kriegel et al. (2011)
- We simulate the directions of elementary magnets based on: https://rajeshrinet.github.io/blog/2014/ising-model/ (last access: May 2024). The procedure for the sampling can be summarized into the following steps: "(1) Prepare an initial configuration of N spins; (2) Flip the spin of a randomly chosen lattice site. (3) Calculate the change in energy dE. (4) If $dE < 0$, accept the move. Otherwise accept the move with probability $\exp^{-dE/T}$. This satisfies the detailed balance condition, ensuring a final equilibrium state. (5) Repeat 2-4.", because with the simplifying assumptions, quick simulation is possible)
- Monte Carlo sampling is based on metropolis algorithm, see: Scherer (2017)
- T-SNE has also been utilized to cluster quarks, gluons, and Higgs in high energy physics, which typically deals with very high dimensional data, see: https://indico.physics.lbl.gov/event/975/contributions/8262/attachments/4079/5490/BOOST_Krupa_2.pdf, last access: Feb 2024

Chapter 8

- Different definitions (uses) of "work" in physics, see: Williams (1999)
- For details on ELIZA, see: https://de.wikipedia.org/wiki/ELIZA (last access Dec 2023)
- Deep learning enabled more performant models, see: Prince (2023)
- The human brain as an experience machine, see: Clark (2023)
- Problems with defining what a word is, see: https://en.wikiversity.org/wiki/Psycholinguistics/What_is_a_Word (last access Dec 2023)
- Response length as a predictor for high scores, see: Nehm and Härtig (2012), Lai and Calandra (2010), Powers (2005)
- Problems with dot product, see: Jurafsky and Martin (2014)
- What are language models, see: Jurafsky and Martin (2014)
- Simple n-gram models produce reasonable text, see: Jurafsky and Martin (2014)
- Sparse estimation problem with language data, see: Rosenfeld (2000)

- Complexity explosion is referred to as the curse of dimensionality (Bishop, 2006), which is a serious problem in many real-world applications: Examples in the training set are sparse compared to the number of possible observations overall. The boon of statistical (language) modelling approaches is that they can more or less tackle this challenge—or, at least, make it less problematic. In fact, for deep learning Sejnowski (2024) mentions the surprise of researchers when it was found that high-dimensional optimization and generalization with highly over-parameterized models is possible. He called it a "blessing of dimensionality" instead.
- Document embeddings are more flexible compared to n-gram models, see: Bengio et al. (2003)
- ANNs can be used to train document embeddings, see: Bengio et al. (2003) and Mikolov et al. (2013)
- Reconstruction of text based on word vectors is sometimes possible, see: Morris et al. (2023)
- Transformer-based LLMs, see: Vaswani et al. (2017). Find excellent resources for how LLMs, particularly transformers, work here: https://e2eml.school/transfor mers (last access Feb 2024), http://jalammar.github.io/illustrated-transformer/ (last access Feb 2024), in Bishop and Bishop (2024), or in Wolfram (2023)
- Architecture choices of transformer LLMs, see: Chang and Bergen (2023), Prince (2023)
- Explanation of the self-attention mechanism: see a detailed explanation how they work in Prince (2023)
- Scaling laws for LLMs, see: Kaplan et al. (2020); for the importance of the quality of the training data, see: ?.
- Encoder and decoder LLMs, see: Prince (2023)
- Emergent capabilities of LLMs with increasing training size, etc., see: Devlin et al. (2018), Wei et al. (2022)
- Prompting, see: White et al. (2023); Performance comparisons with varying prompting strategies, see: Brown et al. (2020); Prompt engineering, see: Bubeck et al. (2023), Kojima et al. (2022), Wang et al. (2023), Yao et al. (2023); prompting "step-by-step reasoning" as "thinking slow", see: Wang et al. (2023), Polverini and Gregorcic (2024), Kahneman (2011); general guidelines on effective prompting strategies are documented in White et al. (2023)
- Automated annotation and coding with generative LLM, see: ?
- Potentials of LLMs in education, see: Kasneci et al. (2023)
- `gensim` module, see: Rehurek and Sojka (2010)
- For an overview of deep semantic models (from: Lenci et al. (2022))
- Asymmetry of analogical reasoning in humans, see: Christian (2021)
- BERT model, see: Devlin et al. (2018)
- Shortcut learning in BERT, see: Mitchell (2023); Example of shortcut learning with BERT, see: Niven and Kao (2019)
- Warrants in arguments are often left implicit, see: Walton and Reed (2005)

- Approximate memory retrieval as a limitation of LLMs, see: https://cacm.acm.org/blogcacm/can-llms-really-reason-and-plan/?utm_source=substack&utm_medium=email (last access: May 2024)
- LLMs between developing a world model and merely mimicking the training data, see: https://www.preposterousuniverse.com/podcast/2024/06/24/280-francois-chollet-on-deep-learning-and-the-meaning-of-intelligence/ (last access: June 2024)
- Monty Hall problem with LLMs, see: Wu et al. (2024); fixes with LLM size, see: Macmillan-Scott and Musolesi (2024)
- Hallucination in textual summarization, see: Hughes (2023)
- Toroidal chess example, see: https://www.preposterousuniverse.com/podcast/2023/11/27/258-solo-ai-thinks-different/ (last access Nov 2023)
- Problem of test data set contamination in LLMs, see: Li et al. (2023)

Chapter 12

- Wordpiece tokenization improves LLM's performance, see: Wu et al. (2016)
- Segmenting language data into elementary discourse units, see: Stede and Neumann (2014)
- Idea of using generative LLMs for educational data augmentation, see: Kieser et al. (2023)
- An accessible introduction to transformer LLMs can be found in Alammar (2018), and a Python implementation from scratch in http://nlp.seas.harvard.edu/annotated-transformer/.
- Energy expenditure of LLMs, see: Dodge et al. (2022), de Vries (2023); inequity in environmental impacts, see: https://themarkup.org/hello-world/2023/07/08/ai-environmental-equity-its-not-easy-being-green (last access: June 2024)
- Information on BERT model, see: Devlin et al. (2018)
- Sometimes simpler model are better, see: Urrutia and Araya (2024)
- Integrated gradients, see: Schrouff et al. (2022)
- Explainable AI, see: Lipton (2018); the company Anthropic (among others) actively seeks to research the "inner workings" of AI models such as LLMs.
- Requirements for Integrated Gradients, see: Sundararajan et al. (2017)
- Tutorial for Integrated Gradients can be found here: https://captum.ai/docs/extension/integrated_gradients (last access: Dec 2023)
- Some of the tasks and solutions can be found here: https://chatgpt.com/share/0fff74d5-a22a-40d2-80ee-21b7a1733707 (May 2024); difficult tasks are marked by *.

Chapter 13

- For reflections about preservation of local and global structures in data, see: McInnes et al. (2020)
- More information and reasoning on HDBSCAN, see: Campello et al. (2013)
- BERTtopic module, see: ?
- PCA for reducing co-occurrence matrix, see: Lebret and Collobert (2013)

Chapter 16

- For another example of using computational grounded theory for science education research, see: Tschisgale et al. (2023)
- More information on a contemporary approach to assessment development in education, see: Wilson (2023)
- For a summary of a Bayesian approach to teaching and learning about and under conditions of uncertainty, see: Rosenberg et al. (2022)

References

Aaronson, S. (2011). Why philosophers should care about computational complexity. arXiv.

Adlakha, V., & Kuo, E. (2023). Critical issues in statistical causal inference for observational physics education research. *Physical Review Physics Education Research, 19*(2).

Alammar, J. (2018). The illustrated transformer [blog post].

Alvarez-Lacalle, E., Dorow, B., Eckmann, J.-P., & Moses, E. (2006). Hierarchical structures induce long-range dynamical correlations in written texts. *Proceedings of the National Academy of Sciences of the United States of America, 103*(21), 7956–7961.

Asendorpf, J. B. (2004). *Psychologie der Persönlichkeit*. Berlin and Heidelberg and New York: Springer.

Baig, M. I., Shuib, L., & Yadegaridehkordi, E. (2020). Big data in education: A state of the art, limitations, and future research directions. *International Journal of Educational Technology in Higher Education, 17*(1).

Ball, P. (2023). Is AI leading to a reproducibility crisis in science? *Nature*.

Bar-Yam, Y. (1997). *Dynamics of complex systems*. Studies in nonlinearity. Reading Mass.: Addison-Wesley.

Batterink, L. J., & Paller, K. (2016). Picking up patterns in language. *Psychological Science Agenda*.

Bender, E. M., Gebru, T., McMillan-Major, A., & Shmitchell, S. (2021). On the dangers of stochastic parrots. *FAccT*, 610–623.

Bengio, Y., Ducharme, R., Vincent, P., & Jauvin, C. (2003). A neural probabilistic language model. *Journal of Machine Learning Research, 3*, 1137–1155.

Berliner, D. C. (2001). Learning about and learning from expert teachers. *International Journal of Educational Research, 35*, 463–482.

Bishop, C. M. (2006). *Pattern recognition and machine learning*. Information science and statistics. New York, NY: Springer Science+Business Media LLC.

Bishop, C. M., & Bishop, H. (2024). *Deep learning: Foundations and concepts*. Cham: Springer.

Blei, D. M., Ng, A. Y., & Jordan, M. I. (2003). Latent dirichlet allocation. *Journal of Machine Learning Research, 3*(4–5), 993–1022.

Bodnar, C., Bruinsma, W. P., Lucic, A., Stanley, M., Brandstetter, J., Garvan, P. (2024). Aurora: A foundation model of the atmosphere. In: *arXiv*.

Bommasani, R., Hudson, D. A., Adeli, E., Altman, R., Arora, S., Arx, S. v., Bernstein, M. S., Bohg, J., Bosselut, A., Brunskill, E., Brynjolfsson, E., Buch, S., Card, D., Castellon, R., Chatterji, N., Chen, A., Creel, K., Davis, J. Q., Demszky, D., Donahue, C., Doumbouya, M., Durmus, E., Ermon, S., Etchemendy, J., Ethayarajh, K., Fei-Fei, L., Finn, C., Gale, T., Gillespie, L., Goel, K., Goodman, N., Grossman, S., Guha, N., Hashimoto, T., Henderson, P., Hewitt, J., Ho, D. E., Hong, J., Hsu, K., Huang, J., Icard, T., Jain, S., Jurafsky, D., Kalluri, P., Karamcheti, S., Keeling, G., Khani, F., Khattab, O., Koh, P. W., Krass, M., Krishna, R., Kuditipudi, R., Kumar, A., Ladhak, F., Lee, M., Lee, T., Leskovec, J., Levent, I., Li, X. L., Li, X., Ma, T., Malik, A., Manning, C. D., Mirchandani, S., Mitchell, E., Munyikwa, Z., Nair, S., Narayan, A., Narayanan, D., Newman, B., Nie, A., Niebles, J. C., Nilforoshan, H., Nyarko, J., Ogut, G., Orr, L., Papadimitriou, I., Park, J. S., Piech, C., Portelance, E., Potts, C., Raghunathan, A., Reich, R., Ren, H., Rong, F., Roohani, Y., Ruiz, C., Ryan, J., Ré, C., Sadigh, D., Sagawa, S., Santhanam, K., Shih, A., Srinivasan, K., Tamkin, A., Taori, R., Thomas, A. W., Tramèr, F., Wang, R. E., Wang, W., Wu, B., Wu, J., Wu, Y., Xie, S. M., Yasunaga, M., You, J., Zaharia, M., Zhang, M., Zhang, T., Zhang, X., Zhang, Y., Zheng, L., Zhou, K., & Liang, P. (2022). On the opportunities and risks of foundation models. arXiv.

Bone, D., Goodwin, M. S., Black, M. P., Lee, C.-C., Audhkhasi, K., & Narayanan, S. (2015). Applying machine learning to facilitate autism diagnostics: Pitfalls and promises. *Journal of Autism and Developmental Disorders, 45*(5), 1121–1136.

Borchers, C., Gala, D. S., Gilburt, B., Oravkin, E., Bounsi, W., Asano, Y. M., & Kirk, H. R. (2022). Looking for a handsome carpenter! debiasing gpt-3 job advertisements. In *Proceedings of the the 4th workshop on gender bias in natural language processing (gebnlp)* (pp. 212–224).

Boroditsky, L. (2011). How language shapes thought. *Scientific American, 304*(2), 63–65.

Bortz, J., & Döring, N. (2002). *Forschungsmethoden und Evaluation*. Berlin, Heidelberg: Springer.

Bostrom, N. (2017). *Superintelligence: Paths, dangers, strategies*. Oxford University Press, Oxford, United Kingdom, reprinted with corrections 2017 edition.

Brazdil, P. B., van Rijn, J. N., Soares, C., & Vanschoren, J. (Eds.). (2022). *Metalearning: Applications to automated machine learning and data mining*. Springer eBook collection. Cham: Springer. second edition edition.

Brown, T. B., Mann, B., Ryder, N., Subbiah, M., Kaplan, J., Dhariwal, P., Neelakantan, A., Shyam, P., Sastry, G., Askell, A., Agarwal, S., Herbert-Voss, A., Krueger, G., Henighan, T., Child, R., Ramesh, A., Ziegler, D. M., Wu, J., Winter, C., Hesse, C., Chen, M., Sigler, E., Litwin, M., Gray, S., Chess, B., Clark, J., Berner, C., McCandlish, S., Radford, A., Sutskever, I., & Amodei, D. (2020). Language models are few-shot learners. arXiv.

Brunton, S. L., & Kutz, J. N. (2019). *Data-Driven Science and Engineering*. Cambridge University Press.

Bubeck, S., Chandrasekaran, V., Eldan, R., Gehrke, J., Horvitz, E., Kamar, E., Lee, P., Lee, Y. T., Li, Y., Lundberg, S., Nori, H., Palangi, H., Ribeiro, M. T., & Zhang, Y. (2023). Sparks of artificial general intelligence: Early experiments with gpt-4. arXiv.

Caliskan, A., Bryson, J. J., & Narayanan, A. (2017). Semantics derived automatically from language corpora contain human-like biases. *Science (New York, N.Y.), 356*(6334), 183–186.

Campello, R. J., Moulavi, D., & Sander, J. (2013). Density-based clustering based on hierarchical density estimates. In J. Pei, V. S. Tseng, L. Cao, H. Motoda, & Xu, G. (Eds.), *Advances in knowledge discovery and data mining* (pp. 160–172). Springer: Berlin, Heidelberg.

Carleo, G., Cirac, I., Cranmer, K., Daudet, L., Schuld, M., Tishby, N., Vogt-Maranto, L., & Zdeborová, L. (2019). Machine learning and the physical sciences. *Reviews of Modern Physics, 91*(4).

Carpenter, D., Geden, M., Rowe, J., Azevedo, R., & Lester, J. (2020). Automated analysis of middle school students' written reflections during game-based learning. In I. I. Bittencourt, M. Cukurova, K. Muldner, R. Luckin, & E. Millán (Eds.), *Artificial intelligence in education* (pp. 67–78). Cham: Springer.

Chalmers, D. J. (2010). The singularity: A philosophical analysis. *Journal of Consciousness Studies, 17*, 7–65.

Chang, T. A., & Bergen, B. K. (2023). Language model behavior: A comprehensive survey. arXiv.

Chen, S., Fang, Y., Shi, G., Sabatini, J., Greenberg, D., Frijters, J., & Graesser, A. C. (2020). Automated disengagement tracking within an intelligent tutoring system. *Frontiers in Artificial Intelligence, 3*, 595627.

Cheuk, T. (2021). Can AI be racist? color-evasiveness in the application of machine learning to science assessments. *Science Education, 105*(5), 825–836.

Chialvo, D. R. (2010). Emergent complex neural dynamics. *Nature Physics, 6*(10), 744–750.

Chollet, F. (2018). *Deep learning with Python*. Safari Tech Books Online. Manning, Shelter Island, NY.

Christian, B. (2021). *The alignment problem: How can machines learn human values?* London: Atlantic Books.

Chung, Gregory K. W. K., & Baker, E. L. (2003). Issues in the reliability and validity of automated scoring of constructed responses. *Automated essay scoring: A cross-disciplinary perspective* (pp. 23–40). Mahwah, NJ, US: Lawrence Erlbaum Associates Publishers.

Clark, A. (2023). *The experience machine: How our minds predict and shape reality*. New York: Knopf Doubleday Publishing Group.

Cranmer, M., Sanchez-Gonzalez, A., Battaglia, P., Xu, R., Cranmer, K., Spergel, D., & Ho, S. (2020). Discovering symbolic models from deep learning with inductive biases. arXiv.

Crokidakis, N., Menezes, M. A. D., & Cajueiro, D. O. (2023). Questions of science: chatting with chatgpt about complex systems. arXiv.

Da Xu, L., & Duan, L. (2019). Big data for cyber physical systems in industry 4.0: A survey. *Enterprise Information Systems, 13*(2), 148–169.

Davis, B., & Sumara, D. (2006). *Complexity and education: Inquiries into learning, teaching, and research*. Mahwah, NJ: Lawrence Erlbaum Associates Publishers.

de Vries, A. (2023). The growing energy footprint of artificial intelligence. *Joule, 7*(10), 2191–2194.

Deacon, T. W. (1997). *The symbolic species: The co-evolution of language and the brain*. New York: W.W. Norton. 1st ed. edition.

Deerwester, S., Dumais, S. T., Furnas, G. W., Landauer, T. K., & Harshman, R. (1990). Indexing by latent semantic analysis.

Degrave, J., Felici, F., Buchli, J., Neunert, M., Tracey, B., Carpanese, F., Ewalds, T., Hafner, R., Abdolmaleki, A., & de las Casas, D., Donner, C., Fritz, L., Galperti, C., Huber, A., Keeling, J., Tsimpoukelli, M., Kay, J., Merle, A., Moret, J.-M., Noury, S., Pesamosca, F., Pfau, D., Sauter, O., Sommariva, C., Coda, S., Duval, B., Fasoli, A., Kohli, P., Kavukcuoglu, K., Hassabis, D., & Riedmiller, M. (2022). Magnetic control of tokamak plasmas through deep reinforcement learning. *Nature, 602*(7897), 414–419.

Dehaene, S. (2021). *How we learn: The new science of education and the brain*. Dublin: Penguin Books.

Devlin, J., Chang, M.-W., Lee, K., & Toutanova, K. (2018). Bert: Pre-training of deep bidirectional transformers for language understanding. arXiv:1810.04805

Docktor, J. L., Dornfeld, J., Frodermann, E., Heller, K., Hsu, L., Jackson, K. A., Mason, A., Ryan, Q. X., & Yang, J. (2016). Assessing student written problem solutions: A problem-solving rubric with application to introductory physics. *Physical Review Physics Education Research, 12*(1), 010130.

Dodge, J., Prewitt, T., Des Combes, R. T., Odmark, E., Schwartz, R., Strubell, E., Luccioni, A. S., Smith, N. A., DeCario, N., & Buchanan, W. (2022). Measuring the carbon intensity of AI in cloud instances: Facct '22. arXiv.

Domingos, P. (2015). *The master algorithm : How the quest for the ultimate learning machine will remake our world*. Basic Books.

Du Sautoy, M. (2020). *The creativity code: Art and innovation in the age of AI*. Harper Collins Publ.

Dunbar, K. (1998). Problem solving. In W. Bechtel & G. Graham (Eds.), *A companion to cognitive science* (pp. 289–298). London, England: Blackwell.

Eccles, J. S., & Wigfield, A. (2020). From expectancy-value theory to situated expectancy-value theory: A developmental, social cognitive, and sociocultural perspective on motivation. *Contemporary Educational Psychology*, 101859.

Elizondo, D. (2006). The linear separability problem: Some testing methods. *IEEE Transactions on Neural Networks, 17*(2), 330–344.

Engel, A., & den van Broeck, C. (2001). *Statistical mechanics of learning.* Cambridge: Cambridge University Press.

Ericsson, K. A. (2003). The acquisition of expert performance as problem solving: Construction and modification of mediating mechanisms through deliberate practice. In J. E. Davidson & R. J. Sternberg (Eds.), *The psychology of problem solving* (pp. 31–83). Cambridge, UK: Cambridge University Press.

Evans, V. (2006). Lexical concepts, cognitive models and meaning-construction. *Cognitive Linguistics, 17*(4).

Fang, L., Lee, G.-G., & Zhai, X. (2023). Using gpt-4 to augment unbalanced data for automatic scoring. arXiv.

Fleener, M. J. (2016). Re-searching methods in educational research: A transdisciplinary approach. In M. Koopmans & D. Stamovlasis (Eds.), *Complex Dynamical Systems in Education.* Springer.

Flicker, F. (2022). *The magick of matter: Crystals, chaos and the wizardry of physics.* London: Profile Books.

Friege, G. (2001). *Wissen und Problemlösen: Eine empirische Untersuchung des wissenszentrierten Problemlösens im Gebiet der Elektrizitätslehre auf der Grundlage des Experten-Novizen-Vergleichs.* Berlin: Logos.

Gay, G. (2015). *Culturally responsive teaching: Theory, research, and practice.* Multicultural education. New York: Teachers College Press.

Gelman, A. (2004). Exploratory data analysis for complex models. *Journal of Computational and Graphical Statistics, 13*(4), 755–779.

Géron, A. (2017). *Hands-on machine learning with Scikit-Learn and TensorFlow: Concepts, tools, and techniques to build intelligent systems.* Beijing and Boston and Farnham and Sebastopol and Tokyo: O'Reilly.

Giraud-Carrier, C., & Provost, F. (2005). Toward a justification of meta-learning: Is the no free lunch theorem a show-stopper? In: *Proceedings of the icml-2005 workshop on meta-learning,* Bonn, Germany.

Goodfellow, I., Bengio, Y., & Courville, A. (2016). *Deep learning.* Cambridge, Massachusetts and London, England: MIT Press.

Gottfredson, L. S. (1998). The general intelligence factor. *Scientific American.*

Graesser, A. C. (2016). Conversations with autotutor help students learn. *International Journal of Artificial Intelligence in Education, 26*(1), 124–132.

Gregorcic, B., & Pendrill, A.-M. (2023). Chatgpt and the frustrated socrates. *Physics Education, 58*(3), 035021.

Grotzer, T. A., & Shane Tutwiler, M. (2014). Simplifying causal complexity: How interactions between modes of causal induction and information availability lead to heuristic-driven reasoning. *Mind, Brain, and Education, 8*(3), 97–114.

Halliday, M. A. K. (1978). *Language as social semiotic: The social interpretation of language and meaning.* London: Arnold.

Handelsman, J., Cantor, N., Carnes, M., Denton, D., Fine, E., Grosz, B., Hinshaw, V., Marrett, C., Rosser, S., Donna, S., & Sheridan, J. (2005). More women in science. *Science (New York, N.Y.), 309,* 1190–1191.

Harris, C., Krajcik, J. S., Pellegrino, J. W., & DeBarger, A. H. (2019). Designing knowledge-in-use assessments to promote deeper learning. *Educational Measurement: Issues and Practice, 38*(2), 53–67.

Hastie, T., Tibshirani, R., & Friedman, J. (2008). *The elements of statistical learning: Data mining, inference, and prediction.* Springer.

Hattie, J., & Timperley, H. (2007). The power of feedback. *Review of Educational Research, 77*(1), 81–112.

Hewitt, J., & Manning, C. D. (2019). A structural probe for finding syntax in word representations. In *Proceedings of naacl-hlt 2019.*

Hey, T., Tansley, S., Tolle, K., & Gray, J. (2009). *The fourth paradigm: Data-intensive scientific discovery.* Microsoft Research.

Hilbert, S., Coors, S., Kraus, E., Bischl, B., Lindl, A., Frei, M., Wild, J., Krauss, S., Goretzko, D., & Stachl, C. (2021). Machine learning for the educational sciences. *Review of Education, 9*(3).

Hilpert, J. C., & Marchand, G. C. (2018). Complex systems research in educational psychology: Aligning theory and method. *Educational Psychologist, 53*(3), 185–202.

Hinton, G., Osindero, S., & Teh, Y.-W. (2006). A fast learning algorithm for deep belief nets. *Neural Computation, 18*, 1527–1554.

Hofmann, T. (2001). Unsupervised learning by probabilistic latent semantic analysis. *Machine Learning, 42*, 177–196.

Hornik, K., Stinchcombe, M., & White, H. (1989). Multilayer feedforward networks are universal approximators. *Neural Networks, 2*(5), 359–366.

Hughes, S. (2023). Cut the bull. . . . detecting hallucinations in large language models: Vectara.

Jumper, J., Evans, R., Pritzel, A., Green, T., Figurnov, M., Ronneberger, O., Tunyasuvunakool, K., Bates, R., Žídek, A., Potapenko, A., Bridgland, A., Meyer, C., Kohl, S. A. A., Ballard, A. J., Cowie, A., Romera-Paredes, B., Nikolov, S., Jain, R., Adler, J., ... Hassabis, D. (2021). Highly accurate protein structure prediction with alphafold. *Nature, 596*(7873), 583–589.

Jurafsky, D., & Martin, J. H. (2014). *Speech and language processing.* Always learning. Pearson Education, Harlow, 2. ed., pearson new internat. ed. edition.

Kahneman, D. (2011). *Thinking, fast and slow.* New York: Farrar Straus and Giroux. first edition edition.

Kantz, H., & Schreiber, T. (2003). *Nonlinear time series analysis (Kantz, Schreiber) (Cambridge 2004) (2nd ed).* Cambridge University Press.

Kaplan, J., McCandlish, S., Henighan, T., Brown, T. B., Chess, B., Child, R., Gray, S., Radford, A., Wu, J., & Amodei, D. (2020). Scaling laws for neural language models. arXiv.

Kapoor, S., & Narayanan, A. (2022). Leakage and the reproducibility crisis in ml-based science.

Kapoor, S., & Narayanan, A. (2023). Leakage and the reproducibility crisis in machine-learning-based science. *Patterns (New York, N.Y.), 4*(9), 100804.

Karniadakis, G. E., Kevrekidis, I. G., Lu, L., Perdikaris, P., Wang, S., & Yang, L. (2021). Physics-informed machine learning. *Nature Reviews Physics, 3*(6), 422–440.

Kasneci, E., Sessler, K., Küchemann, S., Bannert, M., Dementieva, D., Fischer, F., Gasser, U., Groh, G., Günnemann, S., Hüllermeier, E., Krusche, S., Kutyniok, G., Michaeli, T., Nerdel, C., Pfeffer, J., Poquet, O., Sailer, M., Schmidt, A., Seidel, T., ... Kasneci, G. (2023). Chatgpt for good? on opportunities and challenges of large language models for education. *Learning and Individual Differences, 103*, 102274.

Keshavarzi Arshadi, A., Webb, J., Salem, M., Cruz, E., Calad-Thomson, S., Ghadirian, N., Collins, J., Diez-Cecilia, E., Kelly, B., Goodarzi, H., & Yuan, J. S. (2020). Artificial intelligence for covid-19 drug discovery and vaccine development. *Frontiers in Artificial Intelligence, 3*, 65.

Khanum, M., Mahboob, T., Imtiaz, W., Abdul Ghafoor, H., & Sehar, R. (2015). A survey on unsupervised machine learning algorithms for automation, classification and maintenance. *International Journal of Computer Applications, 119*(13), 34–39.

Kieser, F., Wulff, P., Kuhn, J., & Küchemann, S. (2023). Educational data augmentation in physics education research using chatgpt. *Physical Review Physics Education Research, 19*(2), 1–13.

Kirkpatrick, J., Pascanu, R., Rabinowitz, N., Veness, J., Desjardins, G., Rusu, A. A., Milan, K., Quan, J., Ramalho, T., Grabska-Barwinska, A., Hassabis, D., Clopath, C., Kumaran, D., & Hadsell, R. (2017). Overcoming catastrophic forgetting in neural networks. *Proceedings of the National Academy of Sciences of the United States of America, 114*(13), 3521–3526.

Kojima, T., Gu, S. S., Reid, M., Matsuo, Y., & Iwasawa, Y. (2022). Large language models are zero-shot reasoners. In *36th conference on neural information processing systems (neurips 2022).*

Koopmans, M., & Stamovlasis, D. (2016). Introduction to education as a complex dynamical system. In M. Koopmans & D. Stamovlasis (Eds.), *Complex Dynamical Systems in Education*. Springer.

Kotsiantis, S. B. (2007). Supervised machine learning: A review of classification techniques. *Informatica, 31*, 249–268.

Krauss, L. M. (2023). *The known unknowns: The unsolved mysteries of the cosmos*. London: Head of Zeus.

Krenn, M., Pollice, R., Guo, S. Y., Aldeghi, M., Cervera-Lierta, A., Friederich, P., Dos Passos Gomes, G., Häse, F., Jinich, A., Nigam, A., Yao, Z., & Aspuru-Guzik, A. (2022). On scientific understanding with artificial intelligence. *Nature Reviews Physics, 4*(12), 761–769.

Kriegel, H.-P., Kröger, P., Sander, J., & Zimek, A. (2011). Density-based clustering. *WIREs Data Mining and Knowledge Discovery, 1*(3), 231–240.

Krizhevsky, A., Sutskever, I., & Hinton, G. (2012). Imagenet classification with deep convolutional neural networks: Neurips proceedings.

Kubsch, M., Krist, C., & Rosenberg, J. (2022). Distributing epistemic functions and tasks - a framework for augmenting human analytic power with machine learning in science education research. *Journal of Research in Science Teaching*.

Kurzweil, R. (2004). The law of accelerating returns. In C. Teuscher (Ed.), *Alan turing: Life and legacy of a great thinker* (pp. 381–416). Berlin and Heidelberg: Springer.

Lai, G., & Calandra, B. (2010). Examining the effects of computer-based scaffolds on novice teachers' reflective journal writing. *Educational Technology Research and Development, 58*(4), 421–437.

Lebret, R., & Collobert, R. (2013). Word embeddings through Hellinger PCA. arXiv.

LeCun, Y., Bottou, L., Bengio, Y., & Haffner, P. (1998). Gradient-based learning applied to document recognition. *Proceedings of the IEEE, 86*(11), 2278–2324.

Lenat, D., & Marcus, G. (2023). Getting from generative AI to trustworthy AI: What llms might learn from cyc. arXiv.

Lenci, A., Sahlgren, M., Jeuniaux, P., Cuba Gyllensten, A., & Miliani, M. (2022). A comparative evaluation and analysis of three generations of distributional semantic models. *Language Resources and Evaluation, 56*(4), 1269–1313.

Liang, X., Nguyen, D., & Jiang, S. B. (2021). Generalizability issues with deep learning models in medicine and their potential solutions: Illustrated with cone-beam computed tomography (cbct) to computed tomography (ct) image conversion. *Machine Learning: Science and Technology, 2*(1), 015007.

Lipton, Z. C. (2018). The mythos of model interpretability. *Machine Learning*.

Li, Y., Sha, L., Yan, L., Lin, J., Raković, M., Galbraith, K., Lyons, K., Gašević, D., & Chen, G. (2023). Can large language models write reflectively. *Computers and Education: Artificial Intelligence, 4*, 100140.

Liu, Z., & Tegmark, M. (2021). AI poincaré: Machine learning conservation laws from trajectories. arXiv.

Lundberg, S. M., & Lee, S.-I. (2017). A unified approach to interpreting model predictions. In *31st conference on neural information processing systems (nips 2017)*, Long Beach, CA, USA.

Lyubchenko, Y., Shlyakhtenko, L., Potaman, V., & Sinden, R. (2002). Global and local dna structure and dynamics. Single molecule studies with afm. *Microscopy and Microanalysis, 8*(S02), 170–171.

Macmillan-Scott, O., & Musolesi, M. (2024). (ir)rationality and cognitive biases in large language models. arXiv.

Malik, M. M. (2020). A hierarchy of limitations in machine learning. arXiv.

Manning, C. D. (2022). Human language understanding & reasoning. *Daedalus, 151*(2), 127–138.

Marcus, G. (2019). *Rebooting AI: Building artificial intelligence we can trust*. New York: Pantheon Books. first edition edition.

Marsland, S. (2015). *Machine learning: An algorithmic perspective*. Chapman & Hall/CRC machine learning & pattern recognition series. Boca Raton, FL: CRC Press. second edition edition.

Martin, P. P., Kranz, D., Wulff, P., & Graulich, N. (2023). Exploring new depths: Applying machine learning for the analysis of student argumentation in chemistry. *Journal of Research in Science Teaching*, 1–36.

McClelland, D. C. (1973). Testing for competence rather than for 'intelligence.'. *American Psychologist, 28*(1), 1–14.

McInnes, L., Healy, J., & Melville, J. (2020). Umap: Uniform manifold approximation and projection for dimension reduction. arXiv.

McNamara, D., Kintsch, E., Butler Songer, N., & Kintsch, W. (1996). Are good texts always better? interactions of text coherence, background knowledge, and levels of understanding in learning from text. *Cognition and Instruction, 14*(1), 1–43.

Mehrabi, N., Morstatter, F., Saxena, N., Lerman, K., & Galstyan, A. (2022). A survey on bias and fairness in machine learning. arXiv.

Metz, C. (2023). Chatbots may 'hallucinate' more often than many realize. *The New York Times*.

Meurers, D. (2012). Natural language processing and language learning. In C. A. Chapelle (Ed.), *The encyclopedia of applied linguistics*. New York, NY: Wiley.

Micceri, T. (1989). The unicorn, the normal curve, and other improbable creatures. *Psychological Bulletin, 105*(1), 156–166.

Mikolov, T., Sutskever, I., Chen, K., Corrado, G. S., & Dean, J. (2013). Distributed representations of words and phrases and their compositionality. In C. J. Burges, L. Bottou, M. Welling, Z. Ghahramani, & K. Q. Weinberger (Eds.), *Advances in neural information processing systems*. (Vol. 26). Curran Associates Inc.

Mitchell, M. (2020). *Artificial intelligence: A guide for thinking humans*. Pelican Books.

Mitchell, M. (2023). Ai's challenge of understanding the world. *Science, 382*(6671), eadm8175.

Mitchell, M., Palmarini, A. B., & Moskvichev, A. (2023) Comparing humans, gpt-4, and gpt-4v on abstraction and reasoning tasks. arXiv.

Mitchell, T. (1997). *Machine learning*. New York, NY: McGraw-Hill Education.

Mitchell, M., & Krakauer, D. C. (2023). The debate over understanding in AI's large language models. *Proceedings of the National Academy of Sciences of the United States of America, 120*(13), e2215907120.

Molnar, C. (2022). *Interpretable machine learning: A guide for making black box models explainable*. Christoph Molnar, Munich, Germany, second edition edition.

Montemurro, M. A., & Zanette, D. (2010). Towards the quantification of the semantic information encoded in written language. *Advances in Complex Systems, 13*(02), 135–153.

Morris, J. X., Kuleshov, V., Shmatikov, V., & Rush, A. M. (2023). Text embeddings reveal (almost) as much as text. arXiv.

Müllner, D. (2011). Modern hierarchical, agglomerative clustering algorithms. arXiv.

Muthukrishna, M. (2023). *A theory of everyone: The new science of who we are, how we got here, and where we're going*. Cambridge, Massachusetts: The MIT Press.

Nehm, R. H., & Härtig, H. (2012). Human vs. computer diagnosis of students' natural selection knowledge: Testing the efficacy of text analytic software. *Journal of Science Education and Technology, 21*(1), 56–73.

Nisbet, R., Elder, J. F., & Miner, G. (2009). *Handbook of statistical analysis and data mining applications*. Amsterdam and Boston: Academic Press/Elsevier.

Niven, T., & Kao, H.-Y. (2019). Probing neural network comprehension of natural language arguments. In *Proceedings of the 57th annual meeting of the association for computational linguistics*.

Nowak, M. A., Komarova, N. L., & Niyogi, P. (2002). Computational and evolutionary aspects of language. *Nature, 417*(6889), 611–617.

Nunnally, J. C. (1978). *Psychometric theory*. McGraw-Hill series in psychology. New York: McGraw-Hill. 2. ed. edition.

Odden, T. O. B., Marin, A., & Caballero, M. D. (2020). Thematic analysis of 18 years of physics education research conference proceedings using natural language processing. *Physical Review Physics Education Research, 16*(1), 1–25.

OpenAI. (2023). Gpt-4 technical report. arXiv.

Patel, R., & Pavlick, E. (2022). Mapping language models to grounded conceptual spaces. In *Published as a conference paper at iclr 2022*.

Patriarcha, M., Heinsalu, E., & Léonard, J. L. (2020). *Languages in space and time: Models and methods from complex systems theory*. Cambridge University Press.

Pearl, J. (2021). Causally colored reflections on leo breiman's 'statistical modeling: The two cultures' (2001). *Observational Studies, 7*(1), 187–190.

Pezzulo, G., Parr, T., & Friston, K. (2021). The evolution of brain architectures for predictive coding and active inference. *Philosophical Transactions of the Royal Society B*.

Polverini, G., & Gregorcic, B. (2024). Performance of chatgpt on the test of understanding graphs in kinematics. arXiv.

Pontzen, A. (2023). *The universe in a box: Simulations and the quest to code the cosmos*. New York: Penguin Publishing Group.

Powers, D. E. (2005). "wordiness": A selective review of its influence, and suggestions for investigating its relevance in tests requiring extended written responses. *ETS Research Report*.

Prince, S. J. D. (2023). *Understanding deep learning*. MIT Press.

Rauf, I. A. (2021). *Physics of data science and machine learning*. Boca Raton: CRC Press.

Raz, A., Heinrichs, B., Avnoon, N., Eyal, G., & Inbar, Y. (2024). Prediction and explainability in AI: Striking a new balance? *Big Data & Society, 11*(1).

Rehurek, R., & Sojka, P. (2010). Software framework for topic modelling with large corpora. In N. Calzolari, K. Choukri, B. Maegaard, J. Mariani, J. Odijk, S. Piperidis, M. Rosner, D. Tapias (Eds.), *Proceedings of the LREC, 2010* (pp. 45–50).

Reinhold, P., Lind, G., & Friege, G. (1999). Wissenszentriertes problemlösen in physik. *Zeitschrift für Didaktik der Naturwissenschaften, 5*(1), 41–62.

Rosenberg, J. M., Kubsch, M., Wagenmakers, E.-J., & Dogucu, M. (2022). Making sense of uncertainty in the science classroom: A Bayesian approach. *Science & Education, 31*(5), 1239–1262.

Rosenfeld, R. (2000). Two decades of statistical language modeling: Where do we go from here? In *Proceedings of the IEEE*.

Rothchild, I. (2006). Induction, deduction, and the scientific method: An eclectic overview of the practice of science. *SSR*.

Russell, S., & Ermon, S. (2016). Label-free supervision of neural networks with physics and domain knowledge. arXiv.

Russell, S., & Norvig, P. (1995). *Artificial intelligence: A modern approach*. Englewood Cliffs, NJ: Prentice Hall.

Sawyer, K. R. (2002). Emergence in psychology. *Human Development, 45*, 2–28.

Scherer, P. O. J. (2017). *Computational physics: Simulation of classical and quantum systems*. Graduate texts in physics. Cham; Springer. 3rd edition 2017 edition.

Schrouff, J., Baur, S., Hou, S., Mincu, D., Loreaux, E., Blanes, R., Wexler, J., Karthikesalingam, A., & Kim, B. (2022). Best of both worlds: Local and global explanations with human-understandable concepts. arXiv.

Sejnowski, T. J. (2024). *ChatGPT and the future of AI. The deep language revolution*. Cambridge, Massachusetts, London, England: The MIT Press.

Simon, H. A., & Gilmartin, K. J. (1973). A simulation of memory for chess positions. *Cognitive Psychology, 5*, 29–46.

Smola, A. J., & Schölkopf, B. (2003). A tutorial on support vector regression.

Sole, R. V. (2011). *Phase Transitions*. Primers in complex systems. Princeton: Princeton University Press.

Solé, R., Moses, M., & Forrest, S. (2019). Liquid brains, solid brains. *Philosophical transactions of the Royal Society of London. Series B, Biological Sciences, 374*(1774), 20190040.

Stamovlasis, D., & Koopmans, M. (2014). Introduction to the special issue: Nonlinear dynamics in education. *Psychology and Life Science, 18*.

Stede, M., & Neumann, A. (2014). Potsdam commentary corpus 2.0: Annotation for discourse research. In *Proceedings of the Language Resources and Evaluation Conference (LREC), Reykjavik*.

Steyvers, M., & Tenenbaum, J. B. (2005). The large-scale structure of semantic networks: Statistical analyses and a model of semantic growth. *Cognitive Science, 29*, 41–78.

Sundararajan, M., Taly, A., & Yan, Q. (2017). Axiomatic attribution for deep networks. In *Proceedings of the 34th international conference on machine learning*, Sydney, Australia. *PMLR*, 70.

Sutton, R. S., & Barto, A. G. (2015). *Reinforcement learning: An introduction*. MIT Press.

Tadros, T., Krishnan, G. P., Ramyaa, R., & Bazhenov, M. (2022). Sleep-like unsupervised replay reduces catastrophic forgetting in artificial neural networks. *Nature Communications, 13*(1), 7742.

Taleb, N. N. (2020). *Statistical consequences of fat tails: Real World preasymptotics, epistemology, and applications: Papers and commentary*. The Technical Incerto Collection. STEM Academic Press.

Tegmark, M. (2018). *Life 3.0: Being human in the age of artificial intelligence*. New York: Vintage Books. first vintage books edition edition.

Thelen, E., & Smith, L. (1996). *A dynamic systems approach to the development of cognition and action*. MIT Press.

Theobald, E. J., Aikens, M., Eddy, S. L., & Jordt, H. (2019). Beyond linear regression: A reference for analyzing common data types in discipline based education research. *Physical Review Physics Education Research, 15*.

Thomas, R. L., & Uminsky, D. (2022). Reliance on metrics is a fundamental challenge for AI. arXiv.

Tschisgale, P., Wulff, P., & Kubsch, M. (2023). Integrating artificial intelligence-based methods into qualitative research in physics education research: A case for computational grounded theory. *Physical Review Physics Education Research, 19*(020123), 1–24.

Udrescu, S.-M., & Tegmark, M. (2020). AI Feynman: A physics-inspired method for symbolic regression. *Science Advances, 6*.

Ulrich, B. (2010). Gender diversity and nurse-physician relationships. *American Medical Association Journal of Ethics, 12*(1), 41–45.

Urrutia, F., & Araya, R. (2024). Who's the best detective? large language models vs. traditional machine learning in detecting incoherent fourth grade math answers. *Journal of Educational Computing Research, 61*(8), 187–218.

Valiant, L. G. (1984). A theory of the learnable. *Communication of the ACM, 27*.

van Es, E., & Sherin, M. G. (2002). Learning to notice: Scaffolding new teachers' interpretations of classroom interactions. *Journal of Technology and Teacher Education, 10*(4), 571–596.

Vansteelandt, S. (2021). Statistical modelling in the age of data science. *Observational Studies, 7*(1), 217–228.

Vapnik, V. (1996). Structure of statistical learning theory. In A. Gammerman (Ed.), *Computational learning and probabilistic reasoning* (pp. 3–31). Chichester and New York: Wiley.

Vaswani, A., Shazeer, N., Parmar, N., Uszkoreit, J., Jones, L., Gomez, A. N., Kaiser, Ł., & Polosukhin, I. (2017). Attention is all you need: Conference on neural information processing systems. *Advances in Neural Information Processing Systems*, 6000–6010.

Vilalta, R., & Meskhi, M. M. (2022). Transfer of knowledge across tasks. In P. B. Brazdil, J. N. van Rijn, C. Soares, & J. Vanschoren (Eds.), *Metalearning, Springer eBook Collection* (p. 219). Cham: Springer.

Vincent, J. (2019). Deepmind's ai agents conquer human pros at starcraft ii.

Walton, D. (2008). *Informal logic: A pragmatic approach* (2nd ed.). Cambridge MA: Cambridge University Press.

Walton, D., & Reed, C. A. (2005). Argumentation schemes and enthymemes. *Synthese, 145*, 339–370.

Wang, L. (2016). Discovering phase transitions with unsupervised learning. *Physical Review B, 94*(19).

Wang, H., Fu, T., Du, Y., Gao, W., Huang, K., Liu, Z., Chandak, P., Liu, S., van Katwyk, P., Deac, A., Anandkumar, A., Bergen, K., Gomes, C. P., Ho, S., Kohli, P., Lasenby, J., Leskovec, J., Liu, T.-Y., Manrai, A., ... Zitnik, M. (2023). Scientific discovery in the age of artificial intelligence. *Nature,620*(7972), 47–60.

Wang, Q., Ma, Y., Zhao, K., & Tian, Y. (2022). A comprehensive survey of loss functions in machine learning. *Annals of Data Science, 9*(2), 187–212.

Wei, J., Wang, X., Schuurmans, D., Bosma, M., Ichter, B., Xia, F., Chi, E., Le Quoc, & Zhou, D. (2022). Chain-of-thought prompting elicits reasoning in large language models. arXiv.

West, C. G. (2023). AI and the FCI: Can chatgpt project an understanding of introductory physics? arXiv.

White, J., Fu, Q., Hays, S., Sandborn, M., Olea, C., Gilbert, H., Elnashar, A., Spencer-Smith, J., & Schmidt, D. C. (2023). A prompt pattern catalog to enhance prompt engineering with chatgpt. arXiv.

Wickramasinghe, C. S., Amarasinghe, K., Marino, D. L., Rieger, C., & Manic, M. (2021). Explainable unsupervised machine learning for cyber-physical systems. *IEEE Access, 9,* 131824–131843.

Williams, H. T. (1999). Semantics in teaching introductory physics. *American Journal of Physics, 67*(8), 670–680.

Wilson, M. (2023). *Constructing measures: An item response modeling approach.* Taylor & Francis.

Wolfram, S. (2002). *A new kind of science.* Champaign, Ill.: Wolfram Media, 1. edition edition.

Wolfram, S. (2023). *What is ChatGPT doing and why does it work?* Wolfram Media.

Wu, Z., Qiu, L., Ross, A., Akyürek, E., Chen, B., Wang, B., Kim, N., Andreas, J., and Kim, Y. (2024). Reasoning or reciting? exploring the capabilities and limitations of language models through counterfactual tasks. arXiv.

Wu, Y., Schuster, M., Chen, Z., Le V, Q., Norouzi, M., Macherey, W., Krikun, M., Cao, Y., Gao, Q., Macherey, K., Klingner, J., Shah, A., Johnson, M., Liu, X., Kaiser, Ł., Gouws, S., Kato, Y., Kudo, T., Kazawa, H., Stevens, K., Kurian, G., Patil, N., Wang, W., Young, C., Smith, J., Riesa, J., Rudnick, A., Vinyals, O., Corrado, G., Hughes, M., & Dean, J. (2016). Google's neural machine translation system: Bridging the gap between human and machine translation. arXiv.

Wulff, P., Mientus, L., Nowak, A., & Borowski, A. (2022). Utilizing a pretrained language model (bert) to classify preservice physics teachers' written reflections. *International Journal of Artificial Intelligence in Education.*

Wulff, P. (2023). Network analysis of terms in the natural sciences insights from wikipedia through natural language processing and network analysis. *Education and Information Technologies, 28,* 14325–14346.

Wulff, P., Buschhüter, D., Westphal, A., Mientus, L., Nowak, A., & Borowski, A. (2022). Bridging the gap between qualitative and quantitative assessment in science education research with machine learning – a case for pretrained language models-based clustering. *Journal of Science Education and Technology, 31,* 490–513.

Yang, Y.-Y., Chou, C.-N., & Chaudhuri, K. (2022). Understanding rare spurious correlations in neural networks. arXiv.

Yao, S., Yu, D., Zhao, J., Shafran, I., Griffiths, T. L., Cao, Y., & Narasimhan, K. (2023). Tree of thoughts: Deliberate problem solving with large language models. arXiv.

Ye, C., Fu, T., Hao, S., Zhang, Y., Wang, O., Jin, B., Xia, M., Liu, M., Zhou, X., Wu, Q., Guo, Y., Zhu, C., Li, Y.-M., Culver, D. S., Alfreds, S. T., Stearns, F., Sylvester, K. G., Widen, E., McElhinney, D., & Ling, X. (2018). Prediction of incident hypertension within the next year: Prospective study using statewide electronic health records and machine learning. *Journal of Medical Internet Research, 20*(1), e22.

Zador, A. M. (2019). A critique of pure learning and what artificial neural networks can learn from animal brains. *Nature Communications, 10*(1), 3770.

Zanette, D. (2014). Statistical pattern in written language. arXiv:1412.3336.

Zhai, X., Haudek, K., Shi, L., Nehm, R., & Urban-Lurain, M. (2020). From substitution to redefinition: A framework of machine learning-based science assessment. *Journal of Research in Science Teaching, 57*(9), 1430–1459.

Zhu, X., & Goldberg, A. B. (Eds.). (2009). *Introduction to semi-supervised learning.* Cham: Springer.